Restoring Earth, Community, and Soul:

Creating the Social, Economic, and Religious Transformations Required by Global Warming

Andrea Lynn Orcutt

Earth Community Press
Evanston, Illinois

MW01711956

The chart, Wall Street Capitalism vs Main Street Markets on page 167 is reprinted with permission of the publisher. From *Agenda for a New Economy,* copyright© 2009 by David Korten, Berrett-Koehler Publishers, Inc., San Francisco, CA. All rights reserved. www.bkconnection.com

Earth Community Press 2011

This work is licensed under a Creative Commons Attribution-NonCommercial-ShareAlike 3.0 Unported (CC BY-NC-SA 3.0) United States License

You are free:
- to Share — to copy, distribute and transmit the work
- to Remix — to adapt the work

Under the following conditions:
- Attribution — You must attribute the work in the manner specified by the author or licensor (but not in any way that suggests that they endorse you or your use of the work).
- Noncommercial — You may not use this work for commercial purposes.
- Share Alike — If you alter, transform, or build upon this work, you may distribute the resulting work only under the same or similar license to this one.

For further information, see http://creativecommons.org/licenses/by-nc-sa/3.0/

ISBN: 978-0-9832613-0-8

Cover and interior design, illustration, and production by Morgan K Brooke
Additional resources and blog at restoringearth.org

Earth Community Press
Evanston, Illinois

For Jim

For my children and all the children

For elephants, penguins, and trees

For Earth, Our Mother

Restoring Earth, Community, and Soul

Creating the Social, Economic, and Religious Transformations Required by Global Warming

Andrea Lynn Orcutt

Full Table of Contents:

Tamarak

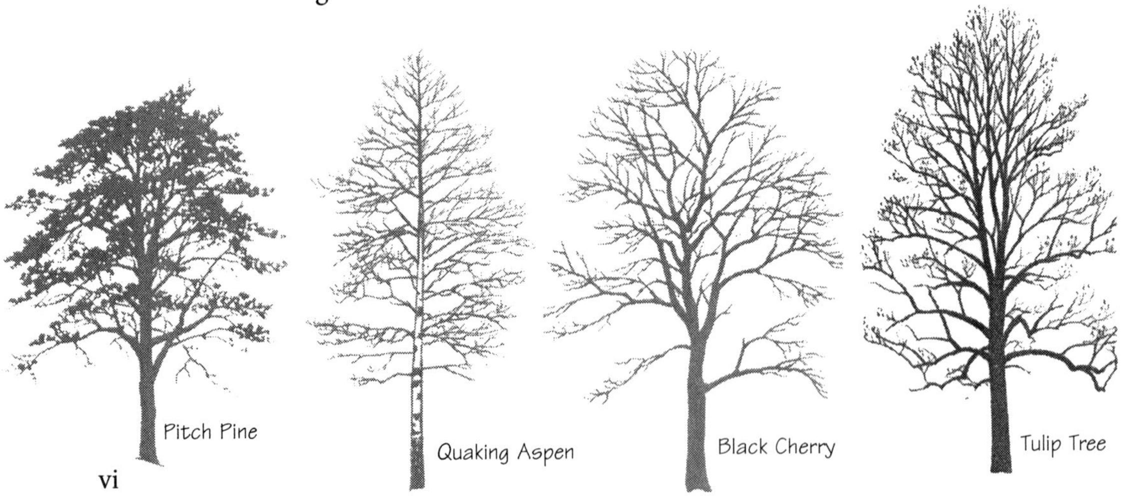
Pitch Pine
Quaking Aspen
Black Cherry
Tulip Tree

Loblolly Pine

Black Willow

Black Locust

Part lll Creating Solutions to Global Warming

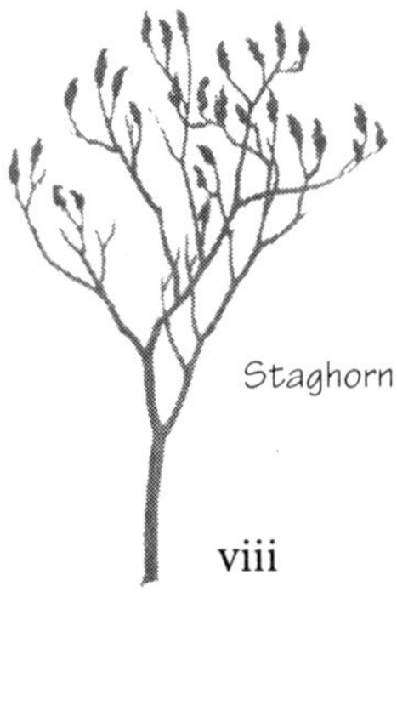

Staghorn Sumac

Flowering Dogwood

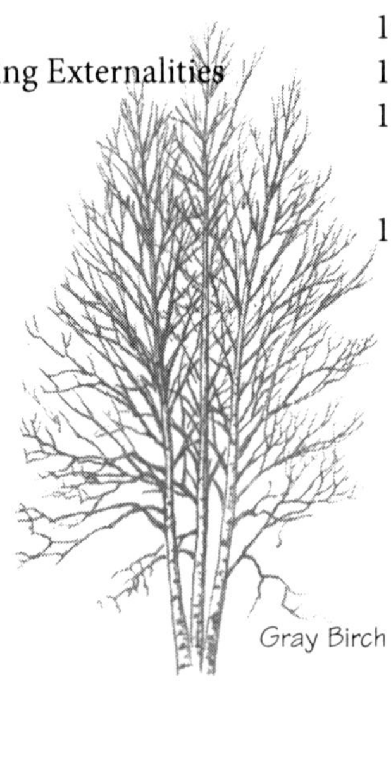

Gray Birch

Hemlock

Osage-Orange

Black Spruce

Sassafras Tree

Bald Cypress

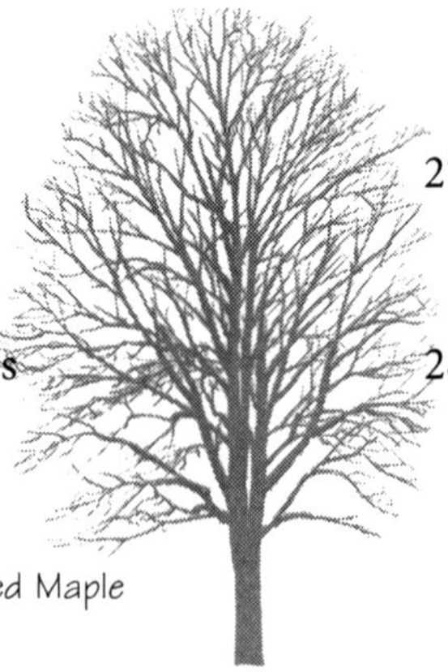
Red Maple

Eastrn Redcedar

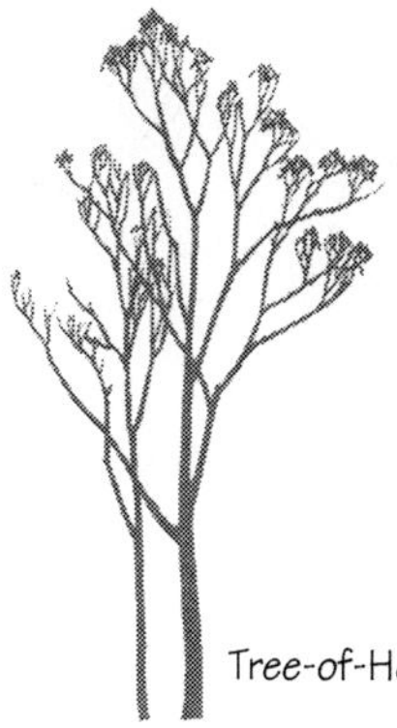
Tree-of-Heaven

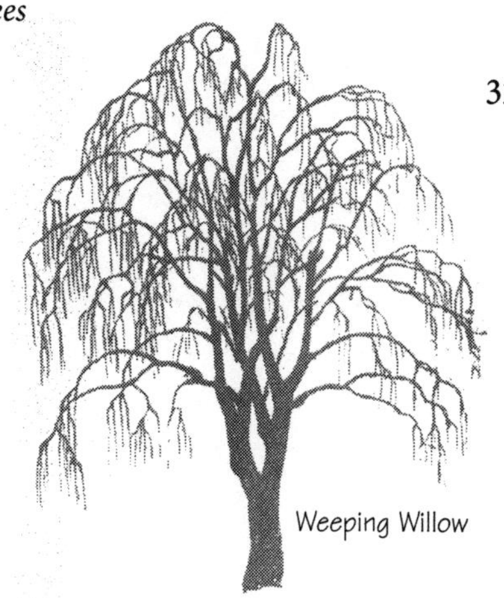
Weeping Willow

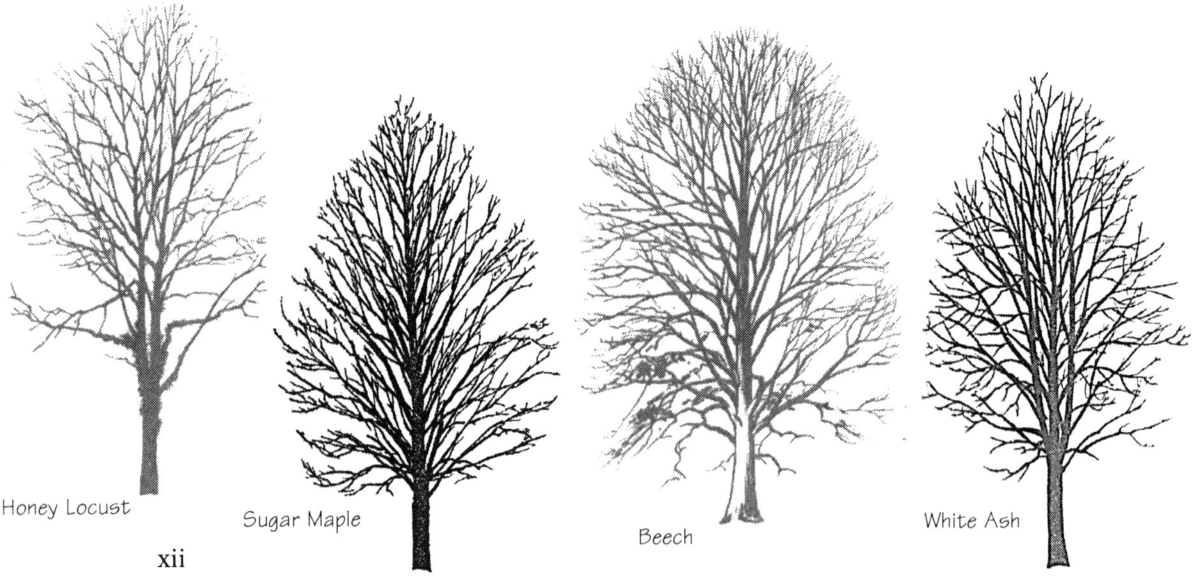

Honey Locust
Sugar Maple
Beech
White Ash

Introduction

Creating a Shared Moral Vision

Indian Elephant Family

Creating a Shared Moral Vision

We are a country in deep denial. Polls in August 2010 found that as many as 45% of Americans believe that global warming "is caused by long-term planetary trends," while only 40% are convinced that "human activity is the main contributor."[1] This is a substantial drop in the number of people who believe that global warming is real. This drop is primarily the result of the efforts to discredit the Intergovernmental Panel on Climate Change's (IPCC's) landmark Fourth Assessment Report, published in early 2007, and worldwide press coverage of the leaked emails from the climate science unit at the University of East Anglia, which the press referred to as "climategate,"[2] all of which was manufactured.[3]

> We stand at a critical moment in Earth's history, a time when humanity must choose its future.
>
> The Earth Charter

Yet even people who accept that global warming is happening are still almost completely in denial about how enormous and urgent our response needs to be. Rajendra K. Pachauri, scientist, economist, and head of the United Nations (U.N.) program on climate change (IPCC), said in January 2005, "Climate change is for real. We have just a small window of opportunity and it is closing rather rapidly. There is not a moment to lose. *We are risking the ability of the human race to survive.*"[4] In 2007, Pachauri said, "If there's no action before 2012, that's too late. What we do in the next two to three years will determine our future. This is the defining moment."[5] James Hansen, Director of the National Aeronautics and Space Administration's (NASA's) Goddard Institute for Space Studies and a leading climate scientist, says, "If we burn even half of Earth's remaining fossil fuels we will destroy the planet as humanity knows it. The added emissions of heat-trapping carbon dioxide will set our Earth irreversibly onto a course toward an ice-free state, a course that will initiate a chain reaction of irreversible and catastrophic climate changes."[6]

Because people generally don't know about this level of risk from global warming, they don't have the required sense of urgency needed to respond adequately and in time. Global warming is affecting the entire world and all aspects of life at once, and the effects are irreversible. If we are to avoid enormous catastrophes, we have at most 10 years to make drastic changes in order to keep the final warming less than an additional 2°C (3.6°F). To do this, the U.S. must cut its emissions by 80% by 2020; however, this level of change, although technically possible, is completely absent from our political discussions.

Our current way of life cannot be sustained. Continuing as we are is no longer an option; nature will change us even if we don't act. That's just the reality. The fate of the world will be decided in the next 10 years, and the U.S. will decide it. We need a global response to global warming, which will never happen until the U.S. takes the lead. The Bush administration has cost us what was probably the most important 8 years. For that matter, the first President Bush, as well as Clinton (and Gore), also did nothing. The crisis of global warming today is fundamentally different than it was 10 years ago. It is exponentially more serious and more difficult.

U.S. policies are a major source of our global ecological problems. With only 5% of the world's population, we create more than 25% of the emissions that cause global warming. Therefore, if the U.S. becomes a major part of the solution, the impact on the world will be decisive. In addition, our history makes Americans particularly well suited for creating a new shared vision and making it a reality. If the U.S. ever had a Manifest Destiny, this is it. Creating this new future requires a wide-open, highly diverse society, with little dependence on the traditional ways in which things were done. It requires a willingness to openly share information and a belief that whatever we commit to, we can accomplish. Americans have a history of innovation and creativity, of storming off into the unknown—it's what we do. Once we commit, no other country could invent this new world as well as we can. The world can't do this unless the U.S. leads, and the U.S. can't do it without the rest of the world. We already know what to do; what is missing is not solutions, but individual and collective commitment.

Our Ecological Crises Are Enormous and Urgent

Addressing global warming is difficult because it is a type of problem that human beings have never faced before. The very nature of the problem is so foreign to our experience because global warming is outside our normal methods, processes, and experiences with problem solving. Never before has the collective impact of individual choices changed the planet. In addition, global warming is a process, and people tend to respond only to *events* (but by the time global warming creates major events, it will be too late to stop it).

To address this problem, we need new ways of thinking and a new consciousness. Since there is no historical precedent for this type of problem solving, we will have to invent something new. Global warming is not just another issue alongside terrorism, the economy, and health care. It is the fundamental crisis of our world today, and solving it will address the other problems because they all have a common cause.

The fundamental issue is the dilemma of scale. The problem of global warming is enormous and quickly getting worse: time is running out. What if we hear that we have now crossed a threshold and the complete

Barred Owl

James Hansen and his team issued a series of papers showing that any atmospheric carbon content greater than 350 ppm is not compatible with "a planet similar to that on which civilization developed and to which life on Earth is adapted." For the past 10,000 years, during which human civilization has developed and flourished, the earth's climate has been unusually stable. Hansen and his team warn that a world of 450 ppm CO_2 is a world that will eventually be largely ice free.[12] And 450 ppm will probably make all the feedback loops mentioned above unstoppable. To emphasize these numbers and what we need to do, ecologist Bill McKibben and others started a campaign at www.350.org.

There is a very real chance that human beings are simply incapable of acting in anticipation of future catastrophe. When we were trying to avoid nuclear war, we were trying to make something not happen, and working on it gave a lot of money to powerful corporations. That is a lot easier than trying to make something happen (responding effectively to global warming), particularly when it is against the interests of those with enormous economic and political power.

> The choice will be made by our generation, but it will affect life on earth for all generations to come.
>
> Lester Brown

Defining Terms: Ecology and Environmentalism

In discussing our global warming crisis, I use the term "ecology." In science, the term "ecology" is the study of the interconnections and interdependencies of organisms and their environments, such as in a swamp, prairie, or forest. However, the term "ecology" is now also used to describe the social issues involved in environmental problems. Norwegian philosopher Arne Naess first used the term "deep ecology" in 1972, and he began the focus on the deep ethical dimensions of the causes of environmental degradation. His deep ecology used a holistic approach and linked environmental damage to social systems and to our values. The *American Heritage Dictionary* gives one definition of "ecology" as "the study of the detrimental effects of modern civilization on the environment, with a view toward prevention or reversal through conservation." Today, "ecology" has grown far beyond conservation.

The terms ecology and environmentalism are not interchangeable. Environmentalism tends to focus on each problem individually, concentrate on solving problems piecemeal, and assume that new technologies and/or new polices will manage our ecological crises without fundamental systems-wide transformations. Environmentalists do not put solutions to global warming in the context of massive social and economic change but generally focus on what individuals or individual cities can do within the current system. This approach ends up treating only the symptoms, not the cause.

In contrast, ecology and ecologists deal with the social, economic, and natural world as one integrated, interdependent whole. Today, the ecological vision of reality goes far beyond concerns with environmental protection. Rather than focus on how we can be a little less destructive, ecology focuses on the understanding of reality we need to have in order to stop all the destruction and begin restoring Earth and creating a sustainable world.

Defining an Ecological Social Theory

In December 2004, the inadequacies of environmentalism lead "The Death of Environmentalism," by Michael Shellenberger and Ted Nordhaus; the other was "Is Environmentalism Dead?" by Adam Werbach.[13] All three authors agreed that we need to reduce emissions by at least 70% as soon as possible, but they felt the environmental movement was simply incapable of addressing a problem of this scale. Environmentalists (like liberals) concentrate on policy and technical fixes that are fragmentary and incremental, making the environmental movement just one more special-interest group. These authors suggest that because of the enormity of the problems, the environmental movement needs to become more generally a part of the progressive movement.

However, even progressives still share many of the same basic economic assumptions that have created the current crisis. Rather than suggesting that environmental concerns should be subsumed under the progressive movement, I think we need to go in the other direction and subsume everything under "ecology" by creating a new language and a new understanding of politics and economics in light of the threats to Earth's ability to support life. We need a new social theory based on an ecological worldview, that is, an "ecological social theory," including an ecological politics and an ecological economics. Ecological social theory is a new conceptual model for defining social interactions and choices.

This ecological social theory is already well developed, but few people recognize it. This book is a kind of primer and uses the work of many writers as well as my own thoughts to define the world we need to create in terms of this new social theory. Parallel conversations are going on in a wide range of fields, all focusing on the transformations we must make to sustain Earth. Some authors focus on our ecological crises, some critique and contrast models for organizing human affairs, some describe movements for a new democracy, and some consider ethical questions or faith issues. They all overlap and generally end up in roughly the same place. They are all describing different aspects of a new conceptual model for understanding our world and clarifying our values. My main interest is in the connections, interactions, and common threads running through all the different viewpoints on these issues. People misunderstand our ecological problems because they can never find the full range of information in one place. When it is all put in one place, it results in a coherent ecological politics interconnected with an ecological economics.

We cannot address global warming without a whole-systems approach, which creates synergies, allowing different goals to be mutually reinforcing. Without a clear, comprehensive theory of the systemic causes of ecological crises, we cannot find effective solutions. People have trouble facing the issue of global warming because they intuitively know that within our current worldview adequate solutions are not possible. But a new way of thinking creates new solutions.

Dragonfly

The ecological worldview is based on two kinds of science: complex systems theory (and systems thinking) and the science of global warming (and other environmental crises). These two give a particular way of understanding the world as well as the concrete requirements needed for us to avoid mass suicide. Global warming and its implications give this new ecological political and economic theory its profound sense of urgency.

> It's all a question of story. We are in trouble just now because we do not have a good story. We are in between stories. The old story, the account of how we fit into it, is no longer effective. Yet we have not learned the new story.
>
> Thomas Berry
> *The Dream of the Earth*

I am particularly interested in also defining these issues in terms of religion and morality. Conservatives have a much too narrow understanding of morality in their emphasis on individual actions. We need to look at systemic morality. Global warming is essentially a values issue, a moral issue, and a justice issue. In fact, how we live on Earth is *the* moral question of the twenty-first century. Any choices we make that result in the destruction of Earth's ability to support life are profoundly immoral. The issue is also religious because global warming forces us to ask fundamental questions: who are we as individuals and as a nation? What are we here for? What is the meaning of our individual and collective lives? It also forces us to ask moral questions of economics and politics.

In addition, the conversation in ecology inherently includes the discussion of the nature of power and authority, genuine democracy, justice, and equality. Ecology is a philosophy of participation and cooperation that is based on the realization of our deep interdependencies; therefore, restoring our democracy is central to the theory.

Redefining Ecological Activism

The urgency of this problem means we need to redefine environmental activism. In every community, concerned individuals and organizations are working hard to contribute solutions to this crisis. The kinds of individual behavioral changes they advocate can change consciousness and create community power and community connections.

However, we can never solve global warming with lifestyle changes. Focusing on recycling, changing light bulbs, lowering the thermostat, and driving less is important in terms of creating community and awareness, but as concrete responses to global warming, they can have little effect. In the summer of 2008, a huge storm caused the river in Cedar Rapids, Iowa, to crest 23 feet above flood level. In concrete terms, a focus on light bulbs is like people in Cedar Rapids filling sandbags the size of sugar packets. The denial of the scale of the problem creates an enormous disconnect between the severity of the problem and our individual responses to it.

Gelbspan calls this focus on limited goals, such as telling people to be more energy efficient, a form of "blaming the victim." He adds, "People are made to feel guilty if they own a gas-guzzler or live in a poorly insulated home."[14] This in turn gives rise to the eco-police, where people feel they need to judge each other. Instead, Gelbspan says, "People should be *outraged* that the government does not require automakers to sell them cars that run on clean fuels, that building codes do not reduce heating and cooling energy requirements by 70 percent, and that government energy policies do not mandate decentralized home-based or regional sources of clean energy."[15] He adds, "Activists compromise. *Nature does not.*"[16]

Activists can still address the needed changes in individual consumption patterns but within the context of ecological political and economic change; that is, we need to shift from seeing ourselves as consumers to being citizens. We need to shift from reducing the harm we do, to actively restoring Earth's natural systems. Personal change does not equal social change. People focus on incremental individual changes because they do not see an alternative form of action. We must recognize that we are actually destroying Earth. What is destroying Earth is our industrial economy—continual economic growth on a finite planet—and globalization of the economy, which focuses all economic activity on production for export, so that goods need to be shipped around the world. Unless we are able to admit and confront the real causes, we will continue on the path to mass suicide.

Creating adequate responses to climate change is not a conflict between political groups with different opinions who need to work out some compromise position. This conflict is between human beings and *nature* (and physics). There is no "middle ground" or compromise in relation to abrupt or runaway warming or the melting of Greenland or West Antarctica. They either happen or they don't—which depends on what we do and how fast we do it. Global warming is getting worse and worse, faster and faster. It is a fundamentally different problem today than it was 10 years ago or even 5 years ago. Most of the research on ice melting has been done only in the past 5 years. If we had capped the CO_2 level in the atmosphere at the level it was at in the year 2000, we would have half the warming we now have. Any delay exponentially increases the amount of warming and the seriousness of the consequences. The longer we wait to act, the higher the temperature that is locked in. *And we are running out of time.*

Social Systems Can Change Quickly

The basic issue is that we need *transformational changes*: the effort must be all-out, not incremental and not transitional. Individuals can only address incremental changes made within our current system by being better consumers. This approach is catastrophically inadequate, so we are now in a desperate race between natural tipping points and political tipping points.

Wolf

The scale of the problem requires that we shift the economy as suddenly and drastically as we did in World War II; that is, we need to move at wartime speed. Al Gore has called on us to get 100% of our electricity from renewable sources within 10 years. Lester Brown, former Director of Worldwatch, now Director of Earth Policy

Institute, says we must reduce our emissions by 80% by 2020. Only then will the scale of our response match the scale of the problem. In Brown's book, *Plan B 3.0: Mobilizing to Save Civilization,* he offers an example of the scale and speed of the needed social change by comparing it with what the U.S. was able to do after Pearl Harbor. Brown says that President Roosevelt, in his State of the Union address on January 6, 1942 (one month after Pearl Harbor), announced very ambitious production goals, saying that the U.S. would produce 60,000 planes—a goal no one thought was possible. With the production and sale of cars banned, the entire automobile industry concentrated on producing airplane engines. From the beginning of 1942 through 1944, the U.S. turned out not 60,000, but 229,600, aircraft, "a fleet so vast it is hard even today to visualize it."[17] In addition, wartime production added 5,000 ships to the 1,000 we had in 1939. Brown then adds, "*This mobilization of resources within a matter of months demonstrates that a country and, indeed, the world can restructure the economy quickly if convinced of the need to do so*."[18]

There's no guarantee that this tremendous shift will kick in before our life support systems unravel irretrievably. Actually, that very fact that there's no guarantee of success is what will draw forth our greatest courage and creativity.

Joanna Macy

Global warming needs this level of response. After Pearl Harbor, people did not organize to encourage people to reduce their use of various goods voluntarily. Instead, the federal government mandated these changes. Nothing else would have created a sufficient response, and Americans responded to the challenge and agreed to the sacrifices because they understood the necessity. We can do it again, if Americans understand the truth. This restructuring would also create thousands of new industries and could stabilize our economy.

Time for Plan B: Cutting Carbon Emissions 80 Percent by 2020 summarizes in eight pages the recommendations in Brown's book, *Plan B 3.0: Mobilizing to Save Civilization*. Cutting CO_2 emissions by 80% by 2020 will take a worldwide mobilization at wartime speed, but with the following changes, it would be possible:

- First, investing in energy efficiency will allow us to keep global energy demand from increasing.
- Then, we can cut carbon emissions by one-third by replacing fossil fuels with renewable energy sources for electricity and heat production.
- A further 14% drop comes from restructuring our transportation systems and reducing coal and oil use in industry.
- Ending net deforestation worldwide can cut CO_2 emissions by another 16%.
- Last, planting trees and managing soils to sequester carbon can absorb 17% of our current emissions.[19]

Brown also says these changes do not require new technologies; rather, what has been missing is leadership. In the U.S., our political system is paralyzed, and current economic interests are so powerful that it is possible that we are simply incapable of responding adequately and in time. Global warming is not making front-page headlines daily because the media (newspapers and TV) are all owned by multinational corporations who control, in particular, what is *not* discussed. These corporations do not want this conversation to take place. Given the economic interests and the inability to get global warming addressed in the media, perhaps human beings are simply incapable, psychologically and politically, of creating the required change in time.

We face a gathering perfect storm in which we will be blindsided with sudden catastrophic changes. In addition to global warming, there are the coming global food shortages. Food supply is not keeping up with population increases; food prices are rising worldwide and in 2008 resulted in food riots in 33 countries. In addition, water tables are dropping worldwide, topsoil is being lost at catastrophic rates, and warming is reducing yield. (For every degree Celsius of warming over the optimum, photosynthesis slows down by 10%.) At the same time, the population keeps increasing. Between 2009 and 2020 we will have added the equivalent of another China to the world population, all of it in developing countries already under severe stress. All of these crises are interconnected, and they each make the others worse. On the other hand, addressing them can create positive interactions, improving all at once. But we are running out of time.

As people come to understand the scale of the changes needed, they will tend to despair and feel that this level of change will just never happen. However, I'd like to point out two things. First, the world we need to create will be a much better world. Second, this level of change is quite possible. We are at a point where massive positive change can happen, and to create it, we don't need everyone; we only need a critical mass. But it begins with a transformation of human consciousness.

Consciousness can cause enormous change: just a shift in what people consider legitimate or in how they view reality can be transformative. Massive positive change can happen if we remember that human potential is extraordinary. In creating a vision of massive change, Paul Ray and Sherry Anderson in *Cultural Creatives* say that enormous numbers of people share these concerns and want to create solutions—in fact, they say we are probably a majority, but we do not recognize each other because there are no "media mirrors."

Massive change is definitely possible. Just as climate systems can change abruptly, social systems can also change suddenly, for better or for worse, as our financial meltdown demonstrates. Any system can hit a threshold, causing the system to suddenly reorganize and jump to an entirely new level—one that cannot revert to the previous level. For example, after apartheid began, Black (and some White) South Africans maintained a tenacious hold on the shared vision of freedom and equality, with little effect

Goldfinch

on the power system. Then, after Nelson Mandela spent 27 years in prison, the White South African power structure agreed to release him. This event happened without an armed revolution, and it forced the entire social system over a threshold. Only *four* years later, official apartheid was over and Mandela was not only *elected* president of South Africa, but the former government voluntarily gave up power. Communism in the Soviet Union crossed a threshold, and the entire country simply imploded. After years of unsuccessful worldwide pressure, the Berlin Wall seemed to collapse overnight. There is no way to predict the timing of any of these tipping points. However, when we think in systems terms, we can see how the cascading feedback loops can suddenly alter an entire system. Coercive power (a revolution) is not necessary for extreme changes to happen.

The Ecological Crisis Is a Religious Crisis

How we live on our planet Earth is unavoidably a religious question. Choosing the values of life and mutual care and overcoming materialism require respect for the mystery in human life and a strong ethical commitment. The issue is religious because the basic questions are, What is our personal and collective purpose in life? What does it mean to be human? What does it mean to be caretakers of Earth?

> In these turbulent and often frightening times, it is important to remind ourselves that we are privileged to live at the most exciting moment in the whole of the human experience.
>
> David Korten

Religion needs to enter the public political debate, but the God we proclaim in a democracy must be capable of working with all religions. Otherwise, we are not making religious statements about meaning and reality but political statements about power and dominance. My concept of ecological social theory focuses on deepening the sense of Earth as sacred, an idea that can both include and transcend all religions. When all religions advocate ecological issues jointly, they become more effective in creating transformation without being seen as politically coercive.

An added problem is that many people see ecology and global warming as "liberal" issues. Therefore, it is a significant problem that liberals resist any use of religious language (even though most of the liberals I know are deeply religious—whether or not they are connected to a church, temple, or mosque). I describe a religious understanding that I feel is necessary for creating change and one that I believe most liberals can understand and embrace.

The primary task of individuals is an inner transformation to being grounded in a spiritual reality, whether that is through organized religion, a personal spirituality, or a practice like yoga. A new spiritual consciousness is already developing around the world.

Although the ecological theory I envision requires a religious dimension, I am one of the millions who over the past decades have left the mainstream church. I left the church because I felt that religion no longer *mattered*, I felt it was irrelevant and boring, and I do not believe that God requires belief in Christ. So what is the point? However, when I wanted to study the religious aspects of the crisis, a seminary accepted me anyway. Once there, I was amazed at how exciting and open the religious conversations were. The seminary did not require acceptance of rigid beliefs, and reading theologians like Marcus Borg and others reconnected me to Christianity.

The social and economic transformations required to live in harmony with Earth will be very difficult. I believe the shift will only be possible if it becomes an interfaith religious movement. Essentially, the needed change in consciousness is a spiritual quest. The threat is to the survival of human beings on Earth. Creating the world anew will again make religion *matter*. It will again be about what is of ultimate importance, as it should be.

To be truly radical is to make hope possible, not despair convincing.

Raymond Williams

Many people are interested in connecting their faith with care for creation. I believe the church (and synagogue, or mosque) could be decisive in creating a world that cares for creation and each other, and I believe there is no substitute for the church's central role. Therefore, one of my goals is to create a religious vision that could encourage some of those who have left the church to return. They can then help those still there to create a church whose primary mission and identity is to restore Earth.

Creating a Better Future

Finally, there *are* alternatives to our current systems, and a better future is possible. Although it is too late to avoid many of the consequences of global warming, it is *not* too late to avoid the most extreme consequences. A great deal of positive potential is embedded in the climate crisis. The new world we need to create will be a much better world, and the more one knows, the more exciting the opportunities are and the less overwhelming the uncertainties.

To solve any problem, one must first correctly define it. Major political fights are never over solutions, they are always over the definition of the problem. Depending on how you define the problem, the solutions follow naturally. Or, by suggesting certain solutions, a definition is implicit but unconscious. We cannot make adequate changes until we address the problem as it really is, in all its dimensions. Progressives tend to define problems in terms of their preferred solutions: organizing responses by individuals. This is focusing on small-scale solutions, which, in the case of global warming, means you end up never addressing the real problem. Obviously, it is always good for people, and cities, to become more energy efficient. But at that scale, it has only minimal impact on global warming.

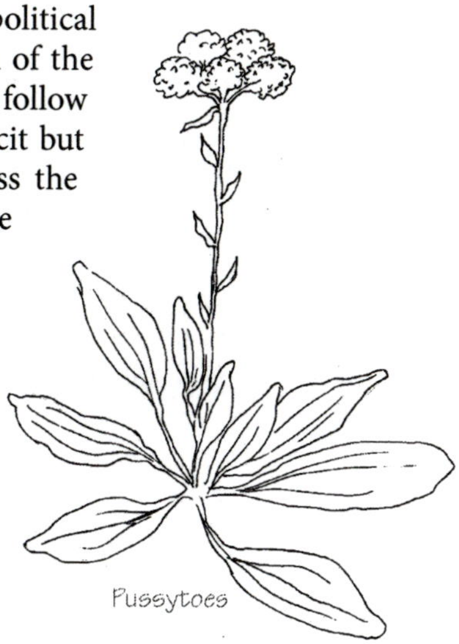
Pussytoes

Yet understanding the changes we need creates a message of hope. The problems now facing us are an unprecedented opportunity. I believe we are capable of making a profoundly positive shift in our world. But because the current mindset dominates our thinking, it is difficult for people to envision alternatives. In order to build the new world, we must first be able to imagine it, which is the goal of this book.

The Organization of the Book

This book goes together like a puzzle—one piece at a time. People have asked me who my audience is, but because I am a generalist, I've never had a single audience in mind. I have a single vision that could be of interest to many different audiences. This book is meant for anyone willing to put the effort into following a complex argument, whether that is individually or in a course on ecology or climate change, or in a seminary course. The book is divided into five parts.

Part I Earth in Crisis

Chapter 1, Causes and Consequences of Global Warming. This chapter gives the background needed to understand global warming and sets the context for the rest of the book. The issue of global warming is inclusive of other ecology problems such as water shortages, biodiversity, soil loss, and public health. This discussion includes the many different feedback loops that are increasing the rate of warming. It also includes a review of the carefully organized, deliberate PR effort to discredit the reality of global warming.

Part II Causes of the Climate Crisis

Chapter 2, Views of Science that Contribute to the Crisis. Much of our understanding of people and the world is created by using metaphors based on our understanding of science. Currently, our worldview is based on mechanistic Newtonian science; however, the new complex systems science allows for new metaphors that create a new, more accurate, way of thinking about the world.

Chapter 3, Views of Society that Contribute to the Crisis. This chapter discusses two worldviews and the two types of politics that result. These two worldviews matter because the social transformations we need are available in only one of them. The chapter uses different ways to contrast these two worldviews. George Lakoff calls the two models an authority-based model and an empathy-based model; David Korten calls them Empire and Earth Community. Finally, I discuss Jeremy Rifkin and his history of empathy.

Chapter 4, Views of Religion that Contribute to the Crisis. This chapter sketches the two worldviews as they are reflected in two understandings of religion. Religion is a key shaper of worldviews, so we need to be very aware of the worldview our religion reflects and sustains. Over time, two contradictory images of God have developed. One could be called a Religion of Earth Community, a religion of compassion, grace, and Spirit, which is the religion Jesus preached. The other is the Religion of Empire, which is imperial, arbitrary, judgmental, and jealous of its power and legitimizes conquest and imperialism.

In a democracy, we need to be conscious of creating legitimate ways to use religious values and language in debating social policy. Democracy depends on holding beliefs openly and working through dialogue for new ideas and solutions. Many liberals believe that religion is inherently coercive and so tend to react negatively to any religious language in democratic debate; however, any belief can be coercive and absolutist, not just religious beliefs. Understanding the nature of beliefs will help us create a democratic debate and still use religious values.

Chapter 5, Views of Economics that Contribute to the Crisis. This chapter discusses our current understanding of economics. The structure of our global economy creates the dynamics and patterns of interaction that have resulted in so much harm to Earth. These same dynamics also create an enormous gap between rich and poor and cause us to lose any genuine democratic control over our lives. Our current economic system has become predatory, ruthlessly competitive, and immoral, so that the drive for profit overrides any concern for either people or Earth.

Chapter 6, Views of Political Power that Contribute to the Crisis. This chapter discusses the ways in which multinational corporations now control the global economy and the American political system, manipulating the decision-making processes so that the powerful corporations can get whatever they want.

Part III Creating Solutions to Global Warming

Chapter 7, Creating a Cooperative World Based on Partnership. This chapter contrasts the type of morality in the two worldviews. One is a morality of authority and obedience and a focus on narrow issues of individual morality, or what has been called "the sufficiency of private virtue." The other is a morality of empathy and compassion, of systemic morality and mutual responsibility and recognition that we are all in this together. Ecological morality would go further and address the whole, the entire world. Lester Brown gives two budgets, the annual amount needed to eradicate poverty worldwide (which will also reduce population) and to restore Earth's natural systems. To fund these two budgets, we must rethink our military budgets.

Chapter 8, Ecological Economics: Putting Earth First. This chapter discusses the field of ecological economics begun by Herman Daly. The basic concept in ecological economics is a redefinition of the "whole." Current economics sees the whole in terms of the global economy. In contrast, ecological economics defines the whole as the natural world, on which we and our economies are profoundly dependent. In this understanding, economics operates as a subset within the limits of nature. The fundamental difference between the two is being able to recognize and accept limits.

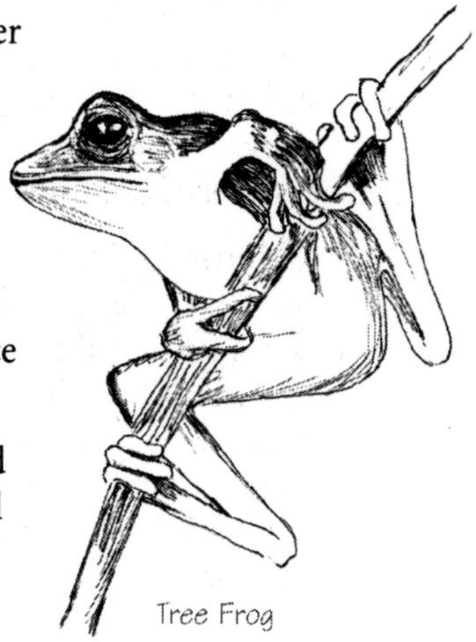

Tree Frog

Daly says that as the world becomes full of us and our stuff, we need to replace our traditional "empty world" economics with a "full world" economics. This questions both globalization and continual economic growth. The fundamental problem is that as the economy grows, it gets bigger—using more resources and creating more waste. However, Earth is as big as it's going to get, and we have already seriously overshot its limits.

Chapter 9, Redesigning Our Banking and Money Systems. This chapter focuses on money and banking. New money in our economy is created by someone going into debt; this has enormous consequences. Our money system does not need to be based on debt, for this is what creates our deep belief in the inevitability of scarcity.

One of the deepest desires underlying shared vision is the desire to be connected, to a larger purpose and to one another.

Peter Senge

Chapter 10, Reconnecting to Spirit, Nature, and the Feminine. This chapter discusses the importance of spirituality because religion is about relationships—in this case, our ongoing relationship with God—rather than a focus on static belief systems. It also discusses our need for a new theology that can again see the natural world as sacred and that can reintroduce the feminine to our images of the divine.

Chapter 11, Sustainability and Reinventing the Church. This chapter discusses the church. I believe the church has become irrelevant to the world, even as it is still essential. We need to reinvent the church, for I see the salvation of the church and the salvation of the planet as necessarily linked.

Part IV Specific Solutions

Chapter 12, Sustainability in Energy Use. This chapter describes how a sustainable energy system would work. The past 20 years of sabotage and procrastination have left us with only one chance: an all-out push to stop the use of fossil fuels. Suggestions considered in the past are now hopelessly out of date. The enormous changes start with conservation and increased energy efficiency. In fact, we can meet *all* our future increases in energy demand by increasing efficiency, and then we need to shift to renewable energy. The fight over energy production is not just over renewable versus nonrenewable energy, it is also over centralization versus decentralization. Rather than use massive power plants, we can create most of our energy by micropower, which is far more resilient and eliminates distribution costs and vulnerabilities.

Chapter 13, Sustainability in Our Food System. This chapter sees our food system as a large contributor to global warming. Raising livestock creates 18% of greenhouse gas emissions, and the food system as a whole creates at least 37%.[20] Therefore, reforming our entire food system could be a major part of the solution to global warming. In addition, our industrial agriculture system is fundamentally unsustainable. Instead, we need to move toward local food economies, stop the use of fertilizers and pesticides that are poisoning our soil, air, and water, and stop monocropping—planting thousands of acres of only one crop.

Chapter 14, Sustainability in Our Use of Materials. This chapter describes how we can drastically reduce the amount of energy and materials used in industry. Industry can accomplish steep reductions by using a process called Zero Waste, in which manufacturers design the production system rather than the product, so that everything is reused in a closed loop. They can also sell a product as a service that is leased and maintained by the company that produces it. This creates incentives to produce goods for durability. And finally, anything that cannot be reused or recycled must be biodegradable.

Part V Conclusion

Chapter 15, Facilitating Self-Organization for Social Transformation. This chapter discusses alternative ways of creating social change. Understanding how our social systems have caused global warming is very learning-intensive, and many democratic processes are available to help people create the change we need. Change in complex systems involves flooding the system with new information (learning) and increasing the number of connections so that the information more easily cascades throughout the system.

Chapter 16, A Shared Moral Vision Can Transform the World. This chapter briefly reviews where we have been and where we need to go.

There are enormous forces of denial aligned against this necessary shift in vision and analytic effort, and to overcome these forces requires a deep philosophical clarification, even religious renewal.

Herman Daly
Beyond Growth

How I Came to Write this Book

I have always been interested in the power of information. That led to a Master's degree (M.L.S.) in library science and a career in libraries. I was director of two different public libraries, both in the suburbs of Chicago. Before that, I worked at the American Library Association headquarters in Chicago. I stopped working to raise a family and over time became intensely interested in our ecological crises. I wanted to be at home to "give my children a good start," but I began to wonder what value a "good start" would be if they would inherit a world in ecological and economic collapse.

After years of a broad, informal education, I entered an M.A. program in political science. From the beginning, I formulated a question that was to define the rest of my work: "What kind of a society must we create to adequately respond to our ecological crises?" I soon could see that political science helped explain the source of the problem but wasn't a source of solutions. I felt that without a radical transformation in values and meaning, an adequate solution would not be possible.

Asking how to transform values and create meaning is a religious question. Therefore, to continue my work using a religious vocabulary, I entered Garrett-Evangelical Theological Seminary, a Methodist Seminary, for an M.A. in theology, again primarily addressing issues in ecology and sustainability. I have also worked

Ostrich

on the Web of Creation (http://www.webofcreation.org), a website on religion and ecology, at the Lutheran School of Theology in Chicago. Three M.A.s and more than 15 years of researching and writing went into this book.

I wrote this book because I had to; it has become an obsession. I've struggled with these ideas for years because I wanted to build up a reasonably complete vision of sustainability for myself and then share it with others. I hope you find this vision compelling and that it moves you to action. I am a generalist, and I can't understand the details unless I can first see the whole. Most books focus on one topic, which makes knowledge highly fragmented and compartmentalized. And that is the fundamental problem: we need a coherent description of the whole and how all the various factors are interconnected.

Throughout this book, I capitalize the word "Earth." I did so at first because I heard that most Native Americans capitalize the word. Then I realized that the English language capitalizes the names of all the other planets: Venus, Jupiter, Mars, yet we still use earth. I think it's a sign of how much we unconsciously discount the importance of our own planet. I also do not say *the* Earth, but rather Earth. We do not say *the* Jupiter, or *the* Mars. To add *the* to *Earth* is another way to distance ourselves and deny our fundamental connection to, and absolute dependence on, Earth.

Zebra

Part I

Earth in Crisis

Polar Bear Family

Chapter 1

The Catastrophe of Global Warming

Climate change is not just another special-interest issue; it is an immediate global crisis of enormous proportions. Global warming has been called "a catastrophe in slow motion;" others have said that there is "an ecological time bomb ticking away" or that we are "sleepwalking into an Apocalypse." Addressing global warming includes other very serious ecological crises—extinction of species, water shortages, new diseases, and pollution of land, water, and soil. Global warming will make all these other problems worse. However, the same solutions address all of them.

> It is an excruciating experience to watch the planet fall apart piece by piece in the face of persistent and pathological denial.
>
> Ross Gelbspan
> *Boiling Point*

The magnitude and urgency of the global warming crisis have been obscured by incomplete information, reports from the media that minimize the situation, and intentional misinformation from corporate and political sources. In particular, there has been a 20-year well-orchestrated, well-funded PR campaign to discredit the science, which I will discuss at the end of the chapter. But first, I will give an overview of the causes and consequences of global warming.

The truth is that these problems are scary and the information can be overwhelming. However, learning the whole truth about what we are doing to Earth's systems is the only way to ground us in reality. Yet, when ecologists present and explain the evidence of global warming, many people dismiss them as "alarmist" or "extremist." Ecologists are alarmists because they are desperately trying to convince people how truly alarming and extreme the ecological crises are.

However, do not lose hope. Very real solutions are already available; we just have to choose to use them.

Background for Understanding Climate Change

Understanding global warming requires some general background. To begin, heat energy from the sun drives Earth's climate when the heat interacts with the oceans and the atmosphere. The sun is the source of all the energy for Earth's climate system, but this heat energy arrives mostly around the equator. This creates circulation patterns in the oceans and atmosphere that transport heat north and south from the equator. Heat is also exchanged between the atmosphere and the ocean's surface, although they retain the sun's heat at different rates.[1]

The oceans cover 71% of Earth's surface and contain 1,100 times more heat than the atmosphere; however, the atmosphere moves heat around more quickly than the oceans do, so that in the end, they each transport nearly equal amounts of the heat that drives Earth's climate. In redistributing heat around the planet, the ocean currents establish the underlying conditions that lead to storms, hurricanes, severe weather, monsoons, El Niños, and rainfall patterns. The differences in temperature and salt content create the circulation patterns in the oceans, while the differences in temperature and air pressure drive the circulation patterns in the atmosphere.

Sunlight (and its heat content) passes through the atmosphere from space and warms Earth's land and ocean surfaces before most of it bounces back into space. However, CO_2, methane, and nitrous oxide molecules in the atmosphere absorb heat and trap it so that it cannot go back out into space. By trapping some of the sun's energy, Earth's temperature is kept constant. The mechanism of greenhouse warming is what makes the atmosphere of our planet livable. In contrast, the moon has no atmosphere and so cannot have a greenhouse effect, so the moon is 33°C (91°F) colder than Earth. Venus, on the other hand, has massive amounts of CO_2 in its atmosphere, which has caused a runaway greenhouse effect. As a result, its surface is so hot 480°C (900°F) that it could melt lead and it gives off a dull red glow.

The science is basic, simple, and well understood; there is nothing controversial about it. Global warming is not a new idea. Svante Arrhenius first described the basic process in 1896. He predicted that a doubling of atmospheric CO_2 concentrations would raise the global mean temperature by approximately 5°C and that this warming would be strongest in the upper latitudes. Many of Arrhenius' predictions are close to what is now being seen.

Since the 1970s, scientists have known that the CO_2 in our atmosphere was increasing because they have kept records of it. It is simply a fact that the amount of CO_2 in the atmosphere is increasing and that because of this increase, the atmosphere is warming. However, there were two basic unknowns. First, scientists didn't know how much or how fast it would warm. Second, they weren't sure what the primary cause was. Over the past two decades, scientists have found that the warming is very fast and that if it goes unchecked it could make the planet uninhabitable. They also demonstrated that the buildup is the result of human actions, primarily our use of fossil fuels. In fact, James Baker,

Emu

head of the U.S. National Oceanic and Atmospheric Administration (NOAA), said of the reality of global warming, "There's no better scientific consensus on any other issue I know—except perhaps Newton's second law of dynamics."[2]

Weather versus Climate

Next, it is important to clarify the difference between weather and climate (as in climate change). Global warming is about the shifting of long-term averages—it is not about today's weather. Weather is short term and local. Weather varies enormously, and it is extremely difficult to predict. Weather is driven by millions of variables, all constantly interacting, creating cascading changes. To try to estimate where all the variables will line up at a particular place and a particular time is extremely difficult.

However, climate change is about long-term averages. In studying climate change, scientists assign each year *one* number. Each number is:

- *Global*—daily temperatures are averaged for the entire globe (so that temperatures at both poles, the equator, and everywhere in between are averaged); and
- *Annual*—daily global average temperatures are averaged over a full year (so that all four seasons are averaged).

> The climate debate is a public issue in which the assets at risk are not specific resources, like oil, fish, or timber, but a life-supporting system.
>
> Paul Hawken, Amory Lovins, and L. Hunter Lovins
> *Natural Capitalism*

Over the course of a full year, over the whole globe, normal regional variations (even unusual ones) average out. The global annual temperature had remained quite stable for millennia. At the turn of the century in 1901, the average global temperature was 13.88°C (56.984°F). Since then, there has been an increase of 0.85°C (1.53°F), and 80% of this gain came over the past two decades.[3] These numbers can get confusing. Scientists use degrees Centigrade (Celsius, °C), but most Americans use degrees Fahrenheit (°F). Newspaper reports in the U.S. sometimes use Fahrenheit and sometimes Centigrade, and sometimes they don't say which they are using.[4]

In addition to tracking current temperatures, scientists also study historical temperatures (paleoclimatology). They do this through a variety of methods, but the most common one involves drilling ice cores deep in the ice in both Antarctica and Greenland (where the ice is 2 miles thick). Over millennia, the passing of each year creates a visible record in the ice, much like in the rings of a tree. Gases are trapped in air bubbles frozen in the ice, and when scientists analyze these gases, they can determine what the atmosphere was like at a particular point in time. People have argued that the warming is just natural variability, but this is exactly what science has been trying to determine, in particular, whether it could be caused by variations in solar output. They have ruled out sun variations and other natural causes.

Because global warming is based on long-term averages, climate scientists never address weather. This means they will never say that a drought in California or the strength of Hurricane Katrina is the result of global warming. These events are weather; they occur in one particular place at one particular time. What climate scientists will say is that global warming will cause stronger hurricanes and more serious droughts.

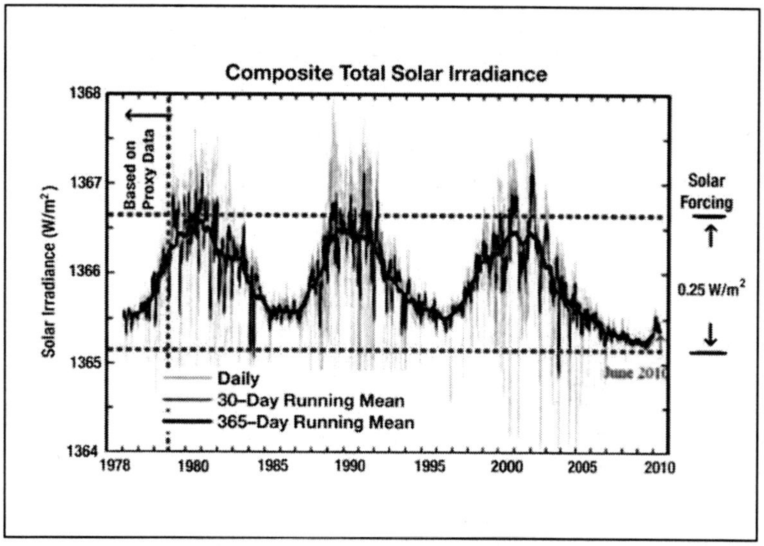

The Decade 2000–2009 Was Hottest on Record

Because 2008 was relatively cool, skeptics in the U.S. argued that global warming has stopped, or even that in fact Earth was cooling! It is true that there were relatively cool temperatures in parts of North America. However, the contiguous 48 states cover only 1.5% of the world area (oceans cover 71%), so temperatures in the U.S. are averaged with the other 98.5% of the world.

The first 6 months of 2010 were the hottest on record, and this was in spite the sun cycle's being at its coolest, called "minimum of solar irradiance."[7] The sun has a cycle of about 11 years during which it gets warmer and cooler. We are now at the coolest part of the sun cycle, in fact, the longest cool phase in a century, yet we are still having the hottest temperatures. James Hansen now says 2012 is likely to set a new record high because the solar irradiance will climb out of the recent long-lasting solar minimum, as seen in the first chart.[8]

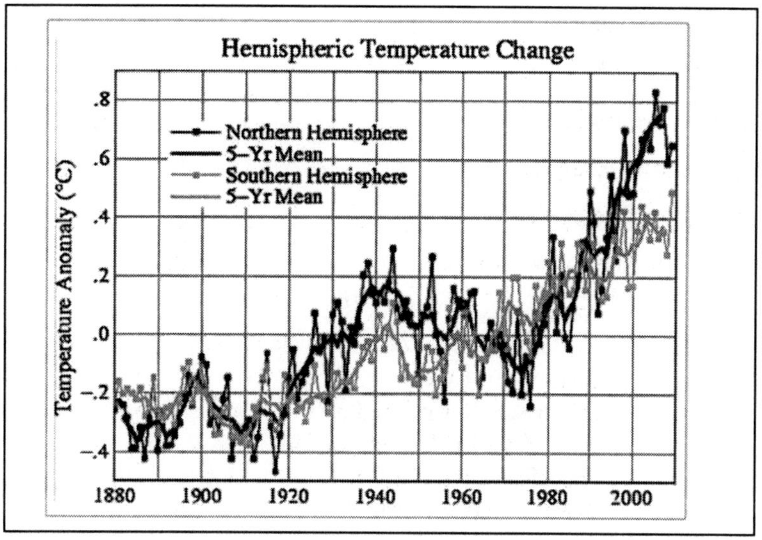

The second chart gives the annual mean temperature change for each hemisphere. Annual and five-year running mean temperature changes relative to the base period 1951–1980 for the northern and southern hemispheres.[9] Before the industrial age, these gases had remained at near stable concentrations for thousands of years. In total, average global temperatures have increased by about 0.8°C (1.5°F) since 1880.

The chart shows a leveling off of temperatures during the 1950s and 1960s. Human activities have increased CO_2 and also have caused increased concentrations of fine reflective particles, or "aerosols," in the atmosphere. This leveling off was the result of the industrialization of World War II and after, which created a great deal of air pollution and aerosols from fossil fuels and other sources, which helped keep the planet cooler. The eruption of Mt. Agung in 1963 also put large quantities of reflective dust into the upper atmosphere. The rapid warming observed since the 1970s has occurred in a period when the increase in greenhouse gases has dominated all other factors.[10]

The United Nations Intergovernmental Panel on Climate Change

The global group coordinating the reporting of the scientific work on climate change is the United Nations Intergovernmental Panel on Climate Change (IPCC), jointly established in 1988 by the World Meteorological Organization and the United Nations Environment Programme (UNEP). This was a year of record heat, burning forests, and severe droughts in the midwestern U.S. The IPCC was formed to provide regular reviews of climate science to member governments in order to inform their governments' policy choices. The IPCC does not conduct research itself; rather, it organizes peer reviews of research done by member scientists and issues summaries of their findings. All these scientists volunteer their time, and their findings represent the largest and most rigorously peer-reviewed scientific collaboration in history. About 2,000 experts have participated in the review process, and approximately 1,000 experts from all over the world have been directly involved in drafting, revising, and finalizing the IPCC reports. This includes climate scientists from 100 countries around the world (50% are from the U.S.).

Governments are in complete denial. They can't deal with [Global Warming]. It's too big, too horrendous. They're terrified of the kind of dialogue they would have to have with the public.

Bill Leiss
Researcher and Professor
of Risk Communications

Since it was established, the IPCC has published four summaries of climate research—in 1990, 1995, 2001, and 2007. The 2007 report is called the Fourth Assessment Report (AR4). The 1995 report established that global warming is real and that humans are causing it, and it was enough to persuade the Europeans and the Japanese to begin work on the Kyoto Protocol. Since then, there has been much more research confirming these conclusions.

The IPCC is organized into three working groups and a task force on national greenhouse gas inventories. Each of the working groups writes a Summary for Policy Makers (usually about 35 pages long), and later they release the full report (often 1,000 pages long). The three working groups are:

- Working Group I—Climate Change 2007: The Physical Science Basis;
- Working Group II—Climate Change 2007: Climate Change Impacts, Adaptation, and Vulnerability; and
- Working Group III—Climate Change 2007: Mitigation of Climate Change.[11]

Each of these working groups has two co-chairs (one from the developed and one from the developing world) and a technical support unit. The assessment summary reports are a state-of-the-art report on the science, written so that they are comprehensive yet able to be understood by non-specialists. These three reports give an overall blueprint that shapes the way the world tackles global warming. The working groups also issue a fourth report that synthesizes the results of all three working group reports. These reports have demonstrated an international scientific consensus that climate change is happening and that human beings are responsible for it.[12]

Global Top Ten Warmest Years

January – December

Year	anomaly °C	anomaly °F
2010	0.62	1.12
2005	0.62	1.12
1998	0.60	1.08
2003	0.58	1.04
2002	0.58	1.04
2009	0.56	1.04
2007	0.55	0.99
2004	0.54	0.97
2001	0.52	0.94

National Oceanic and Atmospheric Administration

NOAA said, "The decadal global land and ocean average temperature anomaly for 2001–2010 was the warmest decade on record for the globe, with a surface global temperature of 0.56°C (1.01°F) above the 20th century average. This surpassed the previous decadal record (1991–2000) value of 0.36°C (0.65°F)."

Working Group I addresses the state of the science in the IPCC Fourth Assessment Report and concludes, "Warming of the climate system is *unequivocal*."[13] These reports use probabilities or percentages to explain their findings, such as unlikely, likely, very likely, etc. The report says, "Most of the observed increase in globally averaged temperatures since the mid-20th century is very likely [this phrase means a 90%–99% chance] due to the observed increase in anthropogenic [caused by humans] greenhouse gas concentrations."[14] The working group is clear that it is human activities, particularly the use of fossil fuels, that have caused global warming.

In other conclusions, Working Group I said, "The atmospheric concentration of carbon dioxide in 2005 exceeds by far the natural range over the last 650,000 years (180 to

300 ppm—parts per million) as determined from ice cores."[15] The global atmospheric concentration of CO_2 has increased from a pre-industrial value of about 280 to 380 ppm.[16] The report also says that scientists have ruled out the alternative explanations for warming such as sunspot activity, varying solar output, or the "urban heat island effect" (the raising of temperatures in cities caused by high building densities and the use of heat-retaining materials such as concrete and asphalt).

The 2007 report of Working Group II presents projections of impacts on policy-relevant sectors, including water, ecosystems, food, coastal systems, industry, settlement and society, and health. It also has sections on major regions, with summaries of the impacts on Africa, Asia, Australia and New Zealand, Europe, Latin America, North America, the polar regions, and small islands. Africa, Asia, the polar regions, and small islands are the most vulnerable regions either because of the degree of climate change projected (polar regions) or because of the vulnerability and lack of resilience of natural and human systems (Africa, Asia, and small islands). The report says that Africa and Asia have the most vulnerable populations. Within the next few decades, these areas will have severely decreased water availability and significant crop reductions. The impacted populations will number in the hundreds of millions.[17] The second and third working group reports are written by social scientists.

> We are taking the reins of the geochemical cycles of the Earth. It's really frightening.
>
> David Archer, an expert in global carbon cycles, University of Chicago, "Alarm over Acidifying Oceans"

The 2007 report of Working Group III focuses on the technological, environmental, economic, and social aspects of mitigation of climate change and on strategies and costs for cutting emissions of the warming gases. The report also makes clear the risks of delay, noting that emissions of greenhouse gases have risen by 70% since 1970 and that if we do not change our behavior, the gases could rise by an additional 90% by 2030. They say that governments are running out of time. This report also makes it clear how much global warming is a matter of equity and demonstrates the degree to which the countries that have contributed the least to the warming problem will be hit the hardest. This disparity leaves the U.S. with an enormous moral problem and eventually perhaps a legal problem. As of early 2009, CO_2 was at 384.9 ppm. During 2008, emissions rose by 2.2 ppm, compared with a rise of 1.8 ppm in 2007.

In reading these reports or articles about them, the terms and numbers they use can be confusing. They use two different numbers. One is CO_2 emissions; the other is total greenhouse gases (sometimes called GHGs), which is CO_2 levels plus methane and nitrous oxide and is called CO_2-eq or CO_2-e (for carbon dioxide equivalent). At the time of the report, CO_2 levels were at 380 ppm, and CO_2-eq was 459 ppm (using the IPCC formula).[18] Newspaper reports may not say which they are using, so some will say emissions are at 380 ppm, and others say 459 ppm.

To keep the warming at 2°C, we must keep CO_2 concentrations at 350 to 400 ppm or keep CO_2-eq (carbon dioxide equivalent) concentrations at 445 to 490 ppm. This will keep the warming to between 2°C and 2.4°C over pre-industrial levels.[19] The report says that even the 2°C warming would subject up to 2 billion people to water shortages by 2050 and threaten the existence of 30% of the world's species.

The final 2007 report, the *Synthesis Report of the IPCC*, summarizes the key findings of the other three volumes and was released in November 2007. U.N. Secretary General Ban Ki-moon described climate change as "the defining challenge of our age." This report created new emphasis and alarm. It says we face abrupt and irreversible impacts and that the window for action is closing. Climate change is occurring faster than the worst-case scenarios of 5 or 6 years ago.[20]

The full reports are published as they were written. However, the summaries are reports by the U.N. to its member countries. For the summaries, scientists first have to agree on a draft of the summary for their working group; then, government representatives from the member countries meet to "edit" the reports. The politicians meet with the scientists and try to cut any language that will threaten their interests. Government politicians edit just the summaries for policymakers; they do not edit the full reports. Conservatives claim that these reports exaggerate the science, but the opposite is true. Scientists spend months summarizing the scientific evidence, but in the end, the summaries release only what receives consensus; therefore, the reports end up with the lowest-common-denominator conclusions.

In addition, the IPCC charter does not allow the scientists to say how dangerous the warming is; the organizers of the IPCC felt that was a political decision. This leads to confusion because the public assumes the IPCC is there to warn us of the danger. For that reason, many scientists say it may be time that the panel's mission and procedures were revised, both because the science is evolving so fast and to limit potential criticisms relating to the behind-the-scenes fights over the language in its reports. Some scientists are suggesting that the very search for consensus may now be distracting us from the need for action.[21]

Many people are questioning whether the IPCC has become irrelevant. Joe Romm of Climate Progress (along with others) has said the Fourth Assessment Report should have been sufficient to jump-start serious action. But it ended up being out of date as soon as it was published for several reasons:

- The IPCC stops taking scientific input a year or two before the year in which the report comes out, and the science is moving fast.
- Most of the climate models had not yet incorporated many if any of the amplifying carbon cycle feedbacks, like the saturation of the sinks or the melting of the permafrost.
- It was published too soon to capture the China-driven explosion in emissions, so its emissions scenarios were completely out of date.

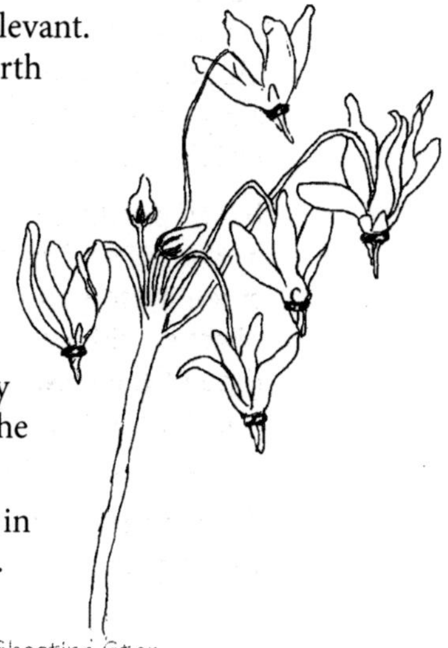

Shooting Star

- The consensus-based process in which every member government signs off on every word in the synthesis reports leads to a least-common-denominator set of statements that further waters down the science.[22]

In particular, the science on ice melting is very recent, and little was included in the Fourth Assessment Report. The IPCC has said it is beginning work on the next report, due out in 2014, which is far too late to be of much use—the science is progressing so quickly and the warming is getting so much worse that the next report, to be of any use, needs to be much sooner. On the other hand, Romm says perhaps the IPCC is no longer useful: what good is yet another report saying that global warming is unequivocal, humans are responsible, and we're headed for extreme, catastrophic changes on Earth. He says the National Academy of Sciences should write a major report and maybe that would get more attention in the U.S.

Kyoto Protocol

Governments first agreed to tackle climate change at the United Nations Framework Convention on Climate Change, informally known as the "Earth Summit," in Rio de Janeiro in 1992. The treaty had non-binding targets but provided for updates (called "protocols") that would set mandatory emission limits. The Kyoto Protocol is the follow-up pact agreed to by governments at a 1997 U.N. conference in Kyoto, Japan. Kyoto is the first binding global agreement to cut greenhouse gases, and the IPCC is the official scientific arm of the Kyoto Protocol.

> The human species is playing dice with the planet, without knowing all the rules of the game.
>
> J. R. McNeill
> *Something New Under the Sun*

Kyoto has had legal force since February 16, 2005, when the minimum number of countries ratified it; 187 nation-states have signed and ratified the Kyoto Protocol. On December 3, 2007, Australia signed the Kyoto Protocol, making the U.S. the only developed country that has not signed the agreement. The Kyoto Protocol included only the 36 developed countries that signed under the principle that richer countries are most to blame and did not address reductions in developing countries.

Countries that ratified the Protocol agreed to reduce greenhouse gases emitted by developed countries to at least 5% below 1990 levels by 2012. The Kyoto agreement called for the U.S. to reduce its emissions by 7%. The targets range from an 8% cut for the European Union (EU) from 1990 levels to a 10% rise for Iceland. U.S. emissions have increased by 25% since 1990; therefore, if we were to sign, we would be required to reduce emissions by 32%. The agreement includes reducing emissions of CO_2, methane, and nitrous oxide.

The U.S. did not sign the Kyoto Protocol. The Bush administration said that Kyoto is too expensive and wrongly omits developing nations from a first round of targets through 2012. They used the fact that the Protocol did not require China to reduce its emissions in this first round as an excuse not to sign it. However, the U.S. has 5% of the world's population and creates at least 25% of the omissions. China has 20% of the world's population and although China has recently caught up to the U.S. in total emissions, they have four times as many people. In addition, a good deal of their emissions result from their manufacturing many of the goods that Americans buy. In any case, the gases causing climate change are cumulative emissions, not current emissions. The industrial revolution—the past 100 years—caused the global warming we have now. The cumulative emissions of the U.S. are more than three times that of any other nation.

Unrestricted emissions of greenhouse gases threaten multiple catastrophes, any one of which justifies action. Together, they represent the gravest threat to humanity imaginable.

Joe Romm
Climateprogress.org

However, China has realized that it can no longer strive for economic growth without taking into account environmental harm. The Chinese government admitted in a January 2007 report that global warming will dramatically affect China's ability to feed its people, for it will cause China's production of wheat, corn, and rice to drop by as much as 37% over the next 50 years.[23] Therefore, China has made a commitment to become a world leader in renewable energy.

Copenhagen Accord

The Copenhagen climate summit was convened to develop a comprehensive international response to the threat of global warming with an agreement to succeed the Kyoto Protocol when it expires in 2012. The meeting was called the COP-15, or the fifteenth 'Conference of the Parties' (or COP) under the United Nations Framework Convention on Climate Change. It met in Copenhagen, Denmark, for two weeks in December 2009. The COP is the highest body of the United Nations Framework Convention on Climate Change and consists of environment ministers from about 190 countries. This second agreement includes developing nations.

The Copenhagen Meeting finished with little accomplished. Some faulted President Obama because he had the power to change the entire debate. Naomi Klein says that if he had come with a "transformative and inspiring commitment to get the US economy off fossil fuels, all the other major emitters would have stepped up. The EU, Japan, China and India had all indicated that they were willing to increase their levels of commitment, but only if the US took the lead." But instead, Obama came with embarrassingly low targets, and the other countries took their cue from him.[24]

After two weeks of negotiations, the only advance was on the issue of financing to help developing countries cope with climate change. Industrialized nations said they would raise $10 billion per year for three years in "fast-start" funding and long-term funds of $100 billion per year by 2020. But the document doesn't say where the annual sum of $100 billion will come from or how the money will be distributed.[25] Another issue was that China did not want independent verification of their

Magellanic Penguins

emission cuts, and the final deal came down to a behind-closed-doors conversation in which they changed "examination and assessment" to "international consultations and analysis."[26]

On the last day, President Barack Obama announced the "Copenhagen Accord," a last-minute deal with the United States, China, India, Brazil, and South Africa, without the remaining 188 members. But after he left, a few nations—including Venezuela, Bolivia, Sudan, and Nicaragua—refused to sign the pact. So the summit's final statement does not adopt the Copenhagen Accord but merely "takes note" of it. The conference failed to produce even a non-binding pact that everyone could live with. Because the document was not adopted unanimously, it has no real legal bearing and may never play a role in future U.N. deliberations.[27] The final document was vague and had few specific numbers—beyond "recognizing the scientific view" that a global temperature rise should be "below 2 degrees." It dropped language from an earlier draft calling for cutting global emissions in half by 2050.[28] One delegate described the watered-down report as a "suicide pact for Africa."[29]

> I'm in the dark as to how close to an edge or transition to a new ocean and climate regime we might be. But I know which way we are walking. We are walking toward the cliff.
>
> Terry Joyce Woods
> Hole Oceanographic Institution

Some blamed the U.N. framework for the minimal outcome. The Kyoto Protocol was "legally binding," yet countries still did little and missed the targets. James Hansen says that global fossil fuel emissions were increasing by 1.5% per year prior to the 1997 Kyoto Protocol and that after Kyoto, emission growth accelerated to 3% per year.

The Group of 77 (G77) block of the least developed countries claimed there was no deal reached. There is now no timeline in the accord to reach a legally binding agreement. The document also discarded the road map to a Kyoto successor, which nations had agreed to at the summit in Bali in 2007.[30]

So the world has still done nothing. The next issue is to understand the causes and consequences of global warming.

The Two Primary Causes of Climate Change

There are two principal causes of global warming. First, we are destroying the carbon sinks on Earth, which sequester (absorb and store) carbon. Second, we are releasing greenhouse gases, especially CO_2, into the atmosphere, primarily through the use of fossil fuels.

As of June 2009, the concentration of CO_2 in our atmosphere is 387 ppm, the highest level in more than 600,000 years and more than 100 ppm higher than the amount at the dawn of the Industrial Revolution.[31] Burning just the oil and gas sitting in known fields will drive atmospheric CO_2 well over 400 ppm and, according to James Hansen, will "ignite a devil's cauldron of melted icecaps, bubbling permafrost, and combustible forests from which there will be no turning back."[32]

Destroying Carbon Sinks

Of all the CO_2 emitted into the atmosphere, only half of it stays there; the rest goes into carbon sinks. There are two major natural carbon sinks: the oceans and the land "biosphere." They are equivalent in size, each absorbing a quarter of all CO_2 emissions. Humans are disturbing these sinks, which can now store less CO_2 than they did earlier and/or are releasing CO_2 that was previously stored there.

When scientists discuss the "point of no return," meaning that we generate runaway global warming, they are most often referring to the loss of carbon sinks. At some point, the world's sinks, forests, oceans, and soil will become so saturated with carbon that they will no longer be able to absorb carbon. At that point, all the carbon we release will go directly into the atmosphere, greatly speeding up the warming. If we reach a point where Earth's carbon sinks become sources of carbon, it is virtually certain that we will enter a time of runaway warming. James Hansen believes that under these circumstances, an eventual sea level rise of 80 feet is a distinct possibility, probably inevitable.[33] So far, the sinks still absorb vast amounts of carbon.

Land-based Sinks

The sinks on land include forests, soils, peat bogs, and the tundra. Plants naturally absorb CO_2 and give off oxygen through the process of photosynthesis.

Forests Forests are called the lungs of the planet. Rapid deforestation alone has been enough to severely disrupt the global CO_2 cycle. Tropical rain forests absorb massive amounts of CO_2; about 70% of all terrestrial carbon moves through forests. With warming, new diseases, and new insects, the ability of forests to absorb carbon may become greatly diminished. Humans have destroyed 50% of Earth's forests, much of it in the past century. In the 1930s, 90% of the forests in the Philippines were intact; today, 10% are still intact. Cutting down trees removes a major absorber of CO_2, and if the wood is burned or left to decay, the carbon it stores is released into the atmosphere. Forest fires will become more and more common as global warming results in more extreme droughts, which will dry out the forests and increase the number and range of fires. Forests are cut down to manufacture wood products and paper, to expand agricultural production, and to increase urban development. In Central and South America, people cut down forests to graze cattle for beef.

Soils Factory farming uses fertilizers and pesticides, which destroys the soil. Normally, soil is teeming with life and so would hold massive amounts of carbon, but our current soils absorb very little. Paul Hawken, Amory Lovins, and L. Hunter Lovins, in *Natural Capitalism*, say, "The earth's 5 billion acres of degraded soils are particularly low in carbon and in need of carbon-restorative vegetative cover. Increasing degraded soil's carbon content to plausible rates could absorb about as much

Quoll

carbon as all human activity emits."[34] This would require an end to industrial agriculture and a return to smaller organic farms. Chapter 14 discusses the sequestration of CO_2 in soil in more detail.

Peat bogs Peat forms when plant material, usually in marshy areas, cannot decay fully because of acidic conditions or because it freezes. Peat bogs are a vast natural reservoir of organic carbon. Peat lands cover 12% of Canada's surface area. By one estimate, the bogs of Europe, Siberia, and North America hold the equivalent of 70 years of global industrial emissions. Because of warming, these bogs are thawing and are now releasing their carbon into rivers in the form of dissolved organic carbon (DOC). Bacteria in the rivers rapidly convert DOC into CO_2, which bubbles into the atmosphere.[35]

There are three numbers you need to really understand global warming, 275, 388, and 350. For all of human history until about 200 years ago, our atmosphere contained 275 parts per million of carbon dioxide. By now the planet has about 388 parts per million CO2 – and this number is rising by about 2 parts per million every year ... the highest safe level of CO2 is 350 parts per million.

350.org

In an article in *New Scientist* called "Peat Bogs Harbour Carbon Time Bomb," Fred Pearce, an environmental and development consultant, says, "The world's peat bogs are hemorrhaging CO_2 into the atmosphere, accelerating global warming. Worse, the process appears to be feeding off itself, as rising atmospheric levels of CO_2 are triggering further releases from the bogs." Billions of tons of carbon could pour into the air from peat bogs in the coming decades.[36] One of the reasons for concern is that the most northerly latitudes (where the peat is) are warming the most. The temperature in the Arctic regions has already risen by 4°C. Pearce concludes, "With these trends, by the middle of the century, DOC emissions from peat bogs and rivers could be as big a source of CO_2 to the atmosphere as burning fossil fuels."[37] To limit the amount of CO_2 released from peat bogs and tundra, we must severely cut our use of greenhouse gases now.

Peat bogs also produce methane. The barren peat lands of Siberia have been massively producing methane since soon after the last Ice Age about 12,000 years ago, far longer than had been previously thought. The part of the soil that thaws in the summer (called the active layer) has become thicker since 1970, and the permafrost has disappeared altogether in some locations. This has led to significant changes and an increase in methane production.

Tundra Like peat bogs, the Arctic tundra areas store carbon. The tundra is a vast treeless zone lying between the ice cap and the timber line of North America and Eurasia and normally having permanently frozen subsoil. Vegetation here is mainly shrubs, mosses, lichens, and grasses. The growing season is only about 50 to 60 days long, so dead plants in the tundra do not decompose because when they die, they are quickly frozen. The tundra therefore stores the remains of plants that have accumulated over thousands of years. Today, global warming is melting the permafrost of the tundra, and every year several feet of tundra are lost. As the tundra melts, the plant mass decomposes, returning CO_2

to the atmosphere.[38]

The tundra now has lakes from the thawing, and they bubble up methane. As the permafrost thaws, it responds differently under different conditions: if it is dry, it releases CO_2, if it is underwater in lakes, it releases methane. The world emits about 7 billion to 8 billion metric tons of carbon a year. Permafrost areas hold 500 billion tons of carbon.[39] The U.N. Working Group I report in June 2007 said, "The thawing of the permafrost will result in the release of this carbon which will create a positive feedback effect to global warming," thereby speeding up the warming.[40] This melting of the permafrost is one of the major positive feedback loops that could greatly accelerate the rate of warming.[41]

Oceans as a Sink

The oceans are a vast reservoir for absorbing and sequestering heat and heat-trapping greenhouse gases. Although the oceans' role in regulating the impact of global warming is critical, scientists still do not understand it well. CO_2 readily dissolves in water, and the oceans provide a huge reservoir for carbon. Across all the oceans, the dissolved CO_2 in water is exchanged with the CO_2 in the atmosphere in a continuous cycle. Around 88 thousand million tons of carbon is released from the surface of the world's oceans each year, while the oceans absorb 90 thousand million tons. Therefore, the net uptake of CO_2 by oceans is estimated to be approximately 2 thousand million tons annually.[42]

The oceans have absorbed one-third of human CO_2 emissions and have become 30% more acidic because the extra CO_2 combines with carbonate ions in seawater, forming carbonic acid. As the oceans become more acidic, they can absorb less CO_2. If the oceans stop taking up CO_2, more will go into the atmosphere, accelerating global warming. This added acidity is also harming the life forms that live in the oceans.

A 2008 study shows that increases in ocean acidity are occurring more than 10 times faster than predicted.[43] Timothy Wootton, of the Department of Ecology and Evolution at the University of Chicago and lead author of the study, said, "It appears that we've crossed a threshold where the ocean can no longer buffer the effects of CO_2 in the atmosphere."[44] Ocean acidification is a very new field of science, and the best ways to conduct the research are still being investigated. Even the term ocean acidification was coined just four years ago. The ocean absorbs about one-third of all human-made CO_2, mainly in the cold regions because CO_2 dissolves easily in cold water (just as a soda in the refrigerator will stay bubbly a lot longer than a soda sitting in the sun).

Red-Tailed Hawk

Adding Greenhouse Gases to the Atmosphere

In addition to the loss of carbon sinks, the release of greenhouse gases causes global warming. Scientists understand the basic physics behind this. Greenhouses gases work by trapping heat from the sun that would normally bounce back into space. By increasing these heat-trapping gases, we have intensified the natural greenhouse effect, altering the chemical composition of Earth's atmosphere and causing the atmosphere to warm.[45] Naturally occurring greenhouse gases include CO_2, methane, nitrous oxide, and water vapor.

Carbon Dioxide

CO_2 accounts for three-fourths of the predicted increase in the greenhouse effect. Carbon is continually exchanged within a closed system that consists of the atmosphere, oceans, biosphere (all the life on Earth), and landmass. Over the short term, carbon is exchanged rapidly between plants and animals through respiration and photosynthesis and between the oceans and the atmosphere through gas exchange. Coal, oil, and natural gas all contain carbon from plants that died hundreds of millions of years ago.[46]

What this means is that, by accident, we are changing the rhythms of nature by which we have lived our lives and planted our crops and written our poetry for 10,000 years.

Bill McKibben

Plants in photosynthesis take CO_2 from the atmosphere and use the carbon they absorb to make up their roots, wood, and leaves. Some of this carbon is then lost—either when the leaves drop or when the plant dies—and becomes soil carbon.

Since the 1950s, the major source of CO_2 added to the atmosphere has been from the fossil fuels people burn to run cars and trucks, heat and light homes and businesses, power factories, and ship goods around the world. The use of fossil fuels is responsible for about 98% of U.S. CO_2 emissions, 24% of methane emissions, and 18% of nitrous oxide emissions. Power plants are responsible for more than one-third of U.S. CO_2 emissions, and yet there are no caps on CO_2 emissions for industry. Cars and light trucks are responsible for one-third of U.S. CO_2 emissions. The third major contributor is raising livestock (see Chapter 13).

Methane

A molecule of methane traps 20 times more heat than a CO_2 molecule, but it lasts only about a decade before it dissipates into CO_2 and other chemicals. CO_2 traps heat for a century. Over the past 200 years, atmospheric methane has more than doubled because of decomposing organic materials in wetlands and swamps and because of human-aided emissions. Humans create methane from the production and transportation of coal, natural gas, and oil; the decomposition of organic wastes in municipal solid waste landfills; raising livestock; coal mining; wastewater treatment; and certain industrial processes.[47] The thawing of peat lands releases methane as well as CO_2. The breakdown of plant material under wet soil conditions releases methane. Of all the sources of methane, natural wetlands and peat bogs are the largest.

Another large potential source of methane comes from decomposing organic matter in the ocean, where sediments have frozen into icy deposits under the seabed, called methane clathrates or hydrates (when gases combine with water, they are called hydrates). In a paper published in the journal *Nature*, researchers said that they believe the methane hydrate that exists in vast quantities beneath the ocean floor constitutes the largest reservoir of organic carbon on Earth.[48] Cold temperatures and high pressure keep this methane stable, but critically pressurized volumes of methane gas exist below many methane hydrate deposits, resulting in a potentially unstable ocean floor that is highly sensitive to changing conditions. Any change in temperature or pressure at these critically pressured gas sites can cause hydrates to convert to methane gas. The highly pressurized gas can cause faults in the ocean floor to break apart, allowing the gas to escape.[49]

One study says that as much as 2,000 gigatons (one gigaton is one billion tons) of methane gas might escape from the ocean into the atmosphere during such events.[50] Because methane is such a powerful greenhouse gas, adding 2,000 gigatons would be equivalent to increasing the total CO_2 in the atmosphere by more than 50 times. A new NASA study has confirmed that such a store of methane has been suddenly released numerous times in the history of Earth. Keeping the temperature of the oceans as low as possible is therefore essential. This also means that the idea of sequestering CO_2 under the ocean floor could be catastrophic.

Nitrous Oxide

A nitrous oxide molecule traps about 200 times more heat than a CO_2 molecule. The 2007 IPCC Working Group I report says the global atmospheric concentration of nitrous oxide increased from a pre-industrial value of about 270 to 319 parts per billion (ppb) in 2005. The growth rate has been fairly constant since 1980. More than one-third of all nitrous oxide emissions are anthropogenic and are primarily the result of agriculture.[51] The gas stays in the atmosphere for a long time, so what we release today could still trap heat more than a century from now.

Nitrogen fertilizers provide nutrients for crops in large non-organic farms, but nitrous oxide is released into the air when fertilizers break down in the soil. Therefore, converting to small organic farms would greatly reduce emissions of this gas. In addition, human and animal waste releases nitrous oxide. Sewage treatment plants can therefore be a major source of this gas. Oceans and bacteria in soils also naturally release nitrous oxide. Coal-fired power stations emit nitrous gas because of their lower combustion efficiency, compared with other fossil fuels.

Butterfly Weed Pod

Water Vapor Is Also a Greenhouse Gas

Water vapor, like CO_2, traps heat and is a potent greenhouse gas. Warming will cause more water to evaporate because of increased temperatures, and it will stay in the air longer because warmer air holds more water. The 2007 Working Group I report says, "Water vapor changes represent the largest feedback affecting climate sensitivity and are becoming better understood. Cloud feedbacks remain the largest source of uncertainty."[52] Clouds are created when water vapor condenses out.

It is possible that when enough of the natural ecosystems are erased, the process might continue autonomously, triggering changes that will be catastrophic for Homo sapiens.

Dr. E. O. Wilson
Conservation International

Clouds are a primary unknown in the science of global warming. It is not clear how they will change as a result of warming and of increased amounts of water vapor in a warmer atmosphere. During the day, clouds can shield Earth from the sun's heat, keeping Earth cooler. At night, clouds can trap the heat rising from the ground, making Earth warmer. Whether the clouds warm or cool Earth also depends on the clouds' height and their depth, color, and density. Researchers now realize how little they know about how many and what sorts of clouds there are. Extra evaporation and water vapor in the atmosphere might create more clouds. However, higher evaporation rates resulting from increased warmth may burn off clouds without the clouds' producing rain. Alternatively, clouds may rain out more quickly, leaving clearer skies (causing more warming), rather than cloudier ones (allowing less warming). Fred Pearce explains:

> Last year, for instance, it emerged that there may be vastly more heat-trapping cirrus clouds in the upper atmosphere than anyone had thought. Some studies suggest that, taken globally, the cooling and warming effects of clouds may largely cancel each other out. But nobody is sure. And small changes in either the area of cloud cover or the types of clouds that form could change things radically. So for modelers of our future climate there are two issues. Will global warming change clouds? And will the changes produce positive or negative feedback on the climate?[53]

According to Pearce, the extent to which clouds control Earth's thermostat may be between two and four times greater than what scientists previously thought.[54]

Adding Aerosols to the Atmosphere

A further major complication is that in addition to pumping gases into the atmosphere, we are also filling the atmosphere with large amounts of microscopic particles, mostly from burning forests, crop waste, and fossil fuels. Pearce says that for more than a decade, scientists have hotly debated the exact nature of the aerosol particles' involvement in global temperature but that most researchers now believe that the dominant effect of these aerosols is to suppress warming by shading the planet.[55] He says, "We are dealing with a coiled spring, with [rising] temperatures being held back by aerosols. If you shut off aerosols, temperatures would increase rapidly, but we don't yet know exactly how coiled the spring is."[56]

The first efforts at directly measuring this "parasol effect" suggest that the spring may be much more tightly coiled than was first thought. In an assessment last year, Nobel prize–winning atmospheric chemist Paul Crutzen argued that aerosols could be disguising between one-half and three-quarters of the present warming. This suggests that the coiled spring is already holding back warming by up to 2°C. This would mean that the climate system is much more sensitive to greenhouse gases than scientists had thought.[57]

This is what has been creating global warming, but what are the consequences?

Consequences of Global Warming

The severity of the consequences of global warming will depend on how much and how fast Earth warms, which in turn depends on how soon we act, that is, how fast we develop the political will to change our social and economic systems.

1) Global warming could create abrupt climate change
2) Global warming threatens human health
3) Global warming endangers millions of species
4) Global warming alters weather patterns
5) Global warming causes ice to melt and seas to rise
6) Global warming causes severe water and food shortages
7) Global warming may create abrupt cooling in Europe
8) Global warming threatens national security

Global Warming Could Create Abrupt Climate Change

Scientists had originally assumed that global warming would be gradual, but they now understand that it can occur abruptly. Paleoclimatic (historical) evidence suggests that such an abrupt climate change could begin in the near future. A report titled *Abrupt Climate Change: Inevitable Surprises,* by the Committee on Abrupt Climate Change of the National Academy of Sciences, states, "Large, abrupt climate changes have repeatedly affected much or all of the earth, locally reaching as much as 10°C change in 10 years. Available evidence suggests that abrupt climate changes are not only possible but likely in the future, potentially with large impacts on ecosystems and societies."[58] The report describes the process: "Technically, an abrupt climate change occurs when the climate system is forced to cross some threshold, triggering a transition to a new state at the rate determined by the climate system itself and faster than the cause. Chaotic processes in the climate system may allow the cause of such an abrupt climate change to be undetectably small."[59] So, the cause might be small, but the speed at which the

Cape May Warbler

system can switch to a new state can be very rapid.

The last period of abrupt climate change was at the end of the Younger Dryas interval, about 11,500 years ago. At that time, global climate shifted dramatically, in many regions by about one-third to one-half of the difference between Ice Age and modern conditions, with much of the change occurring over the course of just a few years.[60] Human civilizations arose after these extreme global Ice Age climate jumps, and the climate has been unusually stable for the 10,000 years since then. The National Academy of Sciences report states, "The only thing we can be sure of is that there will be climatic surprises. Physical, ecological, and human systems are complex, nonlinear, dynamic and imperfectly understood. Climate changes are producing conditions outside the range of recent historical experience and observation, and it is unclear how the systems will interact with and react to the coming climatic changes."[61]

> To wantonly destroy a living species is to silence forever a divine voice.
>
> Thomas Berry
> *Dream of the Earth*

Global Warming Threatens Human Health

Global warming will also be a public health issue because it will increase the spread of disease, create heat stress, and create malnutrition as a result of its impact on agriculture. There will also be an increased risk of new infectious diseases for crops, animals, and humans.

The Working Group II report says that the health status of millions of people will be affected by global warming, particularly those with low adaptive capacity. These impacts include:

- Increases in malnutrition and consequent disorders, with implications for child growth and development;
- Increased deaths, disease, and injury resulting from heat waves, floods, storms, fires, and droughts;
- Increased burden of diseases that cause diarrhea;
- Increased frequency of cardio-respiratory diseases as a result of higher concentrations of ground-level ozone related to climate change; and
- Altered spatial distribution of some infectious disease vectors.[62]

Diseases that are normally confined to southern areas will move farther north because the mosquitoes and other insects follow the higher temperatures. These "vector-borne" diseases include malaria, dengue fever, yellow fever, and encephalitis. Doctors at the Harvard Medical School have linked recent U.S. outbreaks of dengue fever, malaria, Hantavirus, and other diseases to climate change. The range of the mosquito that carries dengue fever is limited by temperatures. Frost kills both adults and larvae. In the past, this has prevented the disease from spreading from the tropics, but warming temperatures are changing that. In the U.S., the mosquito that carries dengue has reached as far north as Chicago. Dengue fever has increased thirtyfold in the past 50 years.[63] People in the north have little immunity to these new diseases.

Diseases such as cholera tend to accompany algae blooms. Some scientists believe that algae blooms could occur more frequently as temperatures warm, particularly in areas with polluted waters. Just as the new diseases introduced by Europeans killed millions of Native Americans because of their lack of resistance, this could greatly affect human health. Malaria is also a mosquito-borne illness that is normally limited by temperatures. Rising temperatures have expanded its range and exposed new populations to infection. IPCC scientists predict that as warmer temperatures continue to spread north and south from the tropics and to higher elevations, malaria-carrying mosquitoes will spread with them. Globally, malaria quadrupled between 1995 and 2000.[64] Scientists project that global warming could put as much as 65% of the world's population at risk of infection by malaria.[65]

Climate-related increases in sea surface temperatures and sea level can lead to higher incidence of other water-borne infectious and toxin-related illnesses, such as cholera and shellfish poisoning. Cholera killed 120,000 people, most of them children, worldwide in 1995.[66] In addition, outbreaks of encephalitis, another illness with strong links to warmer temperatures, have also increased. Since 1987, there have been major U.S. outbreaks in Florida, Mississippi, Louisiana, Texas, Arizona, California, and Colorado.[67]

In addition, crops and farm animals will be subject to a host of new diseases; there is a significant risk that there will occasionally be large die-offs from a new pathogen, creating large, possibly global, food shortages.

Extreme temperatures can also cause the loss of life directly. Untreated heat stress is a very serious medical problem. In July 1995, a heat wave killed more than 700 people in the Chicago area. There are more deaths due to heat than deaths due to cold because people are more sensitive to hot temperatures. The difference of a 5°F decrease from 20°F to 15°F, for example, has a much smaller impact on health than a 5°F increase from 95°F to 100°F. Because the cardiovascular system must work harder to keep the body cool during hot weather, elderly people and people with heart problems are especially vulnerable.[68]

In August 2003, officials estimated that a major heat wave in Europe that summer claimed 35,000 lives. Over the following three years, they revised this estimate: the heat wave killed 52,450 people, making it the deadliest climate-related disaster in Western history.[69] This heat wave was so deadly because it remained hot at night. Never being able to cool down is very dangerous, especially for the elderly, because the human body needs to be cool for at least 3 out of every 24 hours.

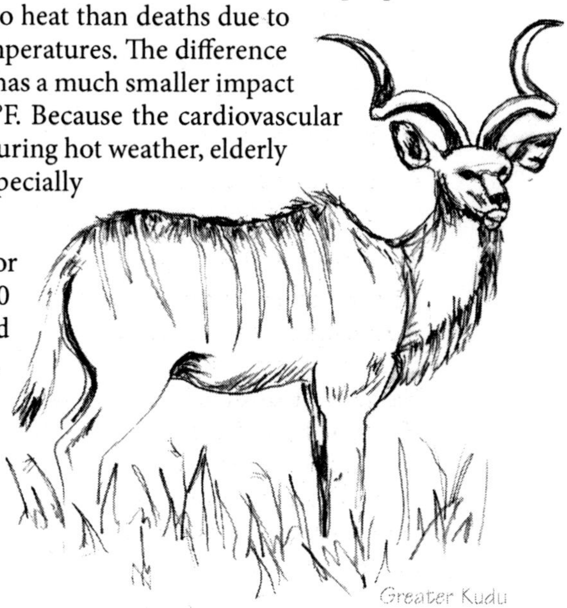

Greater Kudu

Warm temperatures can also increase air and water pollution, as well as the concentration of ozone at ground level. Ozone in the upper atmosphere blocks harmful ultraviolet radiation; however, in the lower atmosphere, it is a harmful pollutant. It damages lung tissue and causes serious problems for people with asthma and other lung diseases. Even modest exposure to ozone can cause healthy individuals to experience chest pains, nausea, and pulmonary congestion.[70] Allergies will also increase as a result of increased pollen levels. The negative impact on crop, animal, and human health will be a major consequence of global warming.

Global Warming Endangers Millions of Species

The primary contributor to the extinction of species has been habitat loss resulting from urban sprawl and deforestation, but now global warming is becoming the major factor. Most plant and animal species cannot adapt to the fast-paced changes in habitat caused by rising temperatures, and the loss of any species will have cascading effects throughout an entire ecosystem. Changes in temperature, rainfall, and plant distribution caused by global warming could affect migratory routes, feeding behavior, and mating habits of many animal species, from insects to elephants.

> A century ago, there must have been some 2 million [chimpanzees] in Africa. Today, there are 150,000 at most.
>
> Jane Goodall
> "When Primates Become Bushmeat"

Massive extinctions have occurred five times during Earth's history; the last one was the extinction of the dinosaurs 65 million years ago. Scientists are calling what is occurring now the *sixth mass extinction*. Stopping the loss of species is about saving the very web of life on Earth. Critics have dismissed or ridiculed people trying to save critical habitat as sentimental "tree-huggers" who want to save the "spotted owls" even if it costs jobs. Most Americans have little idea of the nature or magnitude of the problem. Extinctions have cascading effects, creating the "living dead" because vital interrelationships have been lost; for example, the loss of a pollinator can doom the plant it pollinates, or the loss of a prey species can doom its predator.[71] The sixth mass extinction is not happening because of some external force. It is happening because of us. One scientist has called *Homo sapiens* an "exterminator species."[72]

Although they are uncertain of the exact numbers, most scientists believe the rate of loss is greater now than at any time in the planet's 4.5-billion-year history. In an article in *Nature*, scientist Chris Thomas, professor of conservation biology at the University of Leeds, said that a quarter of all species of plants and land animals—which is more than a *million* species in all—could be driven to extinction.[73] But Richard E. Leakey, author of *The Sixth Extinction*, believes that 50% of Earth's species will vanish within 100 years and that such a dramatic and overwhelming mass extinction threatens the entire complex fabric of life, including *Homo sapiens* (the species responsible for the crisis).[74] Stuart Pimm, professor of ecology at Columbia University, says, "It's not just species on islands or in rain forests or just birds or big charismatic mammals. It's everything and it's everywhere. . . . It is a worldwide epidemic of extinctions."[75] During the 1850s, when Darwin was finishing *The Origin of Species*, the rate of extinction is believed to have been one species every five years. Today, the rate of extinction is estimated at one species every nine minutes.[76]

The problem is not just the loss of species. There is also the loss of the genetic diversity within species, as well as the loss of diversity of different types of ecosystems, which can contribute to or hasten whole species extinction. Preserving the wider gene pool diversity in subdivisions of species, such as subspecies and populations, maintains the raw material needed for the evolution of new species in the future.

The loss of species is also a crisis in ocean species. Oceans cover two-thirds of the planet, and the warmer water temperatures will dramatically change marine life, far more than was originally thought. The warming waters have brought disease-causing organisms that are destroying corals and sea grasses and causing an increase in toxic algae, distemper in seals, and viral disease outbreaks among porpoises and dolphins.[77] As much as 60% of all reefs could be lost by 2030.[78] Coral communities, often referred to as the "rainforests of the oceans," are one of the most diverse ecosystems in the world. One out of every four ocean species so far identified is a reef-dweller, including at least 65% of marine fish species. Coral reefs also protect coastal areas from storms, floods, and erosion, provide habitats for thousands of marine species, and serve as breeding grounds for many species of marine life. The reef itself is actually part of a larger ecosystem of many biological communities.

To study the oceans, a monitoring program called the Continuous Plankton Recorder Survey has been ongoing since 1931. Kenneth Caldeira, of the Lawrence Livermore National Laboratory, said, "We are changing the chemistry of the ocean and we don't know what it's going to do."[79] Chris Reid, Director of the Sir Alistair Hardy Foundation for Ocean Science in England, said, "A regime shift has taken place and the whole ecology of the North Sea has changed quite dramatically. . . . We are seeing a collapse in the system as we knew it."[80]

Global warming affects many other habitats. A drier climate in Central Africa might alter the social behavior and population of large mammals such as African elephants, whose mating patterns are tied to rainfall; a change in rainfall could therefore affect elephant populations. Monarch butterflies from eastern North America migrate to only a few restricted places in mountain forests in central Mexico during the winter. Changes in temperature, rainfall, or distribution of this specialized winter habitat could endanger this butterfly's population.

The massive loss of species could cause cascading effects that could unravel life on Earth.

Chimpanzee

Global Warming Alters Weather Patterns

Global warming will create both increased precipitation (because warm air holds more moisture) and increased evaporation. Where precipitation is greater than evaporation, there will be heavy rains and floods. Where evaporation is greater than precipitation, there will be droughts. The increased warming and the unpredictable changes in weather will greatly affect agriculture. In addition, warming will cause some of the traditional jet stream patterns to shift, distributing precipitation in new patterns. Paul Epstein, a leading climate researcher at Harvard Medical School, commented on the rapid intensification of storms around the world, saying, "We are seeing impacts that were previously projected to occur in 2080."[81]

Computer models show that as the atmosphere warms, the climate not only becomes hotter, it also becomes more unstable and erratic. Scientists know that just warming the air will produce more extreme and destructive rains, even without the addition of more water vapor in the air. However, the increases in water vapor in the atmosphere will further intensify storms. There were many unusual storms and droughts in the summer of 2010. An extreme drought in Russia destroyed 20% of their grain crops, and the President of Russia announced that there would be no further grain exports for the year. The droughts also caused many wildfires, making the air in Moscow hard to breathe. In August, there was concern that the fires would reach the Chernobyl region and spread the contamination there. At the same time, the worst floods in Pakistan in 80 years killed more than 1,600 people and left 2 million homeless.[82] The insurer Munich Re said a natural catastrophe database it runs "shows that the number of extreme weather events like windstorm and floods has tripled since 1980, and the trend is expected to persist."[83]

Hurricanes are not only getting stronger, they also last longer and occur over a wider geographic range.[84] Hurricanes (called typhoons in the Pacific Ocean) will increase in strength because the water in the ocean will be warmer. Nature does not like anything to be out of balance. When the ocean water gets too hot, nature creates a hurricane to dissipate this extra heat energy by converting it to wind energy, cooling the water. The warmer the water, the stronger the cooling winds need to be to restore balance. Out in the ocean, the wind also whips up the cold water from the depths to cool the warmer surface water. However, the Gulf of Mexico is too shallow to have deep, cold water, and so most of the cooling there must be accomplished through strong winds.

The number of Category 4 and 5 hurricanes has *doubled* over the past 30 years.[85] In 2005, there were 28 hurricanes, nearly three times the average annual number of storms over the past 100 years. Because of the high water temperature, 2005 had four Category 5 hurricanes: Emily, Katrina, Rita, and Wilma. Wilma, which hit parts of Mexico's Yucatan Peninsula, was the strongest Atlantic storm in recorded history. Katrina and Rita were the first two Category 5 storms to develop in the same season in the Gulf of Mexico in recorded history; additionally, they were the first to have developed simultaneously.[86] Chris Flavin, President of the Worldwatch Institute, said that Hurricane Katrina, in just over 48 hours, "grew from a low-level Category 1 hurricane to the strongest Atlantic storm ever recorded."[87] In September 2005, Hurricanes Wilma and Rita both broke Katrina's record as the strongest storm ever in that region. It is still unclear whether warming will cause more storms in number, but scientists know that the ones we do have will be more extreme.

The connections between global warming and hurricanes have been politicized. The Bush administration allowed only those government scientists who do not feel hurricanes are connected to global warming to speak to the press. These people are usually meteorologists, who deal with weather, rather than climatologists, who deal with long-term climate. Chris Mooney, a science journalist, has written a book on this controversy titled *Storm World, Hurricanes, Politics and the Battle over Global Warming.*

Damage from hurricanes is growing, bankrupting insurance companies. The Earth Policy Institute describes the extent of the damage: "During the 1960s, worldwide damage from windstorms with economic losses of $1 billion or more totaled just $4 billion. In the 1970s the figure rose to $7 billion, and in the 1980s it topped $24 billion. Next came the 1990s, when hurricane losses soared to $113 billion. Then during the six years from 2000 to 2005, hurricanes left a staggering bill of $273 billion."[88]

Warming is also changing the pattern of El Niños, increasing both their intensity and frequency. El Niños act as another ocean mechanism for expelling stored heat, and they affect weather all over the world. Historically, El Niños occurred approximately every 7 years, but that changed dramatically in the mid-1970s. Atmospheric scientist Kevin Trenberth says, "Clear signs of global surface warming, which began in the late 1970's, is the same time of the more frequent appearances of El Niños."[89] Using fossilized coral reefs, an international team of scientists reconstructed the history of El Niños over the past 130,000 years. Their report says, "The samples indicated that El Niño was never more intense than the events of the last hundred years."[90] And the 1997–1998 El Niño was the most severe on record.[91]

Global Warming Causes Ice to Melt and Seas to Rise

Almost everything frozen on Earth is melting. The 2007 IPCC Working Group I Summary for Policymakers estimates for sea rise created a controversy. They give their best estimate of sea rise as 0.59 meters (m) (23.28 inches), with the possible range as 0.3–0.9 m. This is actually less than the estimated sea rise reported in the 2001 report, so some people are using this reduction in the 2007 estimate to claim the 2001 report was wrong and that sea level rise is not a problem.

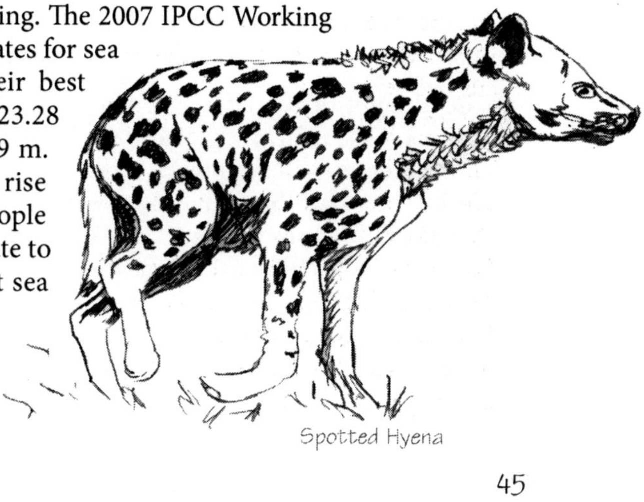

Spotted Hyena

However, the 2001 report included the sea rise from melting ice as well as from

Sometime in the next decade we may pass that tipping point which would put us warmer than temperatures that Greenland can survive. ... What is going on in the Arctic now is the biggest and fastest thing that nature has ever done.

Richard Alley
Professor of geosciences
Pennsylvania State University

warm water expansion. But for the 2007 report, Working Group I decided that their ice sheet models could not model the rise in sea level from dynamic changes because the contribution to sea level rise by the melting of Greenland and Antarctica is too uncertain (but they added that all the uncertainty is on the side of making things worse).[92] Therefore, for the 2007 report, they decided not to even try to estimate sea rise from melting ice. Instead, the 0.59 m refers only to the expected sea level rise because as water warms, it expands.

Most of the research on melting ice is new, and because the 2007 IPCC report had a December 2005 cut-off date for the research to be considered, much of the newer research was not included. One newer report is a study by a team led by Hansen that suggests that the grim reports issued by the IPCC could be absurdly optimistic.[93] But Hansen says this problem with the stability of ice sheets is so critical that it really should be looked at by a panel of our best scientists. Hansen says Congress should ask the National Academy of Sciences to do a study on this and clearly report its conclusions.[94] Hansen reported that the melting in summer 2005 broke all records, and he said that our understanding of what is going on is very new:[95]

> Today's forecasts of sea-level rise use climate models of the ice sheets that say they can only disintegrate over a thousand years or more. But we can now see that the models are almost worthless. They treat the ice sheets like a single block of ice that will slowly melt. But what is happening is much more dynamic. . . . Once the ice starts to melt at the surface, it forms lakes that empty down crevasses to the bottom of the ice. You get rivers of water underneath the ice. And the ice slides towards the ocean.[96]

Scientists now understand from the geological record that ice at the poles does not melt in a gradual, linear fashion but flips suddenly from one state to another. They found from studying the past that ice responds immediately to changes in temperature. That is, they now realize that the ice sheet breaks up suddenly rather than melting slowly.[97] In addition, as pools of water form on the surface, the water is darker than the ice, so it absorbs more heat, and the process feeds on itself.

The amount of sea rise is perhaps the biggest unknown and potentially the most serious consequence of global warming. Hansen explained that once an ice sheet starts to disintegrate, it could reach a tipping point beyond which break-up is "explosively rapid."[98] He said that none of the current climate and ice models predicted this.[99] He continues, "I think sea-level rise is going to be the big issue soon, more even than warming itself. It's hard to say what the world will be like if this happens. It would be another planet."[100] Hansen concludes, "Civilization developed during a period of unusual climate stability, the Holocene, now almost 12,000 years in duration. That period is about to end."[101]

Rising sea levels also contribute to making storms much more destructive because higher sea levels can cause water surges in storms to reach much farther inland. The effect of storm surges increases exponentially with the rise in sea level. Coastal flooding could also cause salt water to flow into freshwater areas where salt is harmful, threatening plants and animals and the quality of drinking water.

Melting in the Arctic

The ice in the arctic normally covers an area the size of the U.S. The Arctic sea ice is the northern hemisphere's major "heat sink," moderating climatic extremes because the ice there covers the cold water, keeping it cold and protecting it from warming. This polar sea ice has helped to keep the climate stable for thousands of years. Losing the sea ice of the Arctic is likely to significantly affect climate worldwide.[102] The Arctic Ocean has been covered with frozen ice all year long for about 14 million years (before that, there was no permanent sea ice cap in the Arctic).[103]

Geophysical Research Letters published a study on the largest ice shelf in the Arctic—the Ward Hunt Ice Shelf—a 150-square-mile, 100-foot-thick mass of ice that has been sitting off the northern Canadian coast for 3,000 years. In September 2003, it broke up into two main parts.[104] With regard to this breakup, Warwick Vincent, professor of biology at Laval University in Quebec, said, "We'd been measuring incremental changes each year. Suddenly in one year, everything changed."[105] The parts themselves were cut through with fissures, and a freshwater lake drained into the sea. In the article, the scientists warned that a "critical threshold" had been breached."[106]

Ice sheets come in two types—land based and marine based. The bottom of a land-based ice sheet (like those in Antarctica) lies mostly above sea level.[107] The marine-based ice in the Arctic is already in the water; therefore, its melting will not make a significant difference to the sea level. However, the white ice in the Arctic reflects 80% of the solar heat energy back into space and absorbs only 20%. Melting ice reverses the process: when ice melts, the dark water has a much lower reflectivity (or albedo) than that of ice. The open water reflects back only 20%, and absorbs 80%, of the sun's heat, creating a feedback loop that greatly speeds up the rate of warming.

The ice in the Arctic normally melts some in summer and builds up again in winter. For the first time on record, in winter 2004–2005, this annual rebound did not occur. Arctic specialists at the U.S. National Snow and Ice Data Center at Colorado University, who have been documenting the gradual loss of polar sea ice since 1978, believe that a more dramatic melt began around 2001. Mark Serreze, one of the scientists at the Center, said, "This will be four Septembers in a row that we've seen a downward trend. The feeling is we are reaching a tipping point or threshold beyond which sea ice will not recover."[108] He added, "The changes we've seen in the Arctic over the past few decades are nothing short of remarkable."[109]

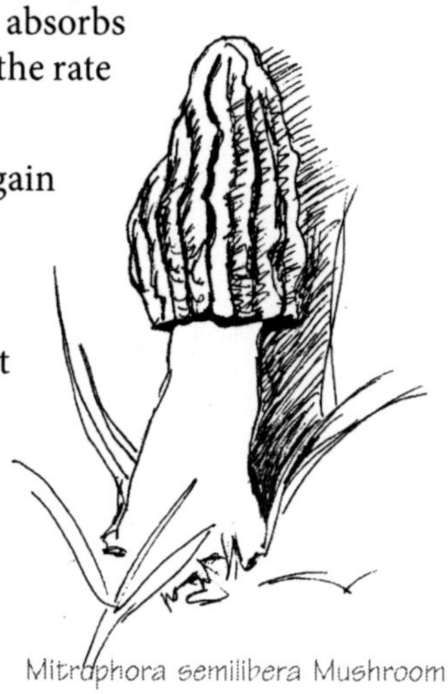

Mitrophora semilibera Mushroom

In September 2007, at the end of the Arctic summer melting, scientists found that 1 million square miles of the Arctic had more open water than scientists have ever seen before. In an article in the May 2007 edition of the *Geophysical Research Letters*, five scientists reported their analysis of the extent of sea ice each summer since 1953. Computer models had predicted a loss of ice of 2.5% per decade. The actual loss was three times greater—7.8% per decade. They said that the melting is 30 years ahead of what the models had predicted.[110] The changes in Arctic ice over the past 28 years have been dramatic. The website Real Climate has photographs of the ice taken in September 1979 and on August 9, 2007.[111] However, the problem is not just the area covered by the ice, but also its thickness. The ice cover is shrinking, and the ice is also losing volume.[112]

The warming in the Arctic also causes the permafrost to melt. A paper published in *Geophysical Research Letters* shows that this warming penetrates 1,500 kilometers (km) inland, covering almost the entire region of continuous permafrost.[113] George Monbiot says that the methane gushers from the new Arctic lakes are now gassing out of some places with such force that they keep the water open in Arctic lakes throughout the winter.[114] Most important, these effects of melting permafrost are not incorporated into any global climate models. Runaway warming in the Arctic alone could flip the entire planet into a new climatic state.

We're changing the Earth in ways that are just inconceivably large."

Ransom Myers
Professor of ocean studies

Melting in Greenland

Greenland is an enormous country about three times the size of Texas, and it contains 12% of the world's ice. Mountains ring the coastline, where people live, and the entire interior (85% of the country's land mass) is a huge valley filled with ice, in some places up to 2 miles thick. This ice sheet lies closer to the equator than the West Antarctic ice sheet at the South Pole does, so it is much more vulnerable to melting.[115] The melting ice in Greenland adds new water that will raise ocean levels, yet it is still uncertain how much of this ice will melt. If all of Greenland melted, it would raise sea level by 21 feet, which would flood some of the world's major population centers, including all of Britain's city ports, all of Holland, all island nations, south Florida, and a good deal of Manhattan.[116]

Two satellites known as the Gravity Recovery and Climate Experiment (GRACE) are monitoring Greenland and Antarctica. They fly in orbit 137 miles apart to determine how the mass of small areas of Earth changes as these masses of ice melt, change to water, and flow into the sea.[117] These new satellite data represent a remarkable advance. Scientists are seeing for the first time the detailed behavior of the ice streams that are draining the Greenland ice sheet.[118] A University of Texas team that is monitoring Greenland with GRACE found that Greenland's ice is melting today at a rate three times faster than it was just five years ago.[119]

Melting in Antarctica

The South Pole is a continent. Since there is land under the ice, any melting will put new water in the ocean. Antarctica has 70% of the world's fresh water. Three large pieces of the Antarctic ice shelf have broken off since 1995. The latest piece, a 500-billion-ton iceberg about the size of Rhode Island, collapsed in only 31 days in the spring of 2002, a speed researchers called "staggering."[120]

Earlier evidence appeared to show that the East Antarctic ice sheet was actually thickening (the warming atmosphere also creates more snow). However, new satellite data show that it is melting, primarily from the West Antarctic ice sheet. This has turned at least 36 cubic miles of ice to fresh water each year between 2002 and 2005.[121] In a *Scientific American* article titled "On Thin Ice," Robert A. Bindschadler and Charles R. Bentley said that dozens of scientific studies conducted over the past 30 years have warned that the ice in West Antarctica, the part lying mainly in the western hemisphere that holds more than 3 million cubic kilometers of fresh water, could melt. It has happened before.[122] This melting would swamp coastal lowlands and force many of the 2 billion inhabitants living on the coast to retreat inland, creating a massive increase in refugees. Like Greenland, the melting of West Antarctica would raise sea levels by about 20 feet, so that if both melted, sea levels would rise by 40 feet.

Melting of Glaciers

The glaciers of the world collectively cover an area the size of South America.[123] Global warming is already melting glaciers all over the world, creating sea level rise and threats of floods, droughts, and water shortages for millions of people. Normally, glaciers release water in dry seasons and collect it in snowy ones and have been referred to as the "world's water towers."[124] According to a report on glaciers by the World Wildlife Fund (WWF), simulations project that a 4°C rise in temperature would eliminate nearly all the world's glaciers.[125]

One study of Alaska's glaciers said that an estimated 24 cubic miles of ice is disappearing annually from Alaskan glaciers, "turning some imposing ice mountains into minor hills and adding to the steady rise in global sea level."[126] Over the past five years, the rate of melting there has almost doubled. After surveying 67 major glaciers using the satellite laser system, researchers at the University of Alaska concluded that the contribution from Alaska's glaciers to the worldwide sea level rise was more than they had expected.[127]

As these Alaskan glaciers melt, they lighten the load on Earth's crust. Tectonic plates, which are mobile pieces of Earth's crust, can then move more freely. In southern Alaska, a tectonic plate under the Pacific Ocean is pushing into the coast. The colliding plates create a great deal of pressure, which builds up

Madagascar Tarsier

and is eventually relieved by earthquakes. The weight of a large glacier on top of these active earthquake areas can help keep things stable. But as the glaciers melt and their load on the plates lessens, there is a greater likelihood of an earthquake happening to relieve the pressure from the colliding tectonic plates underneath.[128]

Glaciers are also melting on all the world's major mountain ranges, including the Rocky Mountains, the Andes, the Alps, Mount Kilimanjaro, and the Himalayas. All seven major rivers in Asia originate in the snow and ice mass of the Himalayas (the Ganga, Indus, Brahmaputra, Salween, Mekong, Yangtze, and Yellow rivers). Drinking water is already scarce in many parts of Asia.[129]

In the second half of 2009, leading up to the Copenhagen meeting, there was a great deal of information denying global warming. One controversy was about the glaciers in the Himalayas. It was found that a comment in the Working Group II report was wrong. It said that the glaciers here could be gone by 2035. Skeptics at a British paper said, "Fresh doubts were cast over controversial global warming theories yesterday after a major climate change argument was discredited." The IPCC report may have used some out-of-date projections or even non-peer-reviewed sources for a pretty minor part of the Working Group II Report (a report written mainly by social scientists on the impacts of global warming; Working Group I is on the science). Even so, there has been an overabundance of new data showing that all glaciers, including those in the Himalayas, are at grave risk. In fact, although they will not melt by 2035, we will know by then whether their eventual melting is inevitable. Himalayan glaciers, by feeding the seven major rivers of Asia, supply fresh water to one-third of the world's population.

> If we continue down the path we are going, we will produce changes greater than any experienced in the past 300 million years.
>
> Kenneth Caldeira
> Lawrence Livermore National Laboratory

On East Africa's Mount Kilimanjaro, the area covered by snow and ice has shrunk by 80% since 1900. Lonnie Thompson, Ohio State University glaciologist, reported that all the snow and ice there is likely to disappear by 2020.[130]

Glaciers in the Bolivian Andes are even more fragile because they are in the tropics. In Bolivia, the glaciers have shrunk by 60% since 1978 and could be gone in 10 years. The total disappearance of the glaciers in South America is now nearly inevitable, and this will lead to severe seasonal water shortages in Bolivia and Peru. Glaciers in Venezuela are already nearly gone.

The number of glaciers in Glacier National Park in the U.S. has dwindled from 150 in 1850 to fewer than 50 today; the park has lost two-thirds of the glaciers and about 75% of the glacier area. WWF estimates that if the present rate of warming continues, there will be no glaciers left in Glacier National Park by 2030.[131]

The ice and permafrost in the Alps are also decreasing. In the summer of 2003, the International Permafrost Association conference, held every 4 years, took place in Zurich, allowing climatologists, civil engineers, and geologists to exchange research data. The outcome of the conference was that scientists concluded that the icy crust of permafrost in the Alps is disappearing. This ice holds stone pillars and rock faces together, and cable car stations and pylons are rooted in it. Melting has already caused several Alpine disasters, one of which killed more than 50 people. Engineers and geologists discovered that part of the problem is that increases in air temperature cause underground temperatures to increase fivefold. In addition to increases in air temperature that warm the ground, the warmer temperatures cause increased evaporation, which is creating thicker snowfalls. The snow insulates the soil and keeps it warm over the winter.[132]

Sea Rise and Island Nations

Sea rise from melting ice threatens island nations and all low-lying coastal areas. Most of the world's largest cities and 30% of the world's population are in coastal areas. Melting could force up to 10 million people to flee in the Philippines and could threaten millions more in Cambodia, Thailand, Egypt, and China, as well as low-lying countries with large, vulnerable coastal populations, such as Indonesia, Pakistan, Gambia, the Maldives, Mozambique, Senegal, Surinam, and many parts of Latin America.[133]

The island countries have organized the Alliance of Small Island States, a group formed in 1990 specifically to lobby on behalf of countries that are particularly vulnerable to climate change.[134] The first island casualty of global warming is Tuvalu, a group of nine islands, with 11,000 people. As sea level has risen, Tuvalu has experienced lowland flooding; saltwater intrusion is affecting its drinking water and food production. Coastal erosion is eating away at all nine islands. President Leo Falcam said, "Climate change is nothing less than a form of slow death."[135]

Lester Brown says this rise in sea level "could force a human migration of almost unimaginable dimensions. It also raises questions about responsibility to other nations and to future generations that humanity has never before faced."[136] He asks, "Will the United Nations be forced to develop a climate-immigrant quota system, allocating the refugees among countries according to the size of their population? Or will the allocation be according to the contribution of individual countries to the climate change that caused the displacement?"[137] If that were the case, the U.S. would be accepting one-fourth of the migrants, for the U.S. creates 25% of the global warming.

Global Warming Causes Severe Water and Food Shortages

Fresh water is a finite resource, and a water crisis of catastrophic proportions is about to explode. How we use water is a large factor in freshwater shortages, but scientists say the single greatest cause of the freshwater shortage in the world is global warming.[138] Only 2.5% of Earth's water is fresh water. The U.N. Commission on Sustainable

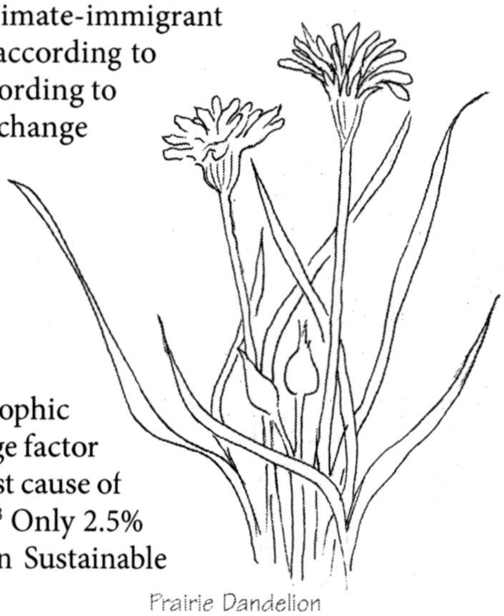

Prairie Dandelion

Development reports that less than 0.01% of Earth's total freshwater resources is accessible for human use. During the past century, world water use has grown at more than twice the rate of population growth. Water shortages are caused by the loss of seasonal water from glaciers, heat and evaporation, pollution, and overuse by industry, people, and agriculture.

An informed global public may be all that is required to bring an end to the climate-destabilizing fossil fuel era.

Dr. Ty Cashman
former California
Secretary of Energy

In 2009, a report titled *World Water Development Report* was released that gave a comprehensive assessment on the world's fresh water. It said that by 2030, nearly one-half the world's population will be living in high water stress. This will create "substantial population displacements" that they call "climate-change refugees." It will also increase conflicts as people and nations compete for water.

There are two principal signs of stress as the demand for fresh water outruns the supply: rivers are running dry, and water tables are falling on every continent. Many major rivers—including the Colorado, Ganges, Indus, Rio Grande, and Yellow—are so over-tapped, they now run dry for part of the year. Freshwater wetlands have shrunk by about one-half worldwide. Underground aquifer depletion is a new problem. Water tables are falling from the over-pumping of groundwater in large portions of the U.S., China, India, Iran, Mexico, the Middle East, North Africa, and Saudi Arabia. India has the highest volume of annual groundwater overdraft of any nation in the world. In most parts of the country, water mining is taking place at twice the rate of natural recharge, causing aquifer water tables to drop by 3 to 10 feet per year.[139]

In *Blue Gold*, Maude Barlow and Tony Clarke say, "There is simply no way to overstate the fresh water crisis on the planet today."[140] Water anywhere on the planet is an integral part of the global hydrologic cycle. Therefore, they state, "Unless we dramatically change our ways, between one-half and two-thirds of humanity will be living with severe freshwater shortages within the next quarter-century."[141] Humans share Earth with other creatures that also need water; therefore, a water shortage is also a crisis for wildlife. Of the 25 biodiversity hotspots designated by Conservation International, 10 are located in regions with water shortages.[142]

Water availability limits food production even more than land availability does, which means that a water shortage can become a food shortage.[143] Agriculture uses by far the largest amount of fresh water (70%). Because farmers produce 40% of the world grain harvest on irrigated land, anything that reduces the irrigation water supply will also reduce the food supply. It takes 1,000 tons of water to produce 1 ton of grain; therefore, exporters of grain are exporting water. The U.S. annual grain exports of 90 million tons of grain represent 90 billion tons of water, an amount that exceeds the 67-billion-ton annual flow of the Missouri River.[144] (see Chapter 13)

The near-simultaneous depletion of aquifers (because of the near-simultaneous shift to electrically powered pumps) means that cutbacks in grain harvests will be occurring in many countries around the same time. In addition, they will be occurring at a time when world population is growing. In effect, governments are currently satisfying the growing demand for food by over-pumping groundwater, creating what Lester Brown calls a "food bubble" economy.

As of November 2007, the capital of Turkey was rationing water. Harvests in the region have dropped by 15% to 30%. By the end of the summer 2007, local reservoirs and dams were holding 5% of their capacity. In Australia, over the past decade, in what they call "The Big Dry," precipitation has decreased by 5%, to 20%. This unprecedented drought over huge parts of Australia is the worst in at least a century, on an already dry continent. Australia is also part of the world's breadbasket, yet crops are failing regularly, and farms are closing down.

Most media reports will focus on one drought at a time; they rarely mention droughts occurring all over the world as a result of global warming.

Global Warming May Cause Abrupt Cooling in Europe

As odd as it sounds, global warming could trigger abrupt cooling in Europe. This severe cooling in Europe would occur while the rest of the world continues to warm. The problem is, again, a tipping point.

Europe is at roughly the same latitude as the northern U.S. and southern Canada. Rome is near the same latitude as Chicago (42°), and London and Paris are even with Duluth and the U.S.-Canada border (49°). However, each of these European cities is warmer than their North American counterparts because Europe is warmed by the huge Atlantic Ocean current, which flows north from the tropics, bringing with it warm water and warm, moist air. About 25% of the heat available to northern Europe comes from the Gulf Stream current. But that current could shut down. If it did, it would cool Europe and also affect the monsoons in India.

The process through which this would happen is clear. When the warm current gets to the North Atlantic near Greenland, it gives up the heat it has transported from the tropics, and the top layer of the ocean evaporates. When sea water evaporates, it leaves the salt behind, so the surface water becomes saltier, making it more dense than the water below. Because it is more dense, the surface water begins to sink, in a process called "flushing." This dropping water pulls more water up from the Gulf Stream to replace it, which keeps the ocean current "conveyer belt" moving. When the water drops, it goes all the way down to the ocean floor, then turns and heads south, where it continues along the bottom of the ocean to Antarctica, then north to the Indian Ocean, where it surfaces and begins its return trip to the North Atlantic. This current moves salt from the North Atlantic to the Indian and Pacific oceans and heat from the tropics to the North Atlantic.[145]

Coral Snake

If the salinity of the top layer is diluted, it would no longer be dense enough to drop, and the flushing would stop. If the process of flushing halted, the current would stop bringing warm air up from the tropics, and Europe would become much colder. Even a slight change in salinity (a few parts per thousand) could tip the balance. The top layer could be diluted with an increase in precipitation (fresh water)—which is one of the characteristics of global warming—or the melting of ice (freshwater ice) in Greenland could also dilute the top layer.

The results of abrupt cooling would be catastrophic. Europe today has more than 650 million people; it has excellent soils and is able to grow its own food. Without the warming from the ocean currents, the European climate would be more like that of Canada, which has good soils but lacks Europe's winter warmth and rainfall. Canada's agriculture supports its 28 million people. In an *Atlantic Monthly* article, "The Great Climate Flip-Flop," William Calvin says, "If Europe had weather like Canada's, it could feed only one out of twenty-three present-day Europeans."[146] He adds that the effect of cooling Europe "looks like a particularly efficient means of committing mass suicide."[147]

It is uncertain how likely this event is, but the point is that it could happen: ice core research shows that it has already happened at least eight times in Earth's recent history. The last two times this cooling happened was during the Younger Dryas, about 12,700 years ago, and again about 8,200 years ago, when it lasted for 1,300 years.[148] Ross Gelbspan, in *Boiling Point*, says that the last time cooling occurred, the entire change took place over about four years.[149] Whether or not this shift occurs, the possibility that it could happen demonstrates the scale of the changes we are making to Earth's systems.

To me the question of the environment is more ominous than that of peace and war. ... I'm more worried about global warming than I am of any major military conflict.

Hans Blix
U.N. weapons inspector

Global Warming Threatens National Security

People are now taking a look at national security issues involved in global warming. The 2007 IPCC Working Group II report on climate change impacts, adaptation, and vulnerability listed the impacts we can expect around the world, including increased famine, drought, heat waves, fires, storms, and infectious-disease outbreaks.[150] The IPCC estimates include the effects of global warming on creating lower agricultural yields in many countries, even as global population continues to grow.[151] This is a humanitarian crisis, but it also is a security crisis, for it will greatly destabilize many areas of the world. Where will all those people go, and how will they be received?

In March and April 2007, interest in the security issues increased dramatically. In March 2007, Senators Chuck Hagel, R-Neb., and Dick Durbin, D-Ill., introduced a bipartisan bill—the Global Climate Change Security Oversight Act—that would require federal intelligence agencies to collaborate on a National Intelligence Estimate to evaluate the security challenges presented by climate change. The National Intelligence Estimate would involve 16 intelligence agencies—including the Central Intelligence Agency (CIA), the Federal Bureau of Investigation (FBI), and various military intelligence groups. Durbin said climate change consequences are "a clear and present danger to the United States" and "a potential threat multiplier for instability around the world."[152] At the Senate hearing on the bill, Durbin said, "Many of the most severe effects of global warming are expected in regions where fragile governments are least capable of responding to them."[153] He described Africa's susceptibility to famine and the flood vulnerability of low-lying coastal areas in the Asia-Pacific region, an area with 58% of the world's population. Disasters in such areas could displace hundreds of millions of people, overburden national militaries, and require an international response.[154]

Developing countries do not have the government and social infrastructure in place to cope with the types of stresses caused by global climate change. When a government can no longer deliver services to its people, ensure domestic order, and protect the nation's borders from invasion and turmoil, the vacuum can be filled by extremism and terrorism. There were food riots in 33 countries in 2009 as a result of food shortages. There is more on failed states in Chapter 7.

The U.N. study also includes the examples of the 1994 genocide in Rwanda, which was intensified by violence over agricultural resources; the violence in Sudan, Chad; the 1970s downfall of Ethiopian Emperor Haile Selassie through his government's inability to respond to food shortages; and the 1974 Nigerian coup, which resulted largely from an insufficient response to famine caused by drought.[155]

An April 2007 report called *National Security and the Threat of Climate Change* was written by 11 retired three-star and four-star admirals and generals from all four services.[156] The report says that climate change presents significant risks to America's national security because climate change could be a "threat multiplier" in already fragile parts of the world. General Gordon R. Sullivan (Ret.), Chairman, Military Advisory Board, and former Chief of Staff, U.S. Army, said, "In the Cold War, there was a concerted effort by all leadership—political and military, national and international—to avoid a potential conflict." However, he continued, "climate change is exactly the opposite. We have a catastrophic event that appears to be inevitable. And the challenge is to stabilize things—to stabilize carbon in the atmosphere. Back then, the challenge was to stop a particular action. Now, the challenge is to inspire a particular action. We have to act if we're to avoid the worst effects."[157]

The risks they describe include rising sea levels that could threaten the livelihoods of more than 1 billion people living within 45 miles of Asia's coastlines. In Africa, recurring heat waves could cause widespread shortages of food and water, leading to large-scale migrations and escalating tensions. Many governments, even some that look stable today, may be unable to deal with these

Longhorn Bull

new stresses. A major concern is the significant spreading of the conditions for vector-borne diseases, such as dengue fever and malaria, and food-borne diseases, such as salmonellosis. The decline in available fresh water in some regions also has an impact on health because adequate supplies of clean water are needed for good health. By 2025, scientists project that 40% of the world's population will be living in countries experiencing significant water shortages. Most countries in the Middle East and northern Africa are already "water scarce," and the International Water Resource Management Institute projects that by 2025, Pakistan, South Africa, and large parts of India and China will also be water scarce. (For the U.S. to be considered "water scarce," it would have to have an 80% decrease in water supply.)[158]

> Journalist Ross Gelbspan has noted how news stories about global warming evoke an 'eerie silence.' This, indeed, is the mother of all silences, because the fossil fuel economy is the mother of all vested interests.
>
> David Edwards
> *Paying the Ultimate Price for Corporate Control of Society*

The UNEP completed an 18-month study of the Sudan in 2007.[159] As an example of the effect of water shortages, the report cites the conflict in Darfur between herders and farmers. Long periods of drought resulted in the loss of both farmland and grazing land to the desert. The failure of their grazing lands compelled the nomadic groups to migrate southward in search of water and herding ground, and that in turn led to conflict with the farming tribes occupying those lands. Rainfall is down by up to 30%, and the desert in northern Sudan has advanced south by 60 miles over the past 40 years, causing tensions between farmers and herders over disappearing pasture and evaporating water holes. As the desert advances, one group displaces another. Competition for land, when added to population growth and tribal, ethnic, and religious differences, turned violent. Darfur provides a case study of how existing marginal situations can be pushed beyond a tipping point by climate-related factors.[160] The UNEP report found that global warming could lead to a drop of up to 70% in crop yields in the most vulnerable areas of the Sahel and in the ecologically fragile belt from Senegal to Sudan.[161]

The climate risks do not need to cause wars over resources. There are increasing calls for "climate justice" because warming is going to have devastating impacts on poor nations. If we created a visible effort to share the world's resources more equitably, the climate crisis could create an unprecedented amount of cooperation. Much will depend on whether the U.S. and the EU work to monopolize most of the world's resources and protect this approach with the military, or whether they agree to help the world share what is available.

The Effort to Discredit the Science and Maintain Public Doubt

As of March 2010, 46% of people in the U.S. believe that climate change is exaggerated, that scientists are unsure about it, or that it is not happening. This is the highest level since polling began 13 years ago. In just two years, there was a strong rise in the number of people denying global warming. Opponents jumped on two events, hacked emails and a mistake about when the glaciers in the Himalayan glaciers might melt.[162]

In 2009, hacked emails of climate scientists from the University of East Anglia started what become known as "climategate," allowing climate change deniers to again claim that climate change is some sort of "worldwide conspiratorial scam." It was on front pages and network news shows. However, since then, five independent investigations have "concluded that the integrity of the science is entirely sound" and that the "deniers' arguments were debunked."[163] But there has been little media coverage of these studies. As Jason Linkins says, "This is, unfortunately, quite common. The right erupts with anger, the media treats the 'controversy' as a legitimate story, and the public hears all about it. We eventually learn that the story was nonsense, but at that point, the media has lost all interest."[164]

People just do not want to believe that global warming is real—it is too overwhelming and would require economic changes they do not believe in, and so they are looking for any excuse to dismiss global warming.

However, the very survival of human beings on Earth is at stake, so it is imperative that as many people as possible know what global warming is really about and why there has been so much confusion. The confusion and controversy over global warming are the result of the deliberate, well-funded, and well-orchestrated campaign of "disinformation" waged by the oil and coal industries. This campaign was designed to be invisible. John Stauber, editor of *PR Watch* and author of *Trust Us... We're Experts*, says, "By design, the PR industry carefully conceals many of its activities. Persuasion, by its definition, is subtle, says one PR executive. The best PR ends up looking like news. You never know when a PR agency is being effective; you'll just find your views slowly shifting."[165]

The technical nature of measuring global warming and the careful language of science have helped PR firms create an effective disinformation campaign. By planting seeds of doubt, anti-regulatory forces achieved confusion, stalemate, and inaction and made any actual evidence refuting global warming unnecessary. This effort used the same tactics (and often the same people) that the tobacco industry used to maintain doubt about the connection between smoking and disease.

Grist, the environmental news website, has a section called "How to Talk to a Climate Skeptic."[166] It includes answers to many of the statements made by skeptics.

Why Conservatives Have Rejected Global Warming

It is astonishing how effective this PR campaign has been. Science was once highly respected as the primary way to establish facts about the physical world. However, people were desperate to deny global warming because of its economic implications, and so in

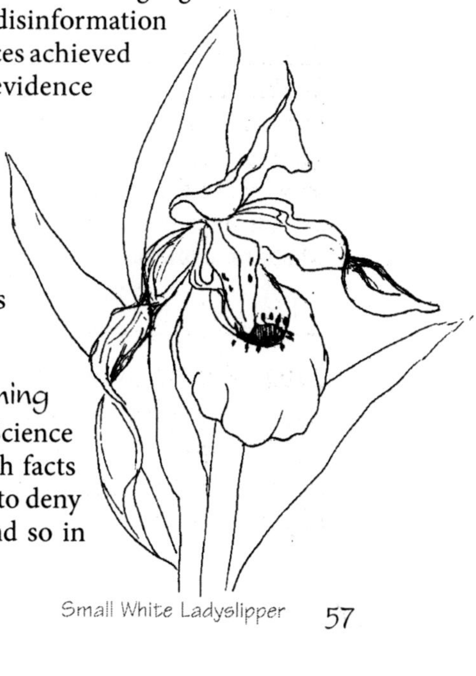

Small White Ladyslipper

order to attack the message, they first had to discredit the messenger: science. The skeptics have said that the science has been "hijacked" by a small group, or that have been some emails between a few scientists who were being argumentative about other scientists, or that have suddenly found a "mistake," or that scientists arguing for the reality of global warming constitute a "liberal special interest group" with an agenda for introducing regulations or government control. They argue that these occurrences reverse decades of research by thousands of scientists all over the world.

> I am a firm believer in the people. If given the truth, they can be depended on to meet any national crises. The great point is to bring them the real facts.
>
> Abraham Lincoln

But the basic science is now settled, and global warming is almost universally accepted by scientists—there is no small group that has "taken over" the science of global warming. Global warming is very real, it is caused by humans, and it is potentially catastrophic. What skeptics are really arguing is that if global warming were real, it would require government regulations and restrictions on economic activity; however, any regulations or restrictions are always wrong, so therefore global warming is wrong. But just because you don't like the required solutions doesn't mean it isn't happening. Just look at the real world: scientists have said that in 30 years the Arctic Sea could be ice free in the summer. The Arctic Sea is about the size of the continental U.S. and the ice there has been frozen year-round for 14 million years. [167] It's melting. And there is a reason for that. And no amount of plots, emails, or mistakes can change that reality.

Conservatives easily accepted this disinformation campaign. Science journalist Chris Mooney gives some of the reasons conservatives reject mainstream science on the issue of climate change. These include 1) the broad and longstanding conservative distrust of academia and "leftwing" campus intellectuals, including scientists; 2) right-wing think tanks, which provide alternative "facts" and alternative "knowledge" tailor-made for conservatives; 3) the growth of a right-wing media, which regularly quotes conservative think-tank "experts;" and 4) the fact that many conservatives do not listen to any other news service beyond the right-wing sources, creating a shielded knowledge environment.[168] I would also add another factor: conservatives tend to focus on individual behavior in isolation. Historically, individual choices have never created global consequences. Unless you can see the interconnections within the whole, global warming does not make sense. Even so, many conservatives have been worried about global warming.

Many evangelicals have begun to take action on global warming. In February 2006, despite some opposition from colleagues, 86 evangelical Christian leaders backed a major initiative to fight global warming, saying, "Millions of people could die in this century because of climate change, most of them our poorest global neighbors." Climate change threatens to condemn millions of people to poverty. Climate disasters between 2000 and 2004 affected 262 million people, 98% of them in the developing world.[169] The leaders called their statement *Climate Change: An Evangelical Call to*

Action.[170] Among the signers of the statement are the presidents of 39 evangelical colleges, such as Duane Lifting, president of Wheaton College, leaders of aid groups and churches, like the Salvation Army, and pastors of mega-churches, including Rick Warren, author of the bestseller *The Purpose-Driven Life*, and Rev. Jim Ball, Executive Director of the Evangelical Environmental Network.[171]

ExxonMobil's Disinformation Campaign

Much of the campaign to discredit the science of global warming has been orchestrated by ExxonMobil. By the end of the 1990s, most oil companies accepted global warming and began the shift to renewable energy. British Petroleum began going into solar, and Royal Dutch Shell began working on wind power. Nevertheless, ExxonMobil decided to continue to fight the reality of global warming. (Paul Krugman called ExxonMobil "an enemy of the planet."[172])

In January 2007, the Union of Concerned Scientists (UCS) published a study called *Smoke, Mirrors and Hot Air: How ExxonMobil Uses Big Tobacco's Tactics to Manufacture Uncertainty on Climate Science.*[173] This report demonstrates that ExxonMobil used the same five tactics that the tobacco industry used so effectively to confuse and control public opinion about the connection between cigarettes and cancer. Using these tactics, ExxonMobil:

- Manufactured uncertainty by raising doubts about even the most indisputable scientific evidence.
- Adopted a strategy of information laundering by using seemingly independent front organizations to publicly further its desired message and thereby confuse the public.
- Promoted scientific spokespeople who misrepresent peer-reviewed scientific findings or cherry-pick facts in their attempts to persuade the media and the public that there is still serious debate among scientists that burning fossil fuels has contributed to global warming and that human-caused warming will have serious consequences.
- *Attempted to shift the focus* away from meaningful action on global warming with misleading charges about the need for "sound science."
- *Used its extraordinary access to the Bush administration* to block federal policies and shape government communications on global warming.[174]

Great Horned Owl

When big tobacco developed their disinformation process, they realized they did not need to prove that their products were safe; they only had to "maintain doubt." The UCS report quoted one internal memo from the Brown and Williamson tobacco company: "Doubt is our product, since it is the best means of competing with the 'body of fact' that exists in the minds of the general public. It is also the means of establishing a controversy."[175]

This is *not* just a legitimate difference of opinion. In fact, while they waged this campaign, the fossil fuel industry knew that global warming was real. As long ago as 1994, Western Fuels (a coal company) stated in their annual report, "There has been a close to universal impulse in the [fossil fuel] trade association community here in Washington to concede the scientific premise of global warming ... while arguing over policy prescriptions that would be the least disruptive to our economy. . . .We have disagreed, and do disagree, with this strategy."[176] They decided they could not concede the science but must create confusion over the science of global warming to forestall debate over what to do about it.[177]

In their effort to create uncertainty and confusion, ExxonMobil helped fund a network of 43 different front organizations (identified in the UCS report) to publish in-house articles by select scientists. The UCS report shows how, by copying the tobacco company's pioneering "information laundering" plan, these organizations tried to:

> raise objections about legitimate climate science research that has withstood rigorous peer review and has been replicated in multiple independent peer-reviewed studies—in other words, to attack research findings that were well established in the scientific community. The network ExxonMobil created masqueraded as a credible scientific alternative, but it publicized discredited studies and cherry-picked information to present misleading conclusions.[178]

This web of organizations, with its overlapping personnel, advisors, and spokespeople, quietly gave the appearance of a broad platform of climate science contrarians or "skeptics." The UCS report said that this manufactured "diversity of the organizations creates an 'echo chamber' that amplifies and sustains scientific disinformation even though many of the assertions have been repeatedly debunked by the real scientific community."[179]

ExxonMobil also funds legitimate established research institutions to deflect criticism. One tactic they use to create doubt is to label peer-reviewed science "junk science" and call for "sound science," "balanced science," and "informed decisions." They also call the science of global warming the work of "ideologues," "cranks," and "incompetents" and argue that it is politically motivated. The attack does not need to be true or even accepted to block action. The objective is only to dismiss or marginalize global warming science as "junk science." However, serious scientists never criticize research using terms like "junk science" (this is only a PR term). Scientists are always specific; they will argue over how scientists collect their data, or how models are created and used, or that the conclusions are not supported by the data; they never label research as "junk science."

> Everyone is entitled to his or her opinions, but no one is entitled to his or her facts.
>
> Sen. Daniel Moynihan

Only *peer-reviewed* scientific journals publish legitimate scientific research and criticism. *None* of the scientists who argue against global warming have published in peer-reviewed scientific journals. The peer-reviewed, highly competitive journals

> practice strict quality control and require other scientists to replicate any results before an article is published, so that acceptance of research ultimately requires that they win over their scientific peers. Building consensus requires repeated testing and retesting of an idea.[181]

Ross Gelbspan says, "From the perspective of an investigative reporter, the central drama underlying this issue is crystal clear. It pits the ability of this planet to sustain civilization versus the survival of one of the largest commercial enterprises in human history. The oil and coal industries together generate more than a trillion dollars a year in revenues. In this battle, their resources are virtually without limit."[182]

In the period from January 2009 to June 2010, the world's top 35 companies and trade associations linked to fossil fuel, mining, and electric utility companies invested more than $500 million "in lobbying and campaign contributions . . . to defeat clean energy legislation," successfully convincing enough U.S. senators to oppose energy reforms.[183]

How the McCain-Lieberman Climate Stewardship Act Was Stopped

The people in these front organizations have published many articles on global warming. For instance, Sallie Baliunas and Willie Soon of the Harvard-Smithsonian Center for Astrophysics (an astrophysicist and a physicist—neither are in any discipline related to climatology) published an article in the journal *Climate Research*, a small, non-peer-reviewed journal. This journal allows any of its 10 editors to approve an article and publish it, and one of the editors (Chris de Freitas) was a known skeptic who approved the publication of their article. Afterward, the publishing of this article created a controversy, and the other editors protested. In fact, three of the 10 editors at *Climate Research* resigned in protest over the publication.[184] Malcolm Hughes, of the Laboratory of Tree-Ring Research at the University of Arizona, said, "The Soon et al. paper is so fundamentally misconceived and contains so many egregious errors that it would take weeks to list and explain them all."[185] In fact, 13 of the scientists cited in their paper took the unprecedented step of publishing a rebuttal explaining that Soon and Baliunas had seriously misinterpreted their research.[186] Nevertheless, it makes no difference that these articles are inaccurate—they are *in print* and can be used repeatedly in arguments by opponents of climate change.

This article was referred to in one of the emails in "climategate." The email objects to this paper in *Climate Research* and calls for scientists to boycott the journal. Skeptics argued that was evidence of scientists manipulating peer review to squeeze out legitimate climate dissenters. Yet

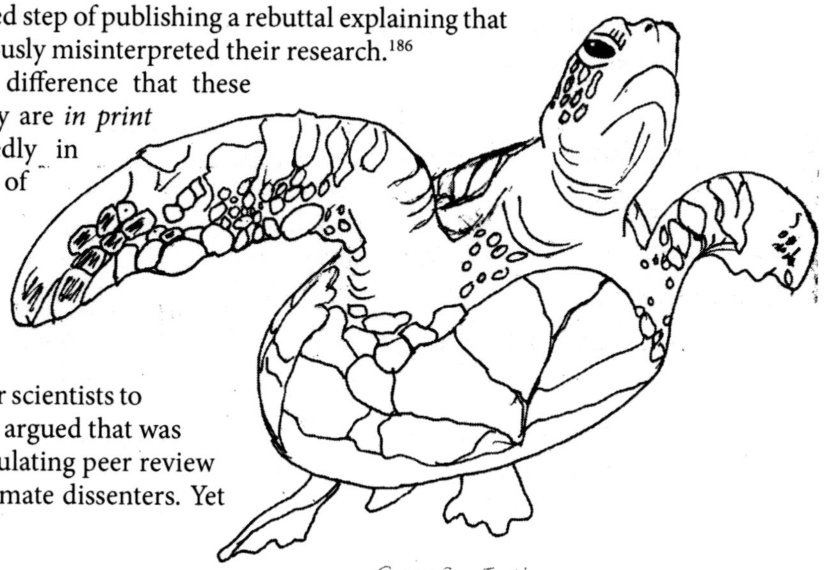

Green Sea Turtle

many subsequent publications set the record straight, which shows that over time, the peer review process corrects inaccurate articles. Scientists later discovered that the paper was funded by the American Petroleum Institute.[187]

It took another article only three days to help defeat the McCain-Lieberman Climate Stewardship Act. As the Senate debated the Act, an article titled "Errors and Defects in Mann et. al. (1998) Proxy Data and Temperature History" was published in *Energy and Environment* (an obscure, non-peer-reviewed journal with a self-proclaimed skeptic editorial board). It was written by Canadians Steven McIntyre (a mining executive) and Ross McKitrick (an economics professor). McKitrick is a senior fellow at the Fraser Institute, a strongly anti-regulatory group, and co-author of the book *The Troubled Science, Policy and Politics of Global Warming*. McIntyre and McKitrick, calling their work an "audit," said that when they took Mann's (a lead author of an IPCC report) data and reproduced his calculations, they arrived at different answers. They did not check their data with Mann before they rushed their paper into print. Once published, it was easy to compare their data with Mann's, which was clearly substantially different from what they had used. Even so, politicians who opposed global warming ideas gave out press releases about this article in Washington for use in the debate: "Important global warming study audited—numerous errors found" and "New research reveals the UN IPCC theory of climate change is flawed."[188]

Long before the human species is extinct, we may know that we are irreversibly en route to extinction. . . . .(and) the opportunity to take the action that would have saved the species has come and gone.

Jack Miles
Global Requiem: The Apocalyptic Moment in Religion, Science, and Art

One day after this article was published, on October 28, 2003, *USA Today* published an op-ed piece by Nick Schulz, who is an editor at an industry and anti-regulatory front-group, Tech Central Station. He said, "An important new paper in the journal *Energy and Environment* upsets a key scientific claim about climate change. If it withstands scrutiny, the collective scientific understanding of recent global warming might need an overhaul."[189] After further attacking Mann's work and procedures, the op-ed piece concludes, "Given the implications of the errors and problems they apparently have unearthed within the Mann study, the two researchers have done a tremendous service to science and the public, which should rely on facts to make informed public policy decisions."[190] This op-ed piece actually says that this single article by two non-scientists threatened the decades-long work of thousands of global warming scientists whose research is published in thousands of articles in respected, peer-reviewed journals.

One day after the *USA Today* op-ed piece appeared, Sen. James Inhofe (R. Oklahoma, an oil state; he is a major opponent of global warming) quoted it on the floor of the U.S. Senate in the anti-regulatory debate over the Climate Stewardship Act. He said, "This is from a publication called *Energy and Environment*. . . . It starts talking about the flaws in the logic that were used by the Mann study [the IPCC report]. The flaws all come out."[191] In these same comments, he also refers to the study by Soon and Baliunas discussed above: "This was the most far-reaching study ever made on climate change .

. . . They came to the conclusion that climate change is not real, that the science is not accurate."[192] The PR campaign was successful. The next day, on October 30, 2003, the McCain-Lieberman Climate Stewardship Act was defeated in the Senate by 43 to 55 votes. (Sen. Inhofe was Chairman of the Senate Committee on Environment and Public Works until the 2006 elections.)

In late 2007, Sen. Inhofe also released a list of more than 400 "prominent scientists" who "disputed man-made global warming claims." By 2009, Inhofe had expanded his list to just under 700 people. The Inhofe list is often cited by climate skeptics; however, the Center for Inquiry conducted a careful study of the list in the summer 2009. The study found that fewer than 10% of the people on Inhofe's list could be identified as climate scientists, that 4% actually favored the IPCC consensus on anthropogenic global warming, and that 80% of the list had no peer-reviewed publications related to climate science.[193]

Media Reporting of Global Warming

The media, although it claims it wants to present a "balanced" view of global warming, aided skeptics in this effort to confuse and mislead. The problem is that "balanced" is not the same as accurate. There is no balanced reporting of scientific fact, only of opinion, but by quoting both sides, the media turned global warming into a matter of opinion. The popular media treated the issue as controversial among scientists and failed to distinguish between scientific opinion and political opinion. Their articles would say "some" scientists believe global warming is happening and "some" do not. Nowhere did they mention that the ratio of the two groups was 1,000 to 1, nor did they mention the oil and coal affiliations of the "skeptics." This allowed skeptics to present a false impression of scientific disagreement. Journalism also likes sensationalism and conflict to sell papers or increase ratings. In addition, the owners of print and TV stations are large, multinational corporations, which are very economically conservative, as are their advertisers, who all have a strong interest in stopping any new regulations and therefore a strong interest in discrediting information about global warming.

The reporting of global warming in the general media is very different from what is in scientific journals. Maxwell T. Boykoff and Jules M. Boykoff analyzed coverage of global warming in the *New York Times*, *Washington Post*, *Wall Street Journal*, and *Los Angeles*

Whooping Crane
One of the most endangered birds in the world

Times between 1988 and 2002. Reporting in these papers did not reflect the scientific consensus. They analyzed a random sample of 636 articles. They found that a majority, 52.7%, gave "roughly equal attention" to the two sides, just 35.3% emphasized the scientific consensus view, 6.2% emphasized the industry-supported view, and only 5.9% focused on the scientific consensus view without bothering to provide the industry/skeptic counterpoint. The authors conclude that the U.S. "prestige-press" has produced "biased coverage of global warming . . . hidden behind the veil of journalistic balance."[194] In contrast, Naomi Oreskes, a historian of science at the University of California, San Diego, published a report in 2004 in the journal *Science*. This report reviewed the peer-reviewed scientific literature for papers on global climate change published between 1993 and 2003. In a random sample of approximately 10% of the literature, or 928 studies, *not one* disagreed with the consensus view that global warming is real and that humans are the cause.[195]

> The integrity of climate research has taken a very public battering . . . . Scientists must now emphasize the science, while acknowledging that they are in a street fight.
>
> Joe Romm
> Climate Progress.org

The Bush Administration Worked to Discredit Global Warming

Democracy cannot exist without the free flow of honest information. However, the Bush administration was part of the effort to manipulate public opinion about global warming. The Bush administration supported only voluntary programs to reduce greenhouse gases, and they worked to stop any other response by the government. President Bush also went back on his pre-election promise to sign the Kyoto Protocol. This has stalled most of the work to stop global warming. The president withdrew the U.S. from the Kyoto talks but said that our withdrawal would not affect the efforts of other countries. Nevertheless, in 2004, the Bush administration was able to seriously weaken the next round of climate talks. When the parties to Kyoto met in Bonn the following May to discuss the next commitment period, they were prohibited from coming out with any action plan at all. As one veteran climate negotiator said, the U.S. left the climate talks "hanging on to a rock face by their fingernails." Ross Gelbspan says, "This is not political conservatism. This is corruption disguised as conservatism."[196]

Robert F. Kennedy, Jr., in his book *Crimes against Nature: How George W. Bush and His Corporate Pals Are Plundering the Country and Hijacking Our Democracy*, says that Bush was the worst environmental governor in America. When he was governor, Texas became the most polluted state in the country, with the highest levels of both air and water pollution and of toxic waste and its uncontrolled release. And it was forty-ninth of the 50 states in per-capita environmental spending. Only Gov. Mike Leavitt of Utah spent less on the environment, and President Bush named Leavitt to be head of the U.S. Environmental Protection Agency (EPA), which put him in charge of all of America's environmental assets and laws.[197]

The Bush administration also tried to control communication to the public and Congress about research done by NASA and NOAA. Taxpayers fund these research institutes, and people have a right to know about their results. James Hansen (Director of the NASA Goddard Institute for Space Studies) testified before the House Committee on Oversight and Government Reform. In his testimony, titled "Political Interference with Government Climate Change Science," he said, "In my more than three decades in government, I have never seen anything approaching the degree to which information flow from scientists to the public has been screened and controlled as it has now."[198] For one thing, he said he "was forced to take down from our web site our routine posting of updated global temperature analysis." Because of this effort to keep the science from the public and even *Congress*, Hansen said there "remains a vast gap between what is understood about global warming by the relevant scientific community, and what is known about global warming by those who need to know, the public and policy-makers."[199]

In 2006, Hansen publicly protested this censorship at NASA, and a month later, the budget for NASA was, by executive action, quietly slashed by 20% retroactively to the beginning of the fiscal year, even though Congress had approved the budget.[200] With all their fixed expenses, "the budget cuts wipe off the books most planned new satellite missions." As he says, "One way to avoid bad news: stop the measurements!"[201]

In 2001, to keep information about global warming from the public and from Congress, President Bush appointed Phillip Cooney chief of staff for the White House Council on Environmental Quality. Here, Cooney edited and changed hundreds of reports on global warming. For the decade before joining the administration, Cooney was a lawyer for the American Petroleum Institute, the oil industry lobby, where he was called the "climate team leader." In 2002, as U.S. government scientists finalized the U.S. government's interagency Climate Change Science Program's strategic plan, Cooney dramatically altered their document, cutting and inserting qualifying words to create an appearance of scientific uncertainty about global warming and its implications and to minimize the extent to which the cause is humans.[202] (See UCS Appendix C for sample edits.) Then in 2005, Rick Piltz, a researcher in the Climate Change Science Program, resigned in protest over Cooney's censorship.[203] On June 10, 2005, two days after the *New York Times* first reported Piltz's revelations, Cooney resigned, and a week later, he announced that he had accepted a high-ranking PR position at ExxonMobil.[204]

The UCS, with the Government Accountability Project (the nation's leading whistleblower protection organization), issued a second report, *Atmosphere of Pressure: Political Interference in Federal Climate Science*. This project surveyed government scientists and found that 150 scientists reported personally experiencing at least 435 incidents of political interference, clearly showing a pattern of abuse.[205] Federal scientists have a constitutional right to speak about their scientific research, and the American public has a right to be informed of the findings of their taxpayer-supported research. This report discusses the ways in which "U.S. federal climate science has been filtered, suppressed, and manipulated in the last five years." A scientist with the U.S. Department of Agriculture said, "To be in denial on climate change is a crime against the Nation."[206]

Tiger Salamander

The ability of scientists to speak out is beginning to change. On March 15, 2007, the House of Representatives overwhelmingly passed the Whistleblower Protection Enhancement Act, which for the first time would grant federal scientists and contractors the right to expose political interference in their research without fear of retribution. The testimony of Hansen, mentioned above, was at hearings on this legislation.

Ross Gelbspan called these efforts of Big Oil and Big Coal "*an unconscionable crime against humanity*."[207] It is hard to imagine or describe how incredibly irresponsible, dangerous, and deliberate this effort to dismiss global warming has been.

Can We Change Fast Enough?

Delaying action to address climate change will increase the environmental and societal consequences as well as the costs. The longer we wait to respond, the harder and more expensive the task will be. These consequences are cumulative, and the problems get worse and worse, faster and faster. In addition, the UCS *Smoke, Mirrors, and Hot Air* report said that positive feedbacks and the inertia of the oceans and ice sheets means that "if we push the climate system hard enough it can obtain a momentum, it can pass tipping points, such that climate changes continue, out of our control. Unless we begin to slow down the human-made climate forcings [greenhouse gases], there is the danger that we will create a different planet, one far outside the range that has existed in the course of human history."[208]

After 20 years of the anti-global-warming PR campaign, how do we convince people that there are so many different nightmarish possibilities, it would take only one of these to throw the world's social systems and economies into chaos? The only way to avoid this chaos is to act now and as fast as humanly possible. Are we incapable, politically and psychologically, of making the changes needed, soon enough? Currently, the U.S. political system can be deadlocked over even minor incremental changes. Elizabeth Kolbert, in her three-part report, "The Climate of Man," in the *New Yorker*, asks,

> As the effects of global warming become more and more apparent, will we react by finally fashioning a global response? Or will we retreat into ever narrower and more destructive forms of self-interest? It may seem impossible to imagine that a technologically advanced society could choose, in essence, to destroy itself, but that is what we are now in the process of doing.[209]

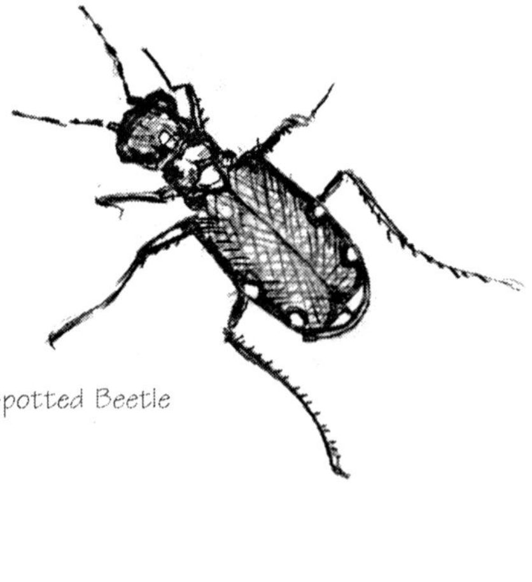

Six-Spotted Beetle

Part II

Causes of the Climate Crisis

Emperor Penguin Family

Chapter 2

Views of Science that Contribute to the Crisis

Global warming clearly demonstrates that the current way our societies and economies are structured must be transformed. Our survival as a species depends on it. Once we understand the enormity of the threat of global warming and have an adequate sense of urgency, we need to concentrate on what we want to create. Peter Senge, in *The Fifth Discipline*, a study of systems thinking, says, "A problem solver tries to make something go away. A creator tries to bring something into being."

For every complex problem there is always a simple solution. And it is wrong.

H. L. Mencken

The first step is to recognize that the world we have now is not the only way the world could work—there are alternatives. What is required is a clear vision of what we want, of the desired result. The future is wide open, but to find the solutions we need, we will need to be passionately curious and creative. Joanna Macy gives her description of the transformations that are already happening:

> What a difference it makes to view our efforts as part of a vast enterprise, a tidal change commensurate to the crisis we face. What is underway, as many have observed, is a revolution that is comparable in magnitude to the agricultural revolution of the late Neolithic and the industrial revolution of the past two centuries. As the industrial-growth society spins out of control comes the third revolution, which is even now given names, like the ecological or sustainability revolution, or the Great Turning. While the first two revolutions, as former EPA administrator William Ruckelshaus reflects, "were gradual, spontaneous, and largely unconscious, this (third) one will have to be a *fully conscious operation.* … If we actually do it, the undertaking will be absolutely unique in humanity's stay on Earth."[1]

This fully conscious transformation has many dimensions. We need to reject the understanding of human nature as selfish and greedy (which justifies selfish and greedy behavior) and instead reclaim human potential for compassion and cooperation (Chapter 3). We will need to create a morality based on a global ethic of care (Chapter 7) and an economics capable of creating well-being for everyone (Chapter 8).

The field of social ecology has been developing these ideas for decades—they now must be put into use and define our behavior. The foundation of the ecological worldview is systems science, and the first step in understanding the creation of a sustainable world is to understand systems science. Much of our understanding of people and the world is created by using metaphors based on our understanding of science. Currently, our worldview is based on mechanistic Newtonian science, where we understand things by looking at them in isolation and where cause and effect are clearly linked. However, the new complex systems science allows for new metaphors that create a new way of thinking about the world. Isolated analysis is replaced by analysis of whole systems—including their relationships and interdependencies. Simple, deterministic cause and effect are replaced by a far more complicated network of interaction in a complex system.

Newtonian/Mechanistic Science: A Focus on the Parts

Newtonian science radically changed the Medieval worldview. In the Middle Ages, people saw life as organic, describing self, society, nature, and the universe with the metaphor of a living organism. But the new science of Newton changed this vision of an organic, living world to a vision of a fragmented, mechanistic world. People used metaphors from this "new" science to redefine life and society in terms of a new metaphor. Rather than imagining the world as a *living organism*, people began to see the world as a *machine*—dead matter moving in a clockwork universe. About the same time, the philosopher Descartes began the development of radical individualism, in which there is a strong separation between subject and object. Then in the nineteenth century, metaphors from Darwin were added to those from Newton, adding concepts of hierarchies, competition, and survival of the fittest.

Essentially reductionism is a watchmaker's view of nature … Reductionism imagines nature as equally capable of being assembled and dissembled.

John Briggs and F. David Peat
Turbulent Mirror: An Illustrated Guide to Chaos Theory and the Science of Wholeness

Using the logic of this scientific revolution, the world is a machine, and behavior is ordered and predictable from analyzing the parts because the machine is static and the output is reliable. This allows for prediction and control. We can simply "fix" any unwanted effect by adjusting the behavior of some of the parts. We assume that we can accurately anticipate any future behavior because we can understand the behavior of the parts and the limited interactions between them. We assume that we can predict how an action will affect the whole, and we engage in complex planning based on the idea that the world is predictable as long as we can correctly quantify and measure the right things.

Newtonian science has functioned powerfully to build the machines of industrialization and to send rockets to the moon. But we have used metaphors based on this science to also describe society, economies, and the natural world, where it is not appropriate. This point of view constrains our ability to understand our problems and limits our ability to envision solutions.

The understanding of causation in this science has had a tremendous impact. In Newtonian physics, causation is always proportional and linear: "for every action there is an equal and opposite reaction." In the Newtonian

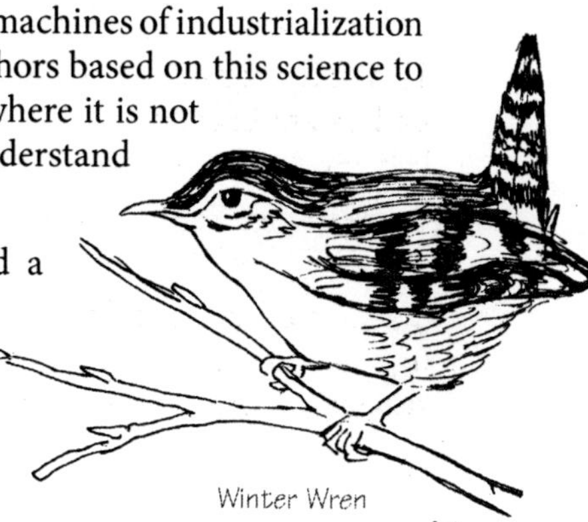

Winter Wren

world, we assume that small changes create small effects and that only a large cause can create a large effect. When considering cause and effect, we normally assume a direct one-to-one link: A causes B, B causes C, etc. In a Newtonian world, causation:

- is clear-cut;
- is sequential;
- creates a response that is proportional to the input;
- goes in one direction: A causes B, which in turn causes C. This is true in a machine, such as a rocket. The parts interact mechanically and reliably.

With a Newtonian understanding of causation, there is no room for surprise or creativity; therefore, this view leaves us with very limited hope. However, causation is understood very differently in a newer kind of science.

A Complex Systems View of the World: Seeing the World Whole

There is an alternative to a Newtonian/mechanistic view of the world in a new field in science called complex systems theory, complex/adaptive systems theory, or the science of complexity. These systems ideas have now begun to influence virtually every discipline and are yielding entirely different images and metaphors from those derived from Newtonian science. This new science is fundamentally about the nature of the whole and about how the whole is created by the interrelationships between the parts. Communities, the natural world, and economies are not machines—they are complex systems.

> The human mind has abstracted from the whole all conscious intelligence and purpose and meaning, and claimed these exclusively for itself, and then projected onto the world a machine.
>
> Richard Tarnas
> *The Passion of the Western Mind*

Complex systems science sees the world as integrated wholes whose properties are of the whole and cannot be reduced to the properties of the parts. This stands in marked contrast to the reductionism of seventeenth-century science. The primary focus is on the *relationships* between the parts, that is, on the nature of the connections. Systems science shifts the focus from objects (parts or individuals) to a focus on relationships and connectedness within the whole; therefore, systems thinking is "contextual thinking." In particular, it sees the whole and the close interconnectedness within the whole; therefore, relationships rather than individuals (parts) are primary. Complex systems are dynamic, not static—they continue to develop and evolve; therefore, systems are always in a process of renewal, change, and transformation. Well-functioning systems also require diversity, which increases resilience. The greater the diversity is, the greater the resilience.

Using concepts from complex systems, our understanding of problems changes and new solutions open up that are appropriate, adequate, and "realistic." A worldview using this systems thinking creates a far more accurate understanding of the dynamics and patterns at work in the world. Complexity science also has a very different understanding of causation and of how change happens. Paradoxically, even the largest changes can be the result of very small inputs. Conversely, very large inputs may only lead to small changes.

The chart at the end of the chapter compares how the world has been understood for centuries using images and metaphors from Newtonian science (on the left side), with new images and metaphors based on complex systems (on the right side). In systems theory, relationships are primary rather than individuals, so that we are interdependent rather than autonomous. It shows that because of the inherent interconnectedness of networks, there are no hierarchies. The world is highly unpredictable, no matter how much we understand the dynamics; therefore, tolerating ambiguity is essential, control is an illusion, and certainty is impossible.

Non-linear Causation and Unpredictability

One fundamental difference in complex systems is a new understanding of causation that is fundamentally different from that of our current Newtonian worldview. In fact, in this view, our general ideas of causation become meaningless. Complex systems exhibit "non-proportional causation." This means that the overall response of the system to a changing input is not only *non-proportional*—it is not meaningful to associate cause and effect between individual parts of the system. Because of the interlocking, interconnected feedback loops in a complex system, small events can be amplified as they cascade through the system, creating a large effect. A "positive" feedback loop amplifies or increases the change, and a "negative" feedback loop dampens or regulates the change.

The "complexity" in complex systems is in the level of interconnections within the system, which creates continual cycles of interactions. Complex systems are dynamic and the interactions are complex, which means that uncertainty grows rapidly, and predictability is very limited.

This close interconnectedness was first understood with the invention of the computer. The invention of the microscope allowed us to see the very small and very near. The invention of the telescope allowed us to see the very large and very far away. The development of the high-speed computer allowed scientists to simulate the behavior of complex systems and to "see" the behavior of a complex system for the first time. With the computer, as with the microscope and the telescope, scientists discovered a completely new level of reality, one that had previously been inaccessible and imagined only by a few.

Understanding complex dynamics of systems began in 1960, when a meteorologist at the Massachusetts Institute of Technology (MIT), Edward Lorenz, discovered "chaos theory" and non-proportional causation. While creating

Wild Rooster

an early computer model of the weather, he re-ran a section of the weather program that he had just run, typing in the same data. When he looked at the results, the printout showed weather predictions for the first few days to be similar to that of the first run, but then for the following days, the predictions were completely different. At first, he thought he had a software problem, then perhaps a hardware problem. He finally realized that the original weather data in the computer were stored using six decimal points. He had retyped the data using the numbers on the printout, which had used only three decimal points to save space. He had assumed that a difference of one part in a thousand would not matter. When he realized that this small change could have such a substantial effect, he knew that the weather would never be predictable for more than the short term.[2] This came to be called the *butterfly effect* because it was "as if a butterfly flapping its wings in Hong Kong could change the weather in New York City days later."[3] This intense interconnectivity, or what is now called "sensitivity to initial conditions," means that the ability to predict specifics in the behavior of complex systems is extremely limited.

> Problems cannot be solved at the same level of consciousness that created them. . . . We shall require a substantially new manner of thinking if humankind is to survive.
>
> Albert Einstein

For most people, non-proportional causation is very counter-intuitive. We are conditioned to intuitively expect even large, fast changes to result in a bell curve outcome. In a so-called normal distribution of events, most results are in the middle, and extreme events are on either end, few in number and not very far from the middle. However, a complex system does not generate our idea of normal outcomes. Instead, when a complex system is in what is known as a "critical state," a state toward which many complex systems normally gravitate, its behavior exhibits what is called a "power law distribution," meaning that the range of outcomes created by some given input can be very slight or enormous or anything in between, and the outcome cannot be predicted. The fundamental difference between a bell curve and a power-law is that the power-law distribution is much more likely to consider very large changes in relation to the probability of small changes than the bell curve will. In other words, the bell curve will not support the wide range of outcomes that a power law will support.

In complex systems, interconnections create a very different kind of causation:

- impulses do not move through the system in a sequential way but ricochet throughout the system, taking many paths at the same time, so that causation quickly becomes too complex to follow;
- since causation goes in all directions, everything can in fact be both the cause and the effect of everything else;
- the responses to an input can be non-proportional; that is, a very small input can have an enormous outcome; therefore,
- the entire concept of linear cause and effect (A causes B, which causes C, ...) breaks down and becomes meaningless.

Examples of Non-proportional Causation

Earthquakes and sand piles are often used as illustrations of this intense interconnectivity of complex systems and the implications for concepts of causation and prediction. In *Ubiquity, Why Catastrophes Happen*, Mark Buchanan describes experiments in which a sand pile is created by a mechanism dropping sand on a table one grain at a time. The pile gradually gets bigger. As the pile grows to a particular slope, however, the entire system of grains (resting on each other) reaches a critical state. It is now impossible to predict the effect that any one grain will have as it is dropped—whether it will cause an avalanche, or how large the avalanche will be if one does occur. When the pile reaches this critical state, it only takes one grain of sand to create an arbitrarily large avalanche. When it will happen depends on where this grain falls and on how all the other grains have stacked up, or on "sensitivity to initial conditions." This sand game, Buchanan says, has been "central to the discovery of the underlying cause of a vast range of tumultuous events."[4]

> Reality is made up of circles, but we see straight lines.
>
> Peter Senge
> *Fifth Discipline*

The sandpile is a good model for Earth's crust in relation to earthquakes. Earthquakes are an example of non-proportional causation and unpredictability. Earth's crust is interconnected in the manner of a complex system and it is in a critical state, which means that every change or shift affects the entire system (in this case, Earth's crust). The magnitude of an earthquake depends not on the size of the first shift in rock but on how this shift reverberates unpredictably throughout the system and on how the rest of the system—the surrounding rock—responds. A small shift in the rock can trigger the release of any amount of energy, anything from a 1.2 to a 9.8 as measured on the Richter scale. The Richter scale measures exponentially growing energy, with a multiplier of 10. It directly implies that earthquakes are governed by a power-law distribution. An earthquake that is a 2 on the Richter scale is 10 times more powerful than one that measures as a 1;

a 3 is 100 times more powerful than a 1;
a 4 is 1,000 times more powerful than a 1; . . . and
a 9 is 100,000,000 times more powerful than a 1.

An earthquake of 3 cannot even be felt. A 9 would release the energy of a thousand nuclear bombs.[5] This is what is meant by a power-law distribution supporting a very wide size range of outcomes. The same input or shift in the rocks triggers an unpredictable result within an enormous range of possible outcomes. The non-linear, chaotic reaction to the initial input determines the magnitude of the response. This means that even when we understand how a system works, we still cannot predict the future state of the system.

Complex systems are inherently so intertwined that no matter how well we understand the basic dynamics between parts, or how the parts themselves work, predicting any particular outcome is, in principle, not possible; therefore, there will always be uncertainty in the system's overall behavior. In a sense, in order to predict the weather long-term, you would need to measure each flap of every butterfly's wings—and do

Honey Bee

it moment by moment. In addition, each flap must be measured to infinite precision; therefore, in practice there is no way to achieve adequate accuracy and thus no way to predict the future state of the system.

Global warming and other ecological problems are about our disturbing multiple interacting complex systems in the physical world that are all individually being greatly perturbed. Systems, in principle and in practice, can undergo abrupt, discontinuous transitions (essentially this means a very large change in the system dynamics resulted from a normal, small input) from one state to another. These events, called structural changes in the system, mean that the system moves directly from one major state to another without transitioning gradually through any intermediate states. This is what climatologists are most concerned about in global warming. It is not just that increased CO_2 in the atmosphere causes gradual warming of the average global temperature, it is that the various complex subsystems of Earth are very sensitive to this warming and we could be setting off a whole series of abrupt, large chain reactions, flipping the larger global climate system into a totally different state, a state within which humanity may not be able to survive. This is why scientists say that change in Earth's climate is regulated by a switch (suddenly) and not by a dial (gradually).

Nature is to be viewed as a dynamical shifting web, not a mechanical, hierarchical pyramid.

John Briggs and F. David Peat
Turbulent Mirror

In *Abrupt Climate Changes: Inevitable Surprises*, the National Academy of Sciences describes why non-proportional climatic responses could cause significant, abrupt changes in Earth's climate, rather than our just experiencing a gradual change in temperature. The report states, "Technically, an abrupt climate change occurs when the climate system is forced to cross some threshold, triggering a transition to a new state at a rate determined by the climate system itself and faster than the cause. Chaotic processes in the climate system may allow the cause of such an abrupt climate change to be undetectably small."[6]

The 2008 financial crisis also demonstrated discontinuous transitions, meaning that because of the close interconnections within the economy, the economy didn't gradually decline over time—it collapsed almost overnight. The unexpected collapse of Russia in the 1980s is another good example, as is the surprising chain of events leading to the election of Nelson Mandela.

This non-proportional change makes clear how much we have misunderstood and underestimated how dangerous global warming really is; however, we also have underestimated how fast societies can cross a threshold and make a positive jump to a new state. That is, understanding non-proportional change allows us to recognize that positive and effective social change is also quite possible.

I will return to this systems understanding of causation in Chapter 7, which discusses morality. Morality is based on one's understanding of causation—specifically whether causation should be viewed from a Newtonian or complex systems perspective, that is, whether a person's poverty is primarily the result of his or her lack of effort or results from social forces. Morality addresses what people think of as the cause of a problem.

We have two different worldviews that have very different understandings of causation; therefore, we have two very different understandings of morality.

The Capacity for Self-Organization

In addition to non-proportional causation, complex systems, through their response to feedback, can "self-organize." Self-organization occurs when widespread patterns of interconnections (relationships) send information cascading throughout the entire system, allowing the system to regulate itself. When new behaviors arise out of these complex dynamics, systems scientists say that the behavior is "emergent." Systems thinking stresses dynamic interaction and change *over time*, rather than simply examining static structures. The effects of change over time are the result of feedback loops. Negative feedback negates the changes; it's a stabilizing force that works to maintain the status quo. Positive feedback loops amplify the effects, which makes the system unstable and open to change.

In any complex system, self-organization emerges spontaneously as the network "learns" how to function in coordinated and seemingly goal-directed ways without a central coordinating authority. This learning takes the form primarily of forming new connections, pathways for new information, energy, or matter. By virtue of these new connections, the system is now capable of new behaviors and typical states. Physicist Ilya Prigogine believes this self-organization or the "sudden appearance of order out of chaos is the rule rather than the exception."[7] We have seen this order emerge repeatedly on the Internet, where no one is coordinating the interactions. The scientist Stuart Kauffman has written extensively about how he thinks this mechanism led to the emergence of life itself on Earth.

Self-organization means that global order (order throughout the whole system) can emerge spontaneously from local behavior without its being imposed by force or coordinated by a central mechanism from above. This is why the full name for this science is complex/adaptive systems theory because systems are capable of adapting to their environment without a centrally administered blueprint for how to do so.

Mechanistic ideas assume that someone or something must be in control. This way of thinking leads people to work toward gaining control and avoiding being controlled. It also leads them to think that only strong central control can direct the behavior of a complex system. But, given the capacity of a complex system to self-organize, we see that in social systems, for example, immense change is possible without strong central control. Moreover, we see that strong central control actually cannot direct the behavior of a complex system with any reliability. Paradoxically, given certain parameters, close interconnections, and sufficient information freely cascading throughout the system, significant change can "emerge" spontaneously. The key to change in this model is not to employ sufficient force from some central point, but rather to change the interconnectedness of the system in a way that causes the system to be

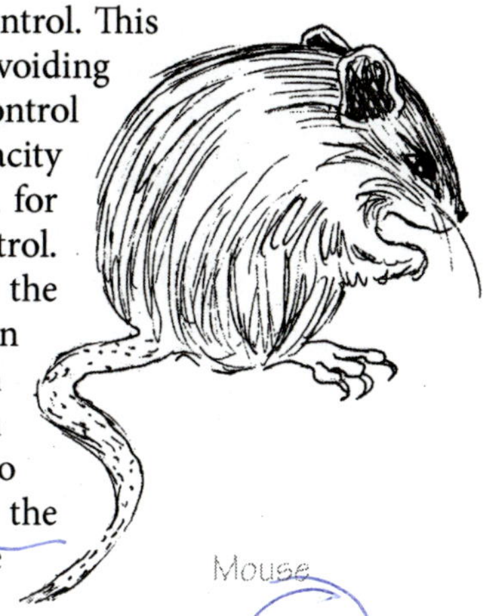

Mouse

capable of new behaviors and adaptations—in other words, to facilitate the ability of the system to further self-organize.

To give another example of emergent change, in *Solidarity: The Origins and Implications of Polish Trade Unions*, Jerzy Wojciechowski describes the level of change created by the Polish movement, Solidarity. He says, "From a few workers involved in a strike at the Lenin shipyard in Gdansk in early August, 1980, the idea grew and spread until by August 31 [Solidarity] had reached 9.5 million card-carrying members."[8] This new movement spread like wildfire, going from a handful of members to almost 10 million in less than *one month*. And that was before the existence of the Internet or cell phones, which today gives us even more potential because we now have a new ability to flood social systems with new information and create new connections between people. The Internet and cell phones allow people to repeatedly distribute messages to multiple recipients. These new technologies create new ways to cooperate, new ways to network. Remember that we do not need everyone; we only need a critical mass working together, which may be a surprisingly small number.

Systems Theory Is the Scientific Formulation of Ecological Awareness

The total-system response to our ecological crises is called "sustainability," in which activities must be able to be sustained over the long term without harm to the system. The focus of ecology is on the world as a whole—humans and all of nature as one interconnected whole.

However, the term sustainability is often misunderstood because it is heard within our current mechanistic worldview and not in terms of nature and living systems. Physicist Fritjof Capra says systems theory is "the scientific formulation of ecological awareness."[9] But, he says, the word sustainability has been "distorted, co-opted and even trivialized" through the use of the term without the ecological context that gives it the proper meaning: "Sustainable community is designed in such a way that its ways of life, business, economy, physical structures, and technologies do not interfere with nature's inherent ability to sustain life."[10] These "principles of sustainability" could also be called principles of ecology or principles of community and include the following:

- an ecosystem generates no waste, one species' waste being another species' food—that is, matter cycles continuously through the web of life;
- the energy driving these ecological cycles flows from the sun;
- diversity gives life resilience and flexibility [but only] as long as the system or network is not fragmented and the diverse information flows freely through all the links of the network;
- life, from its beginning more than three billion years ago, did not take over the planet by combat, but by cooperation, partnership, and networking.[11]

We can't assume the whole is understandable from studying the parts. The "whole" is primary, and nothing can be left out. Causation is very mysterious: enormous change can occur suddenly and without warning. That is true of the sudden melting of the Greenland ice or in the sudden enormous change when Nelson Mandela became president. It turns out that what we know of complex systems science gives a far more accurate understanding of societies and economies than the mechanistic alternative and that Newtonian science in these fields is inappropriate, creates confusion, promotes the wrong kinds of intuition, and causes people to misunderstand problems and solutions. These two worldviews can be applied to society, politics, economics, and even religion. We will only be able to heal Earth and ourselves if we pattern our understanding of the world on self-organizing systems, complex systems, and ecosystems.

Using Metaphors from Science to Understand Society, Economies, and Nature

From Newtonian/Mechanistic Science	*From* Complex Science
Isolated parts, or objects	Interconnections and relationships
The sum of the parts equals the whole	Relationships between parts lead to the essence of the whole
Individual (parts)	Community (the whole)
Reductionist thinking (reducing to the parts) and understanding from the parts only	Holistic thinking (seeing the whole) and understanding from the whole
Behavior is predictable from the parts	Predictability very limited
Hierarchy	Interconnectedness
Simple linear causation and predictability	Non-linear causation and unpredictability
Direct causation, proportional and linear	Causation goes in all directions. Everything is the cause *and* the effect of everything else
Unquestioned acceptance of assumptions and causation	Questioning assumptions *and* causation (thinking about how we think)
Autonomy	Interdependency
Directed control from higher levels	Self-organizing
Pursuit of certainty	Ability to tolerate ambiguity
Logic	Paradox
Objectivity and observation	Creativity and participation
Quantity	Quality
Static, fixed, unchanging landscape	Dynamic interaction over time
Goal and rule-based thinking	Process thinking
The world is chaotic and must be forcibly controlled by an authority	Systems can self-organize, and order can "emerge"

Chapter 3

Views of Society that Contribute to the Crisis

The causes of global collapse run deeper than issues of energy or technology. Much of the crisis lies in our set of assumptions about human nature that govern our behavior. Most of our thinking is unconscious; therefore, we need to be aware of how we unconsciously think about the world around us. In order to create a new world, we first must believe it is possible, which depends largely on how we understand human nature. Human nature can be seen as essentially selfish, materialistic, and competitive, or it can be seen as inherently empathetic, caring and cooperative. Whichever one we believe is the truth becomes a self-fulfilling prophecy. A new understanding of human nature would allow for a politics of community and compassion. Depending on which view of human nature you believe, two very different worldviews naturally follow. We need to understand these two worldviews, for the choice between them is now crucial.

We have no idea of our capacity to create the world anew.

Peter Senge
Presence: Exploring Profound Change in People, Organizations, and Society

The two worldviews described here reflect the differences between the two sciences, as discussed in the last chapter. One science reduces everything to a focus on the parts (in societies, it is individuals), the other focuses on the interdependencies within the whole (communities). One sees a very linear understanding of causation: one person, one action, one effect. The other view sees systemic causation: complex, going in all directions, and unpredictable. The systems view also sees different problems as linked to and as reinforcing one another.

The two worldviews described in this chapter are basically the conservative and progressive worldviews. However, the issue of global warming goes beyond these left/right distinctions and beyond conservative or progressive thinking. We need to start over with a new term: ecology, ecological politics and ecological economics, or "social ecology." But social ecology is more than a set of opinions about the world. It is based on the well-developed science of complex systems and is grounded in the physical reality of what humans are doing to Earth.

Creating a world that can reverse global warming will require a global effort of unprecedented cooperation. It will require a deep sense of sharing resources and a willingness to accept limits on our own behavior and on our economies. It requires that we work for sufficiency and believe that "enough is plenty." Obviously, this will not be possible if we believe that everyone is out for his- or herself and that we must compete in order to survive. The new worldview includes a new understanding of economics, but before those changes are possible, we need to change our understanding of ourselves and our collective life together.

The place to begin is to describe the basic differences between a conservative and a progressive worldview in relation to society

Worldviews

Individuals and cultures always have a "worldview," a lens through which they understand how the world works, what defines moral behavior, and how people should and do relate to each other. The conservative worldview focuses on obedience to authority, strength through force, self-interest, and a morality defined as choices by individuals with a sense of "you're on your own." The liberal/progressive worldview focuses on empathy, relationships in community, systemic morality, and a sense that "we're all in this together."

This chapter explores these two different ways to understand human nature and the two types of politics that result. These two worldviews matter because the social transformations we need are available only from one of them. This chapter uses various ways to contrast these two worldviews:

- First, I compare the conservative worldview as described by George Lakoff using individual psychology and differing understandings of politics. He calls this model an authority-based model. Then, I describe the same worldview using David Korten's Empire worldview, defined by dominance.
- Next, I describe the progressive worldview using Lakoff's empathy-based model and Korten's Earth Community model, defined by partnership.
- Last, I discuss Jeremy Rifkin and his study of how throughout history empathy has been continually expanding to include more and more diverse people.

The shift from the current (conservative) worldview to an ecological one has been given many names. Thomas Berry calls it the Great Work, and David Korten and Joanna Macy call it the Great Turning.

George Lakoff describes the two worldviews using "frames." Lakoff is a linguist and cognitive scientist at the University of California–Berkeley who studies the *unconscious* ways in which people think and how they conceptualize the world. He says people understand the world based on what he calls "frames" (what I've called worldviews), which organize our thoughts. Lakoff describes the two worldviews in terms of the psychology and thinking patterns that create them.

Roseate Spoonbill

The Nature of Frames

Frames are the largely unconscious images or conceptual metaphors that affect the way we reason; they are mental structures that shape the way we see the world.[1] The words we use are defined relative to frames, and people using different frames can have trouble understanding one another. By making the frames people use conscious, we can understand our own thinking better and begin to communicate more effectively. We access frames through language, because all words are defined relative to conceptual frames. Therefore, when you hear a word, its frame (or collection of frames) is activated in your brain.

Frames dictate their own logic. From within any frame, whatever conclusion is reached, people assume it is only "common sense." The frames we use shape the way we think, the goals we seek, the plans we make, the way we act, and how we solve problems, and they dictate what counts as good or bad outcomes of our actions. They define the problem, assign blame, limit possible solutions, and hide relevant issues and causes.[2] In politics, our frames shape our social policies and the institutions we form to carry out policies.[3]

> At its core, the root cause of our collective crisis is two fold: a failure of our values and a failure of our institutions.
>
> Dave Korten
> Life after Capitalism

Lakoff says facts will not change a person's worldview (frames). If the facts do not fit a frame, the frame will stay and the facts will bounce off because frames are not things that can be changed just by someone stating a fact. People can always explain away facts or, if necessary, dismiss them as lies from unreliable or malicious sources. He also says voters do not vote their self-interest; rather, they vote their identity and their values—that is, they vote for people whose worldview and morality (framing) they relate to.

The way to create new ideas is to reframe the way we think and the language we use. Lakoff says that to change our frames is to change everything: "Reframing is social change."[4] But creating new frames requires new language: "Thinking differently requires speaking differently."[5]

When we can make more conscious the frames used in American political arguments, we will be much better at clearly communicating the choices we face and the reasons why the empathy model worldview is now essential if we are to restore Earth.

Political Framing

Lakoff describes the frames used in political discussions by conservatives and by progressives in the U.S. Both sets of frames are based on different metaphorical understandings of idealized family structure. Both frames (worldviews) tend to compare the nation to a family and the government to the parent. These two models of the family assume different concepts of parenting styles, which define how people think about everything: politics, the choice of topics to be concerned about, what are feasible solutions, and the language used. Whichever model people use, they assume that their children will grow up to share their morality system. These two frames define the way in which we communicate ideas and how our perceptions are structured.

He calls the conservative model, the "strict father" or the authority-based model model and the progressive model the "nurturing parent" or empathy-based model. These models simply describe the way people actually think about the world. People are extremely resistant to changing their frames. They often insulate themselves from all information that contradicts their worldview in order to protect their frames. Giving language to the two models can break through the isolation and give people the tools needed to evaluate their own thinking and to communicate their ideas more effectively.

Most people have *both* models, but generally, only one is active, or some people can have active versions of both models but use each in different parts of their life. Lakoff insists that it is important for progressives to understand that conservatives are not being mean spirited; they are committed to acting morally as their worldview defines it. What drives *both models* is the desire to be moral and good, but what constitutes the good differs because the worldviews are so different.

The foundation of both frames is how they understand human nature.

George Lakoff: Conservative Morality as Authority

Conservatives use a traditional worldview that has been the accepted way of thinking for centuries. In this authority-based or "strict father" model, the world is assumed to be a dangerous place because there is evil out there and because the world is very competitive—creating winners and losers. Lakoff says that in this model, children are born bad, in the sense that they just want to do what feels good, not what is right. Therefore, parents need to make them good. There is an absolute right and an absolute wrong, and the father is the moral authority who knows right from wrong. [6]

The father is the authority who teaches his children crucial concepts of self-discipline, self-reliance, and respect for and obedience to legitimate authority. He does this by setting strict rules for their behavior and enforcing the rules through punishment when they do wrong. This punishment will teach them to develop internal discipline so that they do not repeat bad behavior. In this model, morality depends on punishment. The father's exercise of his authority is itself a moral obligation, for in this model, rewarding obedience and punishing disobedience is the only way to produce moral behavior.

The use of discipline in this model not only makes someone moral and good, it makes him or her successful in this difficult, dangerous world. This model assumes that if people are self-disciplined and pursue their own self-interest, they will become self-reliant and prosperous. Therefore, this model links morality with prosperity and defines well-being as wealth. Because life is a struggle for survival in a fundamentally dangerous and difficult world, it is only through individual self-discipline that a child can learn to survive and compete successfully. The pursuit of self-interest by means of self-discipline and self-reliance is moral, and without it, a

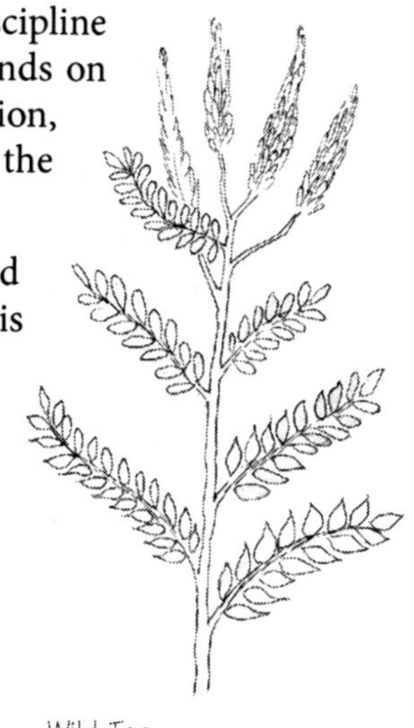

Wild Tea

person cannot take care of himself and become independent. This model focuses on punishment and therefore often includes a commitment to "zero tolerance" and "tough love." In Lakoff's authority-based model:

- The justified reward for acting within this moral system is success. Success means one has learned to be obedient when necessary and to have self-discipline. Earned success is moral. Success is measured by wealth and the ability to consume. This establishes the power of money as legitimate power and a legitimate moral yardstick.
- "Moral" people are better off (more successful) than others are, and deserve to be so. (Calvinism, in which success is a sign of being one of God's elect, also influences this idea.) This creates a strong sense of entitlement.
- The world is a natural hierarchy, and therefore hierarchy is moral. Only the best and the fittest—that is, the "Elect"—should occupy the top of the hierarchy. The best people have authority over others, and their authority is legitimate. This authority is needed to enforce discipline, maintain order, and serve as role models.
- The strict father as the moral authority is responsible for controlling the women in the family. In the moral order, or hierarchy, men are higher than women are; therefore, men are responsible for protecting women, as well as for instilling and monitoring discipline in those lower in the moral order (women and children).
- Competition is necessary to shape children to be the right kind of person. Constraints on competition are therefore immoral. Unearned success (not gotten through competition, etc.) is immoral, for this removes the incentive to become self-disciplined and to obey authority.
- Government social programs "spoil" people by giving them things they have not earned through discipline and self-reliance and therefore do not deserve. Thus, this model opposes any kind of welfare spending. In this view, the poor deserve poverty, and the only explanation for a person's poverty, drug habit, or illegitimate children is insufficient self-discipline and moral weakness. Social systemic causes are irrelevant.
- This model assumes that people behave only because they fear punishment, so a necessary part of learning the rules is learning the punishments for disobedience. Therefore, this model supports capital punishment and other tough-on-crime efforts and supports a strong national defense so that we can control and appropriately punish our enemies.
- This model assumes that one cannot respect the views of one's adversaries and sees a clear-cut us versus them moral dichotomy. There are only heroes and evil-doers.[7]

This models labels social movements for equality (such as feminism) unnatural and therefore counter to the moral order. In addition, an adult neither desires nor requires help from anyone—including his own parents—or the government. If he is sufficiently moral, he will be able to stand on his own and know all he needs to know. Therefore, it logically follows that being "soft" on crime or defining drug use as a health issue instead of a criminal issue threatens the very basis of morality and the social order.

The conservative position fears that without traditional authority structures, life will become immoral and chaotic. When conservatives perceive the world as becoming more and more chaotic, they try to restore order by demanding stricter obedience to rules and to those in power and by emphasizing absolutes. Many conservatives angrily reject "relativism." They want absolutes.

Fear is a powerful activator in this authority-based model and is often used as a tool of distraction and social control. Therefore, the phrase "war on terror" is particularly potent because it activates this entire set of beliefs about the world. This model emphasizes separate, isolated individuals and the importance of power and control, and it creates and operates out of a culture of fear. It is based on the idea that absolute control is possible by the application of power by the right people acting in the right way. The model assumes that we understand the world and that we can create a better world if proper principles are followed and if there are no unnatural constraints—such as any regulatory restraint on competition.

Human kindness has never weakened the stamina or softened the fiber of a free people. A nation does not have to be cruel to be tough.

Franklin D. Roosevelt

In the authority-based model, the nature of relationships is competitive, and people achieve success at the expense of others. It also sees rules, structure, and a hierarchy with a central organizing authority as essential to retaining control. The moral order links dominance to moral authority, that is, dominance becomes moral. Moreover, maintaining this structure, control, and order is a moral duty. Seeing isolated, autonomous individuals competing with each other for their own self-interest, for dominance and control, is considered the only "realistic" view of the world and of human nature. In an authoritarian worldview, winning offers status (rank) in the hierarchy, demonstrating who is more moral and who is on top. Competition is inevitable in a world viewed in terms of hierarchies; one reinforces the other. In addition, morality is strictly expressed through individual behavior—there is no such thing as group morality, only groups of moral and immoral individuals.

Natural Hierarchies

This authority-based model has been influenced by ideas from Charles Darwin, for the model justifies dominance by those who are "naturally" superior. Darwinism had an even more powerful effect on social ideas when Herbert Spencer turned it into Social Darwinism. For Darwin, concepts like the survival of the fittest were simply descriptive. Spencer made them prescriptive: societies ought to behave this way. Therefore, there is no reason to help the poor because they are not meant to survive. Social Darwinism has been discredited, but the ideas still linger.

Morel Mushroom

Ideas about natural hierarchies have a long history. Although they were given scientific legitimacy by Darwin, where survival of the fittest determines who survives, long before Darwin, people believed in concepts of dominance and authority, superior and subordinate, and command and obedience. The Renaissance formalized the idea and called it the "Great Chain of Being." They assumed reality was a hierarchical chain of descending intrinsic value: God was at the top, angels (in nine orders) were next, and then came, in order, the king, men, women, animals, plants, and, finally, rocks. Each link in the chain contained different proportions of spirit and matter. God was completely spirit, and rocks contained only matter.

Progressives need to make explicit what is already inherent in their worldview—vigorous ideas about virtue, morality, character, and freedom, embodied in its central value, empathy.

George Lakoff
Whose Freedom

In this divinely planned hierarchical order, every existing thing in the universe had its "place." God also ranked men by birth and "station in life," and the awareness of their "place" defined everything people thought or did. When life is properly ordered, reason rules the emotions, just as a king rules his subjects, the parent rules the child, and the sun governs the planets.[8] Fear of "disorder" helped reinforce the legitimacy, naturalness, and importance of everyone's (even the king's) staying "in their place." (Shakespeare's King Lear defied God's order when he gave up his throne, throwing all ranks below, even nature, into disorder.)

Even now, although we profess to believe in democracy, we have never completely given up the belief in its opposite: "natural hierarchies." We live with two parallel, mutually exclusive, concepts existing side by side. Murray Bookchin calls this ranking in hierarchical order the "logic of domination"—whether it involves race, gender, class, or ethnicity.[9] Ranking means someone has the authority to give orders and someone else is required to obey. The Great Chain of Being was a way to structure society so that power relations appeared destined by nature or by God. This allows hierarchies of superior and subordinate and of command and obedience to be seen as logical, necessary, and "natural." These ideas justify the naturalness of hierarchical lines of command and obedience found in the military, corporations, government bureaucracies, medicine, the church, and the family. This model assumes that this is simply the way the world "works," the way human nature is, and that to think otherwise is naïve.

David Korten: Empire—the Dominator Model

David Korten takes the same concept and applies it to how societies and countries relate. In *The Great Turning: From Empire to Earth Community*, he describes the worldviews in this country by describing our cultural stories rather than frames. Korten is a co-founder and board chair of the Positive Futures Network, which publishes *Yes! A Journal of Positive Futures*. He is also founder and president of the People-Centered Development Forum.

He says that in any culture, there are stories about the world that people simply accept as a true description of life. Remember, most of our "worldview" is unconscious; it is simply seen as "just the way the world works." We can only create change if we can make our current automatic, unconscious worldview explicit and conscious.

Therefore, telling new stories is a way to communicate new concepts and change the conversation. He took the title of his book from Joanna Macy, who said, "Future generations, if there is a livable world for them, will look back at the epochal transition we are making to a life-sustaining society. And they may well call this the time of the Great Turning."[10] Korten says contemporary cultural politics in the U.S. is mainly a fight over which set of stories is true. New stories can help create a deep sense of alternative realities. For instance, we have a well-known Rags-to-Riches story, but there is no classic American story for the "Cheap Labor Trap."[11] The Rags-to-Riches story tells us that anyone who works hard can succeed, but there is no cultural story about how in reality most people get trapped in low-paying jobs.

Korten calls the conservative worldview or story "The Age of Empire." Empire orders relationships into dominator hierarchies that monopolize power in the hands of elites and suppress the creative potential of everyone else. The alternative, "The Age of Earth Community," orders relationships by partnership networks that distribute power equitably to nurture the well-being and creative potential of each individual and the whole of community. He says each model is within our means, and ultimately it is ours to choose between them.[12]

The Empire worldview refers to the dominator model of organizing human affairs at all levels, whether between nations or family members. The competition inherent in the dominator system creates a hierarchy with only a few at the top; these few become wealthy while the creative potential of everyone else is suppressed and denied. This model will use most of its resources to maintain the systems of domination.[13]

Korten says the Empire worldview sees human possibility as limited. Our natures are inherently flawed and prone to greed and violence and so we need the discipline of competition, the control of a free, unregulated market, and the powers of the police and military to channel individual greed to constructive ends.[14] Leaders in the Empire worldview use fear to control people because "it creates a psychological need for certainty, control, and structured relationships that motivates acquiescence by those below."[15] The Empire story maintains its power because of its ability to control the stories by which we define our possibilities and understand ourselves. It defines the way to prosperity, security, and meaning, all of which affirm the dominator relations of Empire.

Korten describes three of the cultural "stories" of Empire that most people believe. The "Imperial Prosperity Story" says that an eternally growing economy benefits

Giant Panda and Baby

everyone. Growing the economy requires wealthy people who invest in businesses that create jobs. Therefore, it is important to support the wealthy by cutting their taxes and eliminating any regulations that could limit their ability to accumulate wealth. At the same time, welfare programs must be eliminated so that the poor learn the value of working hard at whatever wages the market offers.[16] Economic growth is seen to create opportunity, spread democracy, benefit the poor, and generate a surplus that can be used to protect the environment. The foreign aid system is seen to help deserving poor countries because rich countries promote their development through economic growth.[17] The "Imperial Security Story" says the world is dangerous, filled with criminals, terrorists, and enemies; therefore, major expenditures must be made on the military and the police to maintain order by physical force.[18] The "Imperial Meaning Story" reinforces the other two. It believes righteousness is rewarded with wealth and power and God mandates that the wealthy rule over the poor who, suffer because of their sins or lack of hard work. [19]

Our hearts and minds have to burn with the evolutionary passion that ignites only when we realize that the future literally depends on us.

Peter Senge

Korten says these Empire stories alienate us from the community of life and deny the positive potentials of human nature. They affirm the legitimacy of economic inequality and the use of physical force to maintain imperial order and claim the special righteousness and therefore entitlement of those in power. The Empire worldview believes that this is "just the way people are," but Korten calls this culture of greed and violence our "social pathology."

The U.S. is particularly caught in an Empire mentality. Most Americans assume we are entitled to dominate the world both militarily and economically. The idea that the U.S. should see itself as one nation among equals would be interpreted as a sign of weakness and vulnerability. This worldview and its consequences are now destroying Earth.

George Lakoff: Progressive Morality as Empathy

Lakoff calls the progressive view of politics and morality the "nurturant parent" or empathy-based model. This model is based on relationships and a sense of the whole, as well as on cooperation and reciprocity. This alternative model begins by assuming that although there are dangers and difficulties, the world is good and we can make it better and that people are responsible for working to make it better. Children are born good, and parents can make them better. The "nurturance" in this model is an emphasis on empathy, care, and responsibility. This model emphasizes empowerment rather than domination.

This second model is gender neutral and emphasizes the responsibility of both parents to meet a child's need for loving interactions in order to thrive and live happily. Adults and children find meaning in mutual interaction and care. To be moral, children must learn empathy for others, the capacity for nurturance, cooperation, and maintenance of social ties, which requires the strength, respect, self-discipline, and self-reliance that come through being cared for and caring for others. To raise a socially conscious person, parents must encourage questioning, self-examination, openness, and mutually respectful communication. Therefore, Lakoff says, "This process requires constant communication and negotiation of meaning. It assumes that meanings will be different and that communication will be imperfect. . . . [W]hat keeps the process of communication going is secure attachment, affection, and affectionate behavior, mutual respect, empathy, commitment, clarity of expectations, and trust."[20]

Effective nurturing requires empathy—the ability to feel what someone else is feeling. It also requires responsibility, both for looking out for the well-being of others and for caring for oneself in order to take care of others. That is, this model requires responsibility to oneself as well as to others (social responsibility). Lakoff's nurturant parent model assumes:

- Bonds of affection and earned mutual respect are stronger, more moral, and more legitimate than bonds of dominance.
- Children learn not through reward and punishment, but through their attachments to their parents, which, ideally, are secure and loving attachments. Parents do not maintain moral authority by enforcing rules.
- If children experience and learn abusive behavior, they reproduce that behavior, escalating the creation of a violent society.
- The non-aggressive side of competition is mastery, which develops naturally by nurturing and encouraging a child's efforts. Children learn to be nurturing out of a desire to display their mastery, please their parents, and gain respect.
- Cooperation develops nonhierarchical relationships and an appreciation for interdependence.
- Legitimate moral authority of the parent emerges from their commitment to protect their child and also from their wisdom, good judgment, empathy, and the ability to nurture. We should not gain or maintain authority through dominance.
- A good leader will communicate effectively, encourage and arrange for participation, be honest, and have the wisdom, experience, and strength to help the community succeed.
- This model assumes people who have been helped feel a responsibility to help others, and it is natural for the strong to support the weak.
- This model cannot maintain a strict good-evil dichotomy. Here, moral children are expected to empathize with and be nurturant toward people with different values.[21]

A nurturing parent wants his or her family to be safe, which requires the strength and competency to protect the family, and the parent himself or herself, from harm. This empathy creates a desire for others' well-being. People who experience well-

Cedar Waxwing

being and fulfillment want to see others happy and fulfilled. Correspondingly, unhappy, unfulfilled people often don't want others to be happier than they are, and people who have been harshly judged or made to feel guilty become, in turn, judgmental of others. This model, therefore, wants everyone to be a happy, fulfilled person.

In Lakoff's nurturant-empathy model, the goal is the satisfaction and potential found in non-competitive relationships. The values of this model are mutual care and the well-being of everyone in the community. The central task is to develop secure individuals who have positive relationships with the external world. This model does not depict a hierarchical ladder of command and obedience, but rather communities or "networks" of patterns of interconnections, interdependencies, and relationships. Rather than using punishment to control behavior, the emphasis is on trying to understand the true origins of behavior and creating relationships that can help prevent destructive behavior. This model fosters a democratic personality, whereas the authority-based model rewards the authoritarian personality.

Love is much more demanding than law.

Archbishop Desmond Tutu

Conservative thought, issues, and principles have dominated our cultural discussions. This is largely because conservatives have been extremely effective in presenting their worldview and getting issues defined on their terms. In contrast, progressives have not been good at advocating their own values and principles. Progressives focus on issues one by one and list policies they think we need. But they rarely make clear the underlying values and principles that form the basis of the policies they advocate, that is, they rarely state explicitly their implicit worldview.

Cooperation Instead of Competition

A fundamental assumption of the empathy model is that intense competition does not create positive outcomes. Alfie Kohn's book *No Contest: The Case against Competition* calls competition "mutually exclusive goal attainment," that is, for me to win, you have to lose. He says competition damages self-esteem, depresses empathy, and discourages generosity. Cooperation teaches the value of relationship because it means that the success of any one person is linked to the success of others. In a cooperative world, for me to be right, you do not have to be wrong. This cooperative system leads to mutual assistance and support, which in turn strengthens relationships and the bonds between the individuals in the community.[22] Competition requires our sense of self-worth to be defined by our position on the hierarchical ladder. Instead, Kohn says, we need "to find more successful ways of securing our self-esteem: building an unconditional sense of trust in ourselves that will make it unnecessary to keep demonstrating superiority." That is, "the better I feel about myself, the less I will need to make you lose."[23]

These new concepts of cooperation have appeared even in business. Business consultant Peter Drucker talks of the need for totally new organizations that would eliminate most management layers. Rewards for performance will no longer be promotions to command positions. In fact, he says, "The word 'rank' should disappear entirely." It will be replaced by "assignments."[24] This means there will be no subordinates; there will be

Thinking like an ecosystem

only "associates." This shift, he admits, will raise "tremendous problems of motivation, of reward, and of recognition." Drucker said that now, in knowledge-based organizations, the task of management is not to make everyone a boss, but to make everyone a contributor.[25] Herman Daly and John Cob, Jr., in *For the Common Good*, use the same concept and define community as "a society in which participation is realized."[26]

Those who claim the necessity of competition often point to Darwin and the struggle for survival in nature. But the centrality of cooperation and coevolution in nature now has emerged clearly from new studies in evolution. Biologists no longer see competition as the primary dynamic or mechanism of evolution. For one thing, biologists have found that in evolution not just the organisms adapt and change, but the environment itself also evolves, that is, they "co-evolve."[27] The organisms and the environment are part of one organic whole, in which they both self-organize while continuing to change and evolve. These scientists now identify three mechanisms of evolution, with the selection of random mutations being the least important; the other two are DNA recombination and symbiosis. Both of these depend on cooperation.

Using DNA recombination, bacteria trade genes with each other, which is a far more effective means for evolving than natural selection. Bacteria are prevalent throughout Earth and are indispensable to all living beings; they are not just parasites that cause bacterial infection. They have extreme genetic fluidity and can exchange genes quickly and reversibly. The visiting DNA allows the receiving bacterium to perform functions that its own genes cannot. Therefore, unlike other life forms, all the world's bacteria have access to a single gene pool and therefore to the chemical power of the entire bacterial kingdom. With this worldwide system of communicating, gene-exchanging bacteria have made the planet fertile and habitable for other, larger, life forms. It is also why bacteria have the ability to quickly become resistant to new drugs.

Symbiogenesis

Symbiosis and cooperation are another way organisms evolve. Bacteria have a remarkable capacity to combine their bodies with other organisms, forming alliances that may become permanent. Lynn Margulis developed the idea that cells with nuclei evolved from the merger (symbiosis) of two or more different types of bacterial cells. The theory is called "symbiogenesis" and is now widely accepted. Through symbiotic association, different organisms survive by living in close association or even inside one another. Margulis believes that the primary means of evolution for higher organisms is through symbiotic arrangements that become permanent, thereby creating new forms of life.[28] In *The Web of Life*, Fritjof Capra comments on the significance of Margulis' ideas:

> The theory of symbiogenesis implies a radical shift of perception in evolutionary thought. Whereas the conventional theory sees the unfolding of life as a process in which species only diverge from one another, Lynn Margulis claims that the formation of new composite entities through symbiosis of formerly independent

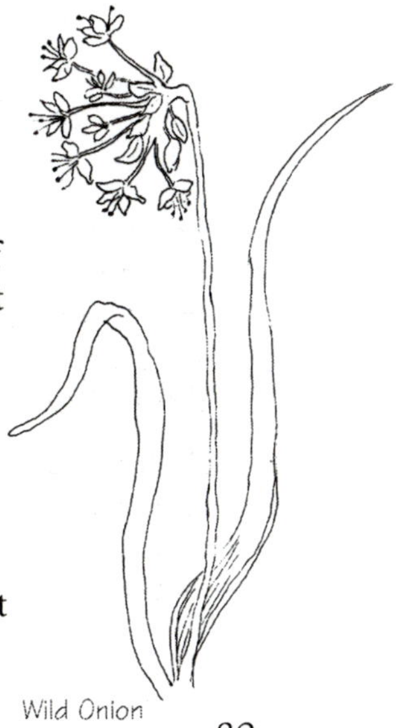

Wild Onion

> organisms has been the more powerful and more important evolutionary force. This new view has forced biologists to recognize the vital importance of cooperation in the evolutionary process. . . . Life is much less a competitive struggle for survival than a triumph of cooperation and creativity.[29]

Or, as Margulis says, "Life did not take over the globe by combat, but by networking."[30]

Darwinists in the nineteenth century talked of the brutal competition in nature, of "Nature, red in tooth and claw." However, Capra says, "In the twentieth century, ecologists have discovered that in the self-organization of ecosystems, cooperation is actually much more important than competition. We constantly observe partnerships, linkages, associations, even species living inside one another depending on one another for survival. Partnership is a key characteristic of life. Self-organization is a collective enterprise."[31] Survival depends more on relationships like those between the bee and the flower than on those between predator and prey. This creates a growing understanding that highly organized systems, even life itself, can emerge spontaneously from networks of interactive and cooperative components. It also means that we can no longer use nature to claim that ruthless competition is only "natural."

> This picture of human beings [as autonomous] is profoundly erroneous. People are constituted by their relationships. We come into being in and through relationships and have no identity apart for them.
>
> Herman Daly and John Cobb Jr.
> *For the Common Good*

The Public/Private Spheres and Hiding Values of Compassion and Care

We could also describe these contrasting worldviews as a conflict between the values that the Western world has defined as "masculine" and those values it has defined as "feminine." In fact, one name given the new worldview is "the return of the feminine." The culture is limited when anything considered "feminine" (such as care, community, nurturing, and cooperation) is marginalized, depreciated, and ignored.

One reason the values of dominance and competition have had such an impact is because they are so visible, while the values of cooperation and mutual care have been hidden. The fact that these values have been hidden has been part of the discussion in feminism.

Feminism now emphasizes the importance of "feminine" values. This is a recent development. In the 1970s, feminists began by assuming the inevitability of a hierarchical worldview, and so they had to insist that, in terms of equality, women were the same as men. Women strongly resisted any admission of difference, for in a hierarchical system, any admission of difference automatically meant the differences are ranked, and women were always ranked inferior to men. But in the 1980s there developed a cultural interest in "celebrating diversity," which was a very radical idea. It meant that differences could be acknowledged *without* ranking them. This allowed feminists to admit that women are different from men but argue that they deserve equality anyway.

Then in the 1990s, feminists (particularly academic feminists in women's study programs) went the next step and said that it was feminine values of empathy, nurturing, and community that are essential and they should play a more defining role in society. They argued that, in our society, it is difficult to value and use care, community, and compassion because our democratic theory hides these values. Democratic theory divides society into public and private spheres, which was a great help in the past in developing and protecting individual rights, but it also hides and isolates the values of care and community.

Because of historical experience, in developing democratic theory, people feared the power of the state (even without a king), which they saw could easily become intrusive and coercive. The construction of public and private spheres confined the legitimate use of state power to the public realm. The private realm is the area in which the state cannot interfere. To create communities of mutual care, we need to become aware of the unacknowledged effects of this conceptual split.

My religion is simple,
my religion is kindness.

Tenzin Gyatso, the Dalai Lama

Feminists in the late 1980s were the first to discuss these contradictions. In the 1960s and 1970s, women believed they had been excluded from the public sphere because of the long history of discrimination against women, which largely relegated women to the private sphere and gave them few opportunities to participate in either business or higher education. Therefore, women believed that ending this exclusion from public life would give women "equality" and end discrimination. Yet, gains in equality did not end discrimination against women. They came to see that equality was not merely a question of access to the public sphere. The division of the public and private spheres was actually a division along gender lines. This division isolated not only women in the private sphere, but also the values traditionally assigned to women, qualities thought of as "feminine." Both race and class complicate this issue; however, for this discussion, I am interested in why our culture emphasizes competition and dominance.

Democratic theory assumed the public sphere needed only the values of self-interest, competition, and economic efficiencies. In turn, we see the values of care, nurturing, cooperation, and compassion as unnecessary in the public sphere because they are preserved in the private sphere. In *Feminism without Illusions: A Critique of Individualism,* Elizabeth FoxGenovese describes this split in values and says, "Historically, the myth of separate spheres promoted not merely a genderspecific division of labor, but a genderspecific division of values. The values of nurture, care, and mutuality that the myth assigned to women in the private sphere made possible the values of autonomy, competition, and abstract justice that it assigned to men in the public sphere. The two sets of values, in other words, sustained and reinforced each other—made each other possible."[32] That is, individualism for men depended on the subordination of women.

Crow

partnership model

It became clear to feminists that the principle of *community* governed the private sphere and the principle of *autonomy* governed the public sphere.[33] Because community values, relationships, and raising children all take place "off stage," we could define the public sphere as "value free." In our political system, "private" community values of care and cooperation appear unnatural and unnecessary, even naïve, in the public sphere. However, a fundamental issue in global warming is that private choices have public consequences; therefore, this split into two spheres has allowed the culture to emphasize autonomy and ignore community. To adequately address global warming, we need to stop fragmenting the world and see it whole. And that whole needs to become a global partnership of mutual care that includes both the public and private spheres.

David Korten: Earth Community—the Partnership Model

David Korten calls the alternative worldview "Earth Community." This worldview uses partnership to organize human potential for creative cooperation, which is then available for the pursuit of the common good. The Earth Community story is one of unrealized human possibility and cooperation. This world of partnership is possible—if we act with the belief that it is possible.[34] Korten says humans are born to connect, learn, and serve. Now, with nature and society near collapse, we must bring the Earth Community world into being by cultivating our capacity for sharing and compassion. This in turn creates the capacity for responsible self-direction and self-organization and thereby the possibility of creating radically democratic organizations and societies.[35]

> The challenge in contemporary America is to create a nurturant society when a significant portion of that society has been raised either by authoritarian or neglectful parents.
>
> George Lakoff
> *Moral Politics*

The Earth Community story supports policies that increase our sense of connection and mutual support. Korten reports that scientists have found that humans are hard wired to cooperate. Scientists using imaging technology to study brain function find that the human brain is wired to reward caring, cooperation, and service. They found that thinking about a person being harmed triggers the same reaction in our brain as when a mother sees distress in her baby's face. At the same time, helping someone triggers the brain's pleasure center, increases our immune system, reduces our heart rate, and prepares us to comfort. Negative emotions do the opposite; they suppress our immune system, increase heart rate, and prepare us to fight or flee.[36] In fact, Korten says, "Developmental psychologists describe the healthy pathway to a fully formed human consciousness as a progression from the self-centered, undifferentiated magical consciousness of the newborn to the fully mature, inclusive, and multidimensional spiritual consciousness of the wise elder."[37]

This alternative story says that when we impose economic growth on developing countries, we transfer control of their economic resources and benefits from the people and institutions of developing countries to the institutions and the wealthy in the donor countries. The Earth Community worldview assumes that "authentic development depends on restoring local, equitably distributed control over local resources and developing the capacities of people to manage those resources efficiently to create a better life for themselves, their families, and their communities."[38] This effort to return to local control applies to communities in both the "developed" and the "developing" world.

The stories of Earth Community help create and legitimize a culture of partnership, which affirms the positive potentials of our human nature. Korten says this story shows that "realizing true prosperity, security, and meaning depends on creating vibrant, caring, interlinked communities that support all persons in realizing their full human possibilities."[39] In fact, membership in a cooperative, caring community is the best indicator of happiness, rather than income (beyond a minimum).

There is no solution short of re-inventing society.

Futurist Willis Harmon
Creative Work

Within each of us, there is a tension between these two concepts. It all depends on what we focus on; when we focus on what divides us, it leads to fear and violent competition, but when we focus on what unites us, it leads to trust and cooperation.[40] Human nature includes a wide range of possibilities. Our capacity for greed, fear, and competition is obvious. We will be able to actualize our capacity for caring and cooperation only after we acknowledge their existence in others. And so Korten says, "The successful negotiation of the transition from Empire to Earth Community will require the creative contribution of every person."[41]

Jeremy Rifkin: Two Ways to View Human Nature

The third author addressing the two worldviews is Jeremy Rifkin, whose book *The Empathic Civilization: The Race to Global Consciousness in a World in Crisis* presents a "new interpretation of the history of civilization," describing the increasing development of human empathy. Rifkin has written 17 bestselling books and is President of the Foundation on Economic Trends, an advisor to the EU, and a senior lecturer at Wharton Business School's Executive Education Program.[42] *The Empathic Civilization* challenges the conventional view of human nature found in our educational systems, business practices, and political culture and seeks to replace that view with a new story that allows humanity to see itself as an extended family living in a shared, interconnected world.

Historians, by and large, write about "social conflict and wars, great heroes and evil wrongdoers, technological progress and the exercise of power, economic injustices and the redress of social grievances." They don't write about happy people, but about the disgruntled, the discontented, the angry, and the rebellious. In fact, Rifkin says much of written history is about the pathology of power. Our collective memory is "measured in terms of crises and calamities, harrowing injustices, and terrifying episodes of brutality inflicted on each other and our fellow creatures." All of which raises the question, "Why have we come to think of life in such dire terms?" The answer he gives is that these tales

Lynx

surprise us. They are unexpected and therefore trigger alarm and heighten our interest. It is because such events are novel and not the norm that they are newsworthy and the stuff of history. Today, our 24-hour cable TV news shows become the chroniclers of these accounts of pathological behavior. But, Rifkin says, if these were the defining elements of human experience, we would have perished as a species long ago.[43]

People do not resist change, they resist being changed.

Peter Senge
The Fifth Discipline

Rifkin says it is the extraordinary evolution of empathic consciousness that is the underlying story of human history, even if it has not been given the serious attention it deserves by our historians.[44] For the most part, the everyday world is lived out in hundreds of small acts of kindness and generosity. Comfort and compassion between people create goodwill, establish the bonds of sociality, and give joy to people's lives. Much of our daily interaction with our fellow human beings is empathic because that is our core nature. Empathy is the means by which we create social life and advance civilization.

Rifkin, like Korten and Lakoff, reviews the recent discoveries in brain science, child development, biology, and cognitive neuroscience that have found that human beings and other species empathize with another's situation as if it were their own. People are social animals and seek intimate participation and companionship with others. If this is true, then we have a chance of creating the transformation we need. Rifkin says that empathic skills emphasize a non-judgmental orientation and tolerance of other perspectives, see reality in terms of layers of complexity, and require living within ambiguous realities where there are no simple formulas or answers, only the constant search for shared meanings and common understandings.[45] Empathy, unlike sympathy, which is passive, includes "an active engagement—the willingness of an observer to become part of another's experience, to share the feeling of that experience."[46] We empathize with a person's joy as well as pain. Rifkin says, "When we say to civilize, we mean to empathize."[47]

Rifkin reviews history to demonstrate that empathy keeps expanding between more and more diverse groups of people. This process happens, he says, because of the convergence of new energy revolutions with new communications revolutions. This convergence creates expanded contact and new economic eras and in the process expands empathy and changes human consciousness—that is, the way the human brain comprehends reality. In this process, empathy expands to larger and larger groups.[48]

The last revolution has allowed empathy to extend to Earth itself, to our species as a whole, and to all the creatures that inhabit the biosphere. Therefore, this biosphere (the natural world) functions like an indivisible organism, and we need to develop a deep empathetic connection to it.

At each communications revolution, as society has become more and more complex, the process creates a greater sense of selfhood, more exposure to diversity, and therefore more empathy. Each communications revolution has expanded the range and depth of human social interaction, and now, for the first time in history, the

Thinking like an ecosystem— social ecology

last revolution is connecting the human race around the world, allowing empathy to expand to a global scale.

However, Rifkin says an enormous problem is created because as empathy and new communication expands, this development occurs alongside increasingly destructive uses of energy, so now we are in a race to see whether our empathy will allow us to avoid our own extinction.[49]

An Ecological Worldview

This chapter has contrasted two worldviews, which in the U.S. are primarily the conservative and progressive worldviews. The conservative view has a long history, and the basic concepts have made sense to people for centuries. It includes the idea that people are basically selfish and competitive and that fighting for dominance and control is the only logical way to live. The progressive view assumes that people are basically good and caring and that these traits need to be nurtured. Cooperation is required for the well-being of everyone, and it is natural for the strong to support the weak. This view can be described in terms of stories of the community of life and human potential, of framing for empathy, of feminine values of care, nurturing, and cooperation, or of the interdependencies in self-organizing complex systems. Progressives believe in the enormous potential in human creativity, ingenuity, and compassion. Whatever terms, frames, or stories are used to clarify the differences, moving to a sense of mutual responsibility for creating the common good is our only hope in solving global warming.

Although we can start with a basically progressive worldview, to address global warming we need to go beyond progressivism, to ecology, ecological politics and ecological economics, or "social ecology." This social ecology is not just an opinion about the world but is based on the well-developed science of complex systems and is grounded in the physical reality of what humans are doing to Earth. Much of both worldviews include assumptions about economics, which are covered in Chapters 5, 6, 8, and 9.

The question now is, Can people who have been taught to obey an authority that has all the answers learn in time to work together to create new answers to completely new problems? In a world of cutthroat competition and every-man-for-himself individualism, it is difficult to address common catastrophic problems. Socially, we need to realize that thinking in terms of mutual care and responsibility is not only realistic, it is now essential for survival. Could social ecology create a social view capable of restoring Earth and ourselves? Or as Jeremy Rifkin says, "The most important question facing humanity is this: Can we reach global empathy in time to avoid the collapse of civilization and save the Earth?"[50]

Common Merganser

Chapter 4

Views of Religion that Contribute to the Crisis

Our social problems are, at heart, a spiritual problem. In this chapter, I want to define a progressive religion, in part to let people know it exists and also to encourage liberals who have left their places of worship to return and work with those still there to help create the religion that God wants of us now. The alternative moral vision of Democrats, liberals, or progressives is grounded in compassion—but we don't say so. Most liberals I know are deeply religious, but they distrust the use of religious language in political debate because religious language has been co-opted by fundamentalists. As David Korten says, "The time has come for the nation's mainstream churches to come out of the closet and speak publicly of values and the spiritual foundations of the progressive agenda and to articulate spiritually grounded stories of human possibility and the world that the living Jesus called us to create."[1]

> The most pressing political issue of the day, both in America and abroad, is a way to integrate the tradition of liberalism with a genuine spirituality.
>
> Ken Wilber
> *Eye of the Spirit*

In creating a sustainable world, the secular fundamentalism of the left is as much of a problem as the religious fundamentalism of the right. We need a third choice that is neither fundamentalism nor secularism. Or, as evangelical Jim Wallis's book title puts it, *God's Politics: Why the Right Gets It Wrong and the Left Doesn't Get It.*

In the U.S., we have two very different Christianities: the fundamentalist version—the only one heard in the public argument—and the mainstream version, which has a very different worldview. Fundamentalism is authoritarian, coercive, and absolute by definition. With this belief system, the power of the church in society is experienced as oppressive. The primary question is not, "Can we mix religion and politics?" but rather, "What *kind* of religion can we legitimately proclaim in public?" Our images of God have great political significance. These images not only define God, but they also define *us* and specify who gets to do the defining. As theologian Marcus Borg says, "Tell me your image of God, and I'll tell you your politics."[2]

Religion is a key definer of worldviews, so we need to be very aware of the worldview our religion reflects and sustains. Religious arguments in our political debates have been limited to a few hot-button symbolic issues focused on controlling individual behavior. Our culture responds by denying that religion has any relevance at all. But the absence of religious values in the public realm is not the only alternative to intolerance. Rather than ignore religion, we need to change the religious voice in political debate, and progressives need to get comfortable with publicly discussing their values. Currently, religion is part of the problem but could be a large part of the solution. However, it is difficult to talk about religion because people have so many automatic assumptions.

In a democracy, we need to consciously create legitimate ways to use religious values and language in debating social policy. We must ask political questions of theological concepts, as well as asking religious questions of political and economic concepts. That is, we need to explicitly evaluate economics as a moral system and determine what moral and value assumptions are implicit in economic choices. Then, we need to take every religious statement and explore the implied political assumptions. What do our religious statements say about power, authority, dominance and submission, superiority and subordination, inclusion and exclusion, cooperation and competition? The religious voice we need is one that will make religious statements about meaning and reality, not political statements about power and dominance. Democracy is not incompatible with religion; it is only incompatible with authoritarian religion.

All decisions finally imply convictions about the holy, or matters of ultimate importance.

David Ray Griffin

If religious institutions take the lead in creating the social change we need, the religion they proclaim in a political argument must be democratic in method (non-authoritarian) and pluralistic in content (capable of working with other religions in a religiously pluralistic world). We need to be able to share our religious vision of an ecologically sustainable world with Christians, Jews, Muslims, Buddhists, Pagans, and others, and we need to listen to the visions of a sacred Earth from other religions. In a religiously pluralistic world, we can make allies of people from other religions who have similar spiritual and compassionate commitments. Creating an ecologically sustainable world depends on the spiritual energy of all religions.

However, I focus on Christianity because it has had the most impact on Western history and culture and because I grew up Christian. In addition, through my seminary training and the books by theologians such as Marcus Borg, I have rediscovered a Christianity that is fundamental to my life. There are many ways to address religion in our ecological crises, but this is my vision for what this change would require. At the seminary, I was in a theology program, not in the ministry program, but this book is my ministry: my calling is to "Preach the Earth."

Instead of dismissing religion, or leaving its expression to a one-sided authoritarian view, we need to understand it differently and claim it as an essential part of our response to an Earth in crisis. We need to create a public theology for a religiously pluralistic world because our current understanding of religion limits our capacity to transform the world,

Ibis

and I believe we can never create sufficient transformational change unless creating a sustainable world becomes part of our religious identity.

Open and Closed Belief Systems

Understanding the nature of beliefs will help us create a democratic debate and still use religious values. Democracy depends on holding beliefs openly and working through dialogue for new ideas and solutions. Many liberals believe that religion is inherently coercive and so tend to be hostile to any religious language in democratic debate. However, we read a book in the seminary titled *The Open and Closed Mind,* by Milton Rokeach, a psychologist, that clarifies how religious ideas function. Rokeach says that any belief (political, religious, or economic) can be absolutist and coercive and can claim to have the One True Answer. There are many forms of intolerance, not just religious or ethnic intolerance. He says there is "intolerance among Marxists, liberals, Freudians, Unitarians, academics, and art critics."[3] The relative openness or closedness of a mind can involve any ideology, religion, philosophy, or scientific viewpoint. Therefore, Rokeach says, *how* you believe is as important as *what* you believe.[4]

> Interwoven in all these pressures to restructure institutions is a change in the locus of authority. This shift, from external authority (e.g. the Church, moral and ethical rules, the Expert) to internal authority (trust in oneself), has been accelerating over the past several decades.
>
> Willis Harmon
> *Creative Work*

Belief systems have a structure independent of their content. For instance, the content of a belief system can be "liberal" while the structure can be rigid, fundamentalist, and absolute. Advocating "politically correct" behavior has a content of respect for diversity and social liberalism, but the structure of the communication of these views has sometimes involved verbal abuse, absolutism, and tactics that stop dialogue. Rokeach says that we also have beliefs about people who have beliefs. In closed systems where authority is seen to be absolute, he says, "Sharp distinctions are made between people: as faithful and unfaithful, orthodox and heretical, loyal and subversive, American and un-American, and friend and enemy. Those who disagree may be rejected as enemies of God, country, mankind, the working class, science, or art. And those who agree may be accepted, but only so long as, and on the condition that, they continue to agree."[5]

Openness depends not on what you believe, but on how you deal with beliefs different from your own. The "openness" of a belief system is primarily determined by the nature of a person's *disbelief* system. In public debate, openness has to do with respect, relationship, and behavior. For people to legitimately argue values in public policy, we need to be aware of how difficult it is to have open and compassionate beliefs and at the same time argue them openly and compassionately. The disbelief systems of conservatives and liberals differ radically. Conservatives are much more absolute about the ideas in their disbelief system being absolutely wrong and not even open to discussion. It means they also know little about the ideas they reject.

Rokeach says closed belief systems are a defense against anxiety, especially the anxiety caused by ambiguity and uncertainty. "The closed mind, through fear of the new, is a passive mind. It achieves its systematic character 'for free,' through the external authority's efforts rather than its own. ... [The closed mind] cannot integrate new beliefs into a new system because it cannot remember them. The reason it cannot remember them is because there is a dynamic unwillingness to 'play along,' to 'entertain' strange belief systems."[6]

Rokeach's research is similar to George Lakoff's morality models because the distinguishing characteristic of both is the role of authority. Authority is also the primary barrier in using religious language in political debates in a democracy. Rokeach defines authority as "any source to whom we look for information about the universe, or to check information we already have." In a closed belief system, people do not question information if it comes from a trusted source. They often cannot see any distinction between information received about the world and information received about the *source* (the one giving the information). In open systems, this dual character of content and source is distinguishable. That is, in an open system, people do not necessarily believe a statement just because a trusted authority figure says it. However, people with a closed system never question information if it comes from a trusted authority, and they protect the information they take in by deliberately avoiding contact with alternative people, events, or books that threaten the validity of their ideology or support competing ideologies.[7]

Within a closed belief system it is very difficult to deal emotionally with the ambiguity and uncertainty that will come with global warming. How can we help people (including ourselves) deal emotionally with the uncertainty and enormous change that is sure to come? People are drawn to fundamentalism when they base their sense of security on a need for order, rules, predictability, certainty, and obedience to legitimate authority. This authority, by definition, can understand and predict the future, can resist change, and can explain away any uncertainty. But in the future, that will simply no longer be possible. Joanna Macy says, "Fundamentalism rears its head in all religions now. It's a reaction against the radical uncertainty of this moment in history. In such times, we tend to revert to the security of rock-bound belief and vent our anxieties in scapegoating others. The temptation to take refuge in our own self-righteousness is strong."[8]

Siberian Tiger
On Ten most endangered list (ICF)

With the coming chaos of climate change, how can we find ways to help people resist absolutism as a defense? Absolutes may allow people to feel more secure—but they cannot deal with the reality of what is occurring to Earth. In addition, without ambiguity, there is no possibility, creativity, or curiosity. When everything is absolute and static, there is no room for surprise or invention.

Open and Closed Christianity

Because liberals are suspicious of religious language in political debates, the culture has the impression that liberals do not base their ideas on principles or values. Fundamentalists have been able to define Christianity as their kind of Christianity—and few have challenged them. The culture and liberals need to be aware that Christianity comes in two different types and that difference matters in a democracy. There is an open Christianity, but most liberals only know about the closed one.

Fundamentalist religious beliefs have reflected, supported, and influenced the economic and political systems of hierarchy, patriarchy, and strict obedience to rules and the power of the wealthy as legitimate. In our current cultural understanding of religion, God is the ultimate strict father, giving rules and laws that must be obeyed, and punishing any disobedience (eternally if necessary). When we describe God as all-powerful and all-good, and when we glorify God's absolute power, we see the use of domination and power by men as legitimate, "normal," even morally required.

Over time, two contradictory images of God have developed. One could be called a Religion of Earth Community, a religion of compassion, grace, and Spirit, which is the religion Jesus preached. The other is the Religion of Empire, which is imperial, arbitrary, judgmental, and jealous of its power and legitimizes conquest and imperialism. This authority-based model describes not what Jesus taught, but what the church has taught about Jesus. When Emperor Constantine made Christianity the religion of the Roman Empire in the fourth century, Christianity became the opposite of what Jesus taught. It literally became the religion of Empire, one with an imperial, monarchical, Emperor God, demanding obedience and harshly judging any disobedience. This God justified the Crusades and the Inquisition and allowed the Holocaust to happen. (The Church and Empire will be discussed further in Chapter 11.)

> The collective psyche seems to be in the grip of a powerful archetypal dynamic in which the long-alienated modern mind is breaking through . . . to rediscover its intimate relationship with nature and the larger cosmos.
>
> Richard Tarnas
> *The Passion of the Western Mind*

People have tried to have a religion that is both compassionate and authoritarian, but the two are mutually exclusive. (The satiric newspaper *The Onion* ran a front-page story with the headline "God Diagnosed Bi-polar.") All major world religions (Christianity, Judaism, Islam) struggle with these two concepts of religion: one as obedience to absolutes and the other as relationship to Spirit/Mystery and to one another.

Both approaches to Christianity have roots in antiquity, yet both are modern interpretations resulting from responses to the Enlightenment and the beginning of modern science in the 1600s. Prior to the Enlightenment, metaphorical meaning was most important. The original Greek language of the Christian Bible does not distinguish between the words *faith, trust*, and *belief*. Greek uses the same word for all three. Translators of the original Biblical Greek have to choose which of these three

English words the writers meant in each instance. Understanding faith as "belief" was not the original meaning in Christianity. Today, mainstream Christians understand this word not as belief in specific statements but as radical trust in divinity. In addition, trust is an emotional connection, defining a relationship with Spirit. Belief is a mental abstraction and automatically requires an acceptance/rejection choice, which in turn creates a judgment about the choice.

Theology today in mainstream seminaries is very different from the fundamentalism the culture hears. It is not concerned with what you must believe. Dr. David Rhoads, professor of New Testament at the Lutheran School of Theology in Chicago, says that theologians are no longer writing for all time, but for their own time, place, and perspective.[9] They no longer assume objectivity, universality, or the existence of one meaning or truth. For the past 30 years, mainline theologians have no longer been writing dogmatics or what everyone is required to believe. Instead, they acknowledge that truth is relative to the writer's perspective. Theologians today often explicitly write from a point of view, such as a commitment to liberation of the poor (liberation theology), of women (feminist theology), or of Earth (eco-theology). And they deliberately ask, Who does my theology benefit? Who does it dismiss? Who does it privilege? and Who does it marginalize and why? This approach is highly decentralized and assumes a multitude of voices and perspectives.[9]

Religion in the Two Worldviews

Just as the two worldviews are reflected in two different understandings of politics, economics, and morality, the two are also reflected in two understandings of religion. Theologian Marcus Borg, in *The Emerging Christian Way*, gives his own names to these two kinds of faith systems, calling them the "belief-centered paradigm" and the "transformation-centered paradigm." He says belief-centered Christianity includes fundamentalists, many conservative-evangelicals, and some mainstream Christians. This type of Christianity believes Christianity is the only way to salvation. It is a religion of requirements and rewards and emphasizes sin, guilt, and forgiveness. It declares the Bible to be a divine product that we must interpret literally, factually, and absolutely.[10] Borg says this belief-centered Christianity is the single biggest reason for the decline of mainstream denominations over the past 40 years. For many people, this requirement of literal belief became "an intellectual and moral stumbling block." Borg asks, "Is 'correct belief' what God most wants from us?"[11]

Interpreting the Bible Metaphorically

The alternative to this belief-centered religious worldview is what Borg calls the "transformation-centered paradigm." In this understanding of religion, we know our relationship with God through the words of Jesus, by which we are changed and transformed.[12] Most mainstream Christians have this kind of religion, although that is not generally understood in our society

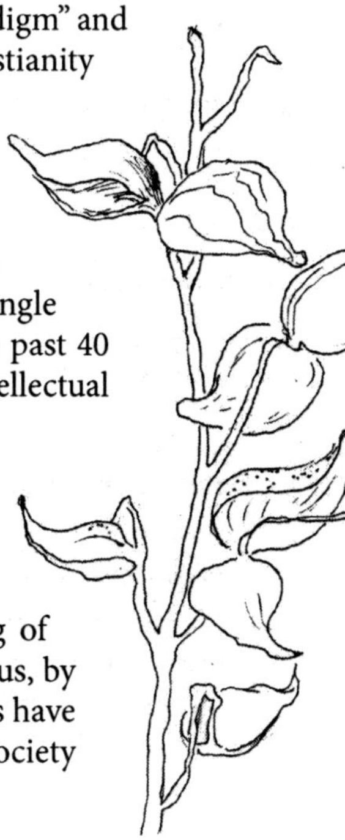

Common Milkweed Pod

and culture. This paradigm does not interpret the Bible literally but metaphorically and historically. People do not believe metaphors literally, because they are about story and meaning. This approach to the Bible emphasizes the underlying meaning of biblical stories and sees the primary stories of the Bible as stories of transformation, of the self and of the world.[13] This form of Christianity is not in conflict with democracy because it is not authoritarian and does not require an acceptance/rejection choice. Borg says the main message in the Bible is a protest of injustices in political and economic systems.[14] Moses, the prophets, Jesus, Paul, and the Book of Revelation all protest against human systems of domination and advocate a vision of what life can be when transformed by being in relationship with God. This transformation-centered Christianity does not deny an afterlife, he says, but recognizes that life after death is a relatively minor theme in the Bible. Salvation is about this process of transformation here and now, not about going to Heaven. This Christianity leaves the afterlife up to God and instead focuses on our responsibility for transforming this life.

Rather than sin and guilt being the central dynamic of the Christian life, the central dynamic becomes relationship—with God, the world, and each other.

Marcus Borg
The God We Never Knew

Historically, concepts of Heaven and Hell were emphasized because they gave enormous power and money to the Church. The Church could say who was going to Heaven and who was condemned to Hell, but a person could increase their chances of salvation by buying indulgences from the Church. Martin Luther ended this form of church income, but the emphasis on attaining salvation remained.

Transformation-centered Christianity accepts other religions as valid because it sees culture as shaping all consciousness and knowledge; therefore, all knowledge (even scientific knowledge) is historically conditioned and relative. All religions are shaped by the culture in which they began, including Christianity. This means, Borg says, that the great religions of the world are those that have endured; they have stood the test of time because they are authentic cultural responses to the experience of the sacred.[15]

Mainstream theologians interpret the Christian Bible not only metaphorically but also historically. Scholarship over the past 50 years has uncovered a great deal about what life was actually like at the time of Jesus. This new knowledge places the Bible in a new context. Jesus was a first-century Jewish rabbi who taught love, compassion, and political justice. (Jesus was never a Christian, for Christianity developed after he died.) The challenge he offered was in conflict with the accepted cultural vision of his time, although this conflict was not new in Judaism. Borg says, "The struggle between Jesus and the wisdom of his time was not a struggle between a new religion (Christianity) and an old religion (Judaism), but a struggle between two ways [worldviews] of being religious. . . . a way of being religious that depended upon observance of externals (the way of conventional wisdom) and a way of being religious that depended upon inner transformation."[16] In fact, he says this conflict is found in all major religions and cultures. The conflict is the same as the two conflicting worldviews.

Interpreting the Bible Historically

The historical context of the Christian Bible begins with the Roman Empire annexation of Palestine in the year 63 B.C. and the imposition of land and crop taxes. The Torah required a system of tithing to support the temple, the priests, and the poor. Many farmers could not pay both Rome and the Temple without risking the loss of their land (and the loss of land was tantamount to a death sentence).[17] The Jews were fighting for their survival. The Roman tax was not optional; however, tithing, although required by divine revelation, had no brutal enforcement. This was a religious crisis because it created a large number of "nonobservant" Jews. Therefore, the Jewish social vision became centered on the holiness or purity system in which a person's birth and behavior (including tithing to the Temple) determined their "purity." Those who followed the purity system were "the righteous," and those who did not were "the sinners."[18]

The Jewish renewal movements—the Essenes, the Pharisees, and what was called "the fourth philosophy"—all intensified the politics of holiness or purity. This politics of holiness came to dominate the Jewish social world at this time.[19] This purity system created a Jewish society with sharp social boundaries between "pure and impure, righteous and sinner, whole and not whole, male and female, rich and poor, Jew and Gentile."[20] Jesus began his ministry in this world shaped by the Jewish politics of holiness and their intense conflict with the Roman Empire (which led to war in A.D. 66–70 when the Romans destroyed the Temple and much of Jerusalem).

> Early Christianity is far more diverse than nearly anyone ecxpected brfore the Nag Hammadi discoveries [in 1945]
>
> Elaine Pagels
> *The Gnostic Gospels*

The purity system governed with whom you could share a meal, as well as what you could eat and how you should prepare it. Sharing a meal meant acceptance, and no decent person would eat with an outcast. However, Jesus preached an inclusive community by having an open table fellowship, eating with sinners of all types, outcasts, lepers, and prostitutes, and in so doing, he challenged the entire purity system.[21] When interpreted historically within the context of the crises of first-century Judaism, the understanding of Jesus's ministry changes.

First, Borg says, Jesus "was a 'mediator of the sacred,' one of those persons in human history to whom the Spirit was an experiential reality." Second, Jesus made compassion rather than authority a political imperative, and he called for social and political transformation. Borg calls this "alternative wisdom" of Jesus a wisdom that sees religious life as one of a relationship with the Spirit of God, rather than one of requirements and rewards.[22] With this understanding, Jesus had a vivid sense that reality was ultimately gracious and compassionate, rather than judgmental and oppressive. Living in relationship to Spirit as Jesus preached shifts the focus from rigid belief systems to "being in relationship to the same Spirit that Jesus knew." Therefore, Borg says, Christian life is not about believing or being good, but "about a relationship with God that involves us in a journey of transformation."[23]

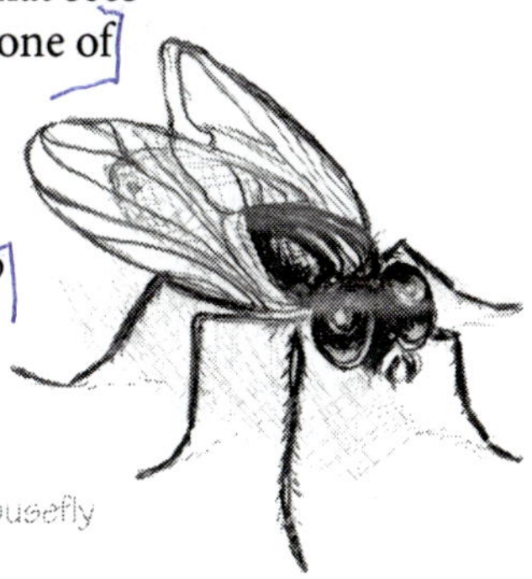

Housefly

In addition to this focus on our relationship with God, Jesus preached that compassion was the "central moral quality of a life centered in God." For Jesus, Borg says, "compassion was more than a quality of God as an individual virtue: it was a social paradigm, the core value for life in community. To put it boldly: compassion for Jesus was political. He directly and repeatedly challenged the dominant sociopolitical paradigm of his social world and advocated instead what might be called a politics of compassion."[24] Jesus drew large crowds, and these crowds and his message were very threatening to the Roman authorities, so much so that the Romans killed him for being a political subversive.

Borg describes our traditional or "popular image" of Jesus as an image of "a divine or semi-divine figure whose purpose was to die for the sins of the world, and whose life and death open up the possibility of eternal life." This image, he says, "is simply not historically true."[25] This is not what Jesus preached. New historical research shows that the three similar "synoptic gospels"—Matthew, Mark, and Luke—describe a very different Jesus than the Gospel of John does. In these three gospels, Jesus did not proclaim himself as God.[26] Scholars and mainstream theologians now believe that Jesus never actually said the statements attributed to him in the Gospel of John. The Gospel of John is a beautiful piece of religious writing written at the very end of the first century. However, Borg says, "instead of being a picture of the historical Jesus. . .[it] is about the risen living Christ of Christian experience. . . . John's gospel is the church's memory transfigured."[27]

In contrast, the words of the historical Jesus were not about speaking of himself as the Son of God whose historical purpose was to die for our sins. His message was not about believing in him. Rather, he was "a spirit person, subversive sage, social prophet, and movement founder who invited his followers and hearers into a transforming relationship with the same Spirit that he himself knew, and into a community whose social vision was shaped by the core value of compassion."[28] Therefore, Borg says, we do not need to base our religion on "Jesus accomplishing something that objectively changes the relationship between God and us, but upon Jesus revealing something that is true."[29] That is, it is about living in relationship with God and acting with compassion. This version of Christianity does not believe that God requires or expects that everyone be Christian.

The decision to focus on Christ as savior rather than on the message Jesus preached was made primarily for political reasons. Elaine Pagels, in *The Gnostic Gospels,* says that the New Testament accounts could support a range of interpretations, but orthodox Christians in the second century emphasized a literal view of the resurrection because it established the legitimate authority of the "men who claimed the right to exercise exclusive leadership over the churches, based on their being the successors of the apostle Peter. From the second century, the doctrine has served to validate the apostolic succession of bishops, the basis for papal authority to this day."[31] Even now, she says, most issues defined as heresy are issues concerning "insubordination to clerical authority."[32]

Seminaries and religion courses in nonsectarian colleges teach this transformative Christianity. Most mainstream church members hear this message of compassion and community, rather than a message of judgment and Hell. Even so, most people in our culture, even those who have heard the message, do not realize how widespread this changing emphasis in Christianity has become.[33] With it, we can create a transformation-centered religion based on Jesus's teachings about compassion, community, and the reality of the Spirit. For me the life and teachings of Jesus most embody what it means to trust your life to God and to act with others as God intended. This transformation unites the "inner" and the "outer" as one: once in relationship with the compassionate God, we cannot do otherwise but to act compassionately.

Creating a Public Theology

Ecology as a systems concept and the dying Earth as a concrete reality offer a new set of categories and a new language that can create an integrated social theory and an explicit theology. We need to rethink the relationship between spirituality and the social order by remaking religion, and with it, politics and society. To do so, we need to return theological reflection to the public domain. This leads theologian David Ray Griffin to ask the central question, "Is it possible to move toward a form of theology that would, through its intrinsic merits, evoke a sufficiently broad response that the values implicit in it would replace the values of the modern worldview that have dominated public policy?"[34] If this is possible, then it becomes legitimate to use compassion and sustainability as a basis for political choices rather than using cost/benefit analysis, the "value-free" workings of the market, or other economic tradeoffs.

A new public theology could combine the political democratic process and religious values. I assume that only this combination can create a society capable of making an adequate response to global warming, and I believe it is the only message worth preaching. If life is truly interdependent, then we can see it as an organic whole, where politics, economics, and religion flow together and define one another. In doing so, a genuinely religious view of the world emerges—one capable of creating sustainable political and economic policies and one energized by an experience of the Holy.

People remain in denial about the severity of the crisis and the solutions needed because they assume the only alternative to denial is despair. Within our current worldview, that's true. The answers—the vision and the hope—lie outside the current contexts and assumptions. I believe that only theological thinking can empower people to participate in the needed changes.

Theology is a solution to real-world problems only if it can move into the public sphere. Yet theology can be public only if it can be accepted in a religiously pluralistic world. Fundamentalists focus on a rigid belief system, and the expression of a particular belief system automatically becomes "private" (and marginalized). The public theology we need would not be defined in terms of belief systems, but in terms of the idea that reality is sacred and that the foremost religious value and therefore political value is compassion for others, ourselves, and Earth, with the understanding that we are our brothers' (and sisters') keepers.

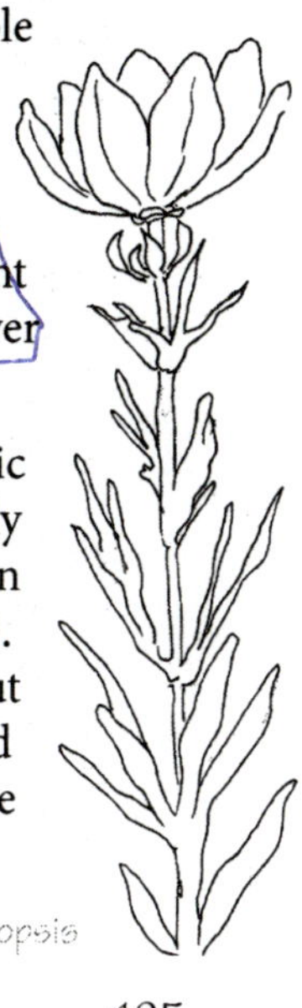

Prairie coreopsis

Chapter 5

Views of Economics that Contribute to the Crisis

We are risking the ability of humans and other creatures to live on Earth. Yet so many people continue to deny this information. They do so because they can't deal with the implications. Part of the problem is that even most progressives and environmentalists only propose responses to global warming that leave our current economic system fundamentally unchanged. But the real solutions are outside left/right distinctions and outside the arguments over capitalism and socialism, for the destruction of Earth occurs in both. Neither accepts that Earth is finite or that we must create economies that can function within nature-imposed limits.

Our worldviews have an economic component. This chapter and the next describe our current economic worldview, which is now destroying Earth.

Capitalism or Socialism—a False Choice

> Under capitalism, man exploits man, under communism it's just the opposite.
>
> John Kenneth Galbraith

In the U.S., any attempt to modify our economic system is met with charges of socialism. The conversation is always limited to only two choices: if it is not our current capitalism, then it is socialism. However, ecologists say that, from the viewpoint of Earth, there is little difference between capitalism and socialism: both are sources of the ecological crisis. Both socialism and capitalism require continual economic growth. Both are materialistic ideologies based on similar concepts of centralization, industrialization, technology, and bureaucracy. For both, production and consumption are the measure of welfare and the defining goals of society, and both define success as the production of more goods. They differ only in what they see as central to an economy and therefore who should get most of the wealth: capital or workers (although in practice, both create elites).

Economist Herman Daly and theologian John Cobb, Jr., in *For the Common Good*, also describe capitalism and socialism in similar terms in relation to sustainability: "Whatever their ideological differences both systems are fully committed to large-scale, factory-style energy and capital-intensive, specialized production units that are hierarchically managed. They also rely heavily on nonrenewable resources and tend to exploit renewable resources and waste absorption capacities at nonsustainable rates."[1] Therefore, the old opposition of capitalism and socialism is now irrelevant.

This alternative has been developing for decades, but now it is imperative that we redesign our economy in terms of ecological economics. The primary difference is that the alternative view acknowledges limits rather than claiming and acting as though there are none. This new vision, called ecological economics, has been developing since the 1970s; it offers an economics that puts Earth and people at the center. I will discuss this new theory in Chapter 8. This chapter reviews where we are now.

It is the current structure of the global economy that creates the dynamics and patterns of interaction that have resulted in so much harm to Earth. These same dynamics also create an enormous gap between rich and poor and cause us to lose any genuine democratic control over our lives. Our current economic system has become predatory, ruthlessly competitive, and immoral. The drive for profit overrides any concern for either people or Earth. David Korten calls it a global suicide economy. Currently, we measure success only in terms of profit and view any ethical considerations as irrelevant. Both conservatives and progressives believe in our current economic system, called neo-liberalism.

Neo-liberalism: Our Current Brand of Capitalism

Capitalism changed greatly in the 1980s. President Reagan and his "Reaganomics" began this shift to an economic philosophy called neo-liberalism ("neo" meaning new). The term "liberalism" can refer to political, economic, or even religious ideas; however, both conservatives and liberals accept economic liberalism. Today we also call neo-liberalism "globalization," free-market economics, Reaganomics, or just "capitalism," although this capitalism has become something very different from what it was before 1980.

The original liberal school of economics (before the "neo") began when Adam Smith published *The Wealth of Nations* in 1776. His economics advocated the abolition of government intervention in economic matters: no restrictions on manufacturing, no barriers to commerce, and no tariffs. He said that free trade was the best way for a nation's economy to develop. These ideas were "liberal" in the sense of having no controls.[2] The objective (then as now) was to free economic life from social and political control. This created a new type of economy in which corporations created prices for the exchange of goods and labor while ignoring any effects on society or the environment. Deregulated markets operate independent of all social needs or values.[3]

This kind of economics was dominant in the U.S. through the 1800s and early 1900s (and created the Gilded Age). Then, during the Great Depression, economist John Maynard Keynes developed a new theory. He said that full employment is necessary for capitalism to grow and that only governments and central banks, by intervening, can increase employment. These ideas influenced President Roosevelt's New Deal,

Peacock Butterfly

and at that time, most people accepted the belief that government should advance the common good.[4]

This new understanding of economics continued through the 1970s. Economist Susan George says that, from 1945 through the 1970s, "At least in the Western countries, at that time, everyone was a Keynesian, a social democrat or a social-Christian democrat or some shade of Marxist. The idea that the market should be allowed to make major social and political decisions; the idea that the State should voluntarily reduce its role in the economy, or that corporations should be given total freedom, that trade unions should be curbed and citizens given much less rather than more social protection—such ideas were utterly foreign to the spirit of the time."[5]

I hope we shall crush in its birth the aristocracy of our monied corporations which dare already to challenge our government to a trial of strength, and bid defiance to the laws of our country.

Thomas Jefferson

Economist Paul Krugman also describes this time and says, "American corporations bore little resemblance to today's hard-nosed institutions. Indeed, by modern standards they were Socialist republics."[6] Executives at this time did not think only of maximizing stock prices. They assumed they were serving multiple constituencies, including their employees. Krugman says that after World War II, the U.S. standard of living doubled, but then growth faltered because of the equalizing effect of this kind of economics, and then the corporate raiders arrived.[7]

Economic liberalism returned in the 1980s, but it was now called neo-liberalism. George says that people again began to believe that the "market mechanism should be allowed to direct the fate of human beings. The economy should dictate its rules to society, not the other way around."[8] The corporate raiders wanted companies to get leaner and meaner and changed the incentives. They replaced much of a company's stock with debt and forced management to show a profit or go bankrupt. There became a sense that whatever it takes to increase profit is just good business. This kind of thinking caused corporations to became more and more ruthless. Executive pay was greatly increased and CEOs became very wealthy, but this led many executives to manipulate the short-term information available to outsiders to allow the company to seem more successful than it was.[9]

The only measure used is money. If it makes more money, it's good; if not, it's bad. Choosing maximum profit was defended on the grounds that it was an accurate measure of economic efficiency. This economic theory sees self-interest as a legitimate measure of choices and greed as a positive force.

Neo-liberalism is not a neutral process with respect to our society or politics. The free market is a deliberate form of social organization, and the government must impose a structure to attain certain goals. Neo-liberalism is a political concept enforced by the state through contracts and by allowing an extreme form of deregulation. This allows neo-liberals to ignore the unequal power dynamics in which the economy functions. In the philosophy of neo-liberalism, the unrestricted operation of a market is a good and an overriding ethic in itself that can legitimately guide all human actions. It is the dominant social value and operates without regard to all other existing ethical beliefs. The market and market values are seen as semi-sacred and a moral necessity so that market choices legitimize all outcomes. Therefore, this economic system operates in an ethical vacuum. But that vacuum then becomes itself the moral value of capitalism—that only what can be measured in monetary terms matters. Whatever the outcomes of market transactions are, neo-liberals assume that they are automatically justified because "the Market" chose them.

However, the selective effects of this form of capitalism are cumulative, creating more and more wealth for the already wealthy and requiring that we ignore the needs of society, workers, and the environment. In neo-liberalism, corporations want to reduce the role of the government, including cutting public funding of social services like education and health care. They want to reduce the safety net for the poor and even money for maintenance of roads, bridges, water supply. They believe the state should sell state-owned enterprises, goods, and services to private investors. In addition, they eliminate any sense of the public good and replace it with "personal responsibility." They pressure the poorest people in a society to find solutions to their lack of health care and education all by themselves and then blame them if they fail, for being "lazy."[10] This focus on increasing profit means wage cuts, less money for social services, and fewer taxes on the rich.

Proponents of this type of economics have been able to make their ideas seem natural, normal, and inevitable. In fact, many people now think that the only alternative to this type of economics is socialism. Our culture rejects socialism, so we assume that neo-liberalism must be the only possible economic and social order. Margaret Thatcher, who became Britain's Prime Minister in 1979, called neo-liberalism TINA, for "There Is No Alternative."[11]

The central value of neo-liberalism is competition—between nations, regions, firms, and individuals—because it creates winners and losers. The theory says that the competitive process is the most efficient way to allocate resources. Because neo-liberals understand competition to be good, they assume the results are always legitimate, particularly in the distribution of income and wealth. The theory claims a privileged social status for the entrepreneur and believes that entrepreneurs should control the economy. Neo-liberals want to drastically reduce the public sector because it cannot function in terms of competition—for profits or market share—and often acts to assist the undeserving losers of competition.

Since the 1980s, neo-liberal economics has fundamentally changed the nature of American politics. The institutions of neo-liberalism lack any transparency or democratic accountability. The theory sees democracy as an encumbrance because neo-liberalism is designed for "winners," not for voters (who include

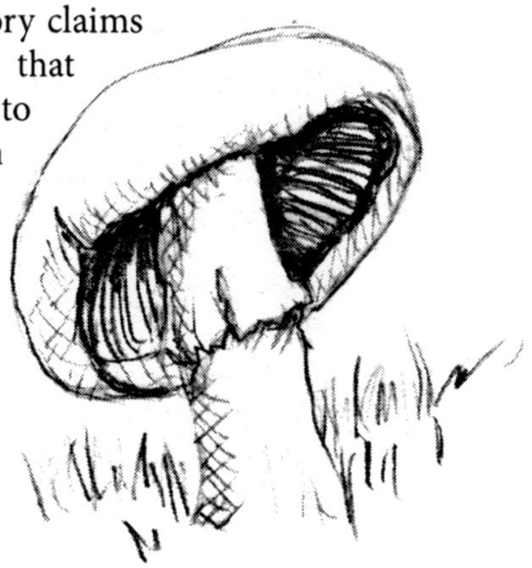

Horse Mushroom

"winners" and "losers").[12] The "winners" of this competition assume that their success is proof of their entitlement and of the unworthiness of the losers. (This is the opposite of the biblical verse "To whom much is given, much will be required."[13]) George describes this understanding of economics as follows:

> For the neo-liberal, the market is so wise and so good that like God, the Invisible Hand can bring good out of apparent evil. Thus Thatcher once said in a speech, "It is our job to glory in inequality and see that talents and abilities are given vent and expression for the benefit of us all." In other words, don't worry about those who might be left behind in the competitive struggle. People are unequal by nature, but this is good because the contributions of the well-born, the best-educated, the toughest, will eventually benefit everyone. Nothing in particular is owed to the weak, the poorly educated, what happens to them is their own fault, never the fault of society. [14]

This ideology is now destroying Earth. It is both inaccurate and immoral.

No Place for Land in Neo-liberal Economics

One reason economics does not measure the destruction of Earth is because markets have no mechanism to put a price on what is happening to nature. By economic definition, land is property; therefore, land only has worth as a commodity, or as something to support the production of various goods and services (such as in agriculture). Nature does not exist, as such, because we cannot buy or sell the natural environment. Therefore, it becomes invisible and exists outside economic theory. Herman Daly and John Cobb, Jr., in *For the Common Good*, say that nature has "disappeared from view, and economics as a discipline floats free from the physical world."[15]

Over time, the place of "land" in economics has shifted.

- In an agricultural society, the "means of production" was land and labor. Economic theory ignored the reality of the wealthy landowner's power or the poverty of the person who labored on the land.
- Later economic theory shifted focus to labor, "earned income," and to the *rent* of the land, rather than the land itself. This rent was "unearned income." This theory devalued the wealth of the landlord class. Daly and Cobb say that this concept of rent "was an ideological weapon in the power struggle between the rising capitalist and the entrenched landlord."[16]
- With the rise of industry, the location of wealth shifted again and pushed the originally dominant class of landowners to third place in economic theory, allowing capital to become the dominant class, followed by organized industrial labor.[17] This shift occurred because land was no longer the limiting or most critical input to production as it is in an agricultural society. Industry generally did not need much land, and so money for funding industrial growth became the limiting factor.

Current economic theory assumes that capital can substitute for land. The central concept of capitalism is that as long as capital is sufficient, there can be no shortages (such as the loss of healthy land). By treating land as capital, economists can depreciate the fertility of the land, just as they do other forms of capital. Economists continue to focus on capital and labor and ignore land, so that Daly and Cobb say, "The goal of increasing capital can proceed without attention to what is physically happening to the land."[18] They say this is particularly true because "above all, land is conceived as inert. It has no principle of life within itself. How anyone who has seen wilderness could think of land in this way is hard to imagine. But a priori modes of thought exercise immense control. Land is matter and matter is passive, ergo only human labor can give form to land. Land not formed by human beings is 'undeveloped,' passively awaiting the actualization of its potentialities."[19]

The modern conservative is engaged in one of man's oldest exercises in moral philosophy; that is, the search for a superior moral justification for selfishness.

John Kenneth Galbraith

Because of the view that land does not have an independent existence, our current economic theory deals only with the flow of money and ignores or misstates what is happening to Earth. For example, in our economic system, trees have no value until they become lumber and can be bought or sold. To compound this, in capitalism, scarcity creates value, so as trees become more scarce, they increase in value and therefore are cut down faster.

Why Our Financial System Collapsed: Deregulation Run Amok

The belief in neo-liberal economics—the belief that markets are self-adjusting and that the role of government should be minimal—caused the financial meltdown that began in 2006. We created a culture of greed and self-interest and then assumed the same culture would "self-regulate." Robert Weissman and Harvey Rosenfield in "Sold Out: How Wall Street and Washington Betrayed America," show that, from 1998–2008, Wall Street firms, including investment firms, commercial banks, hedge funds, real estate companies, and insurance conglomerates, paid $1.725 billion in political contributions and spent another $3.4 billion on lobbyists, all devoted to undercutting federal regulation. In 2007, nearly 3,000 officially registered federal lobbyists worked for the industry.[20] They explain that "for the last three decades, financial regulators, Congress and the executive branch have steadily eroded the regulatory system that restrained the financial sector from acting on its own worst tendencies."[21]

This non-regulatory ideology began with the rise of neo-liberalism, or Reaganomics. The belief in free markets with no restrictions led President Reagan in 1987 to remove Paul Volcker as chair of the Federal Reserve Board because Volcker believed that financial markets needed regulations. Reagan replaced him with Alan Greenspan, who did not believe in regulations. This belief in deregulation led to many changes. In November 1999, after a $300 million lobbying effort by the banking and financial industries, led by Sen. Phil Gramm,

Snail

Congress passed the Financial Services Modernization Act of 1999.[22] This repealed the Glass-Steagall Act, enacted after the Great Depression to separate commercial and investment banks. This meant abandoning antitrust principles and represented the last serious obstacle to the financial crisis that unfolded seven years later.[23]

The resulting flood of mergers changed an entire culture. Joseph E. Stiglitz, a Nobel laureate economist, in an article titled "Capitalist Fools," explains that "commercial banks are not supposed to be high-risk ventures; they are supposed to manage other people's money very conservatively. Investment banks, on the other hand, have traditionally managed rich people's money—people who can take bigger risks in order to get bigger returns. When repeal of Glass-Steagall brought investment and commercial banks together, the investment-bank culture came out on top."[24] Even commercial banks focused on high leverage and big risk taking to get high returns in order to remain competitive.[25] It also allowed for a frenzy of consolidation of banking power in the hands of Wall Street, making them "too big to fail." The concept that some banks were "too big to fail" began at the end of the 1980s, when the Savings and Loans collapsed and Citibank lost 50% of its share price. Negotiations were conducted behind closed doors, and "too big to fail" became standard policy.[26] Banks weren't nationalized, but bank risk was. The government now protects losses in spite of excessive risk-taking or bad management.[27]

The whole aim of practical politics is to keep the populace alarmed by menacing it with an endless series of hobgoblins, all of them imaginary.

H. L. Mencken

During the Clinton administration, the Commodity Futures Trading Commission, which has jurisdiction over futures, options, and other derivatives, wanted to regulate financial derivatives. Treasury Secretary Robert Rubin and, especially, Alan Greenspan stopped them. Then in 2000, Congress passed the Commodities Futures Modernization Act, which specifically exempted financial derivatives from regulation.[28] In April 2004, in the interest of deregulation, the Securities and Exchange Commission (SEC), in a meeting with only five people, decided to allow big investment banks to increase their debt-to-capital ratio. Instead of restricting them to loan at a rate of 10 or 12 times their assets, it was increased to 30 or higher.[29] Merrill Lynch used leverage as high as 40 to 1.[30] In agreeing to this measure, the SEC argued for the virtues of self-regulation. Stiglitz says self-regulation is preposterous, and in any case, it can never identify systemic risks.[31] Having allowed banks to assume much more risk in the businesses they pursue, the regulators now allowed them to become even riskier by increasing their leverage substantially.

Even in a deregulated environment, the banking regulators had the authority to crack down on predatory lending abuses (like the sub-prime mortgages). But since they did nothing, the states tried to enforce consumer protection laws against predatory lenders, but they were stopped. Eliot Spitzer said that in 2003, "During the height of the predatory lending crisis, the Office of the Comptroller of the Currency [OCC] invoked a clause from the 1863 National Bank Act to issue formal opinions preempting all state predatory lending laws, thereby rendering them inoperative. The OCC also promulgated new rules that prevented states from enforcing any of their own consumer protection laws against national banks."[32]

In addition, a major role contributor to the financial crisis was the use of the unregulated derivatives like credit default swaps. Warren Buffett called derivatives "financial weapons of mass destruction."[33] These instruments were originally created to help manage risk—but they were used to gamble. You could bet, for instance, on whether or not a package of mortgages to which you had no intrinsic exposure would rise in value. It was equivalent to being able to take out fire insurance on your neighbor's house. However, with this complicated intertwining of enormous bets, no one could be sure of the financial position of others—or even of themselves—so, when trouble began to surface, the credit markets froze.[34] In Congressional hearings in the early 1990s, the trading of derivatives was challenged as being an illegal form of gambling. Then in 1998, the head of the Commodity Futures Trading Commission, Brooksley Born, called for regulating derivatives. But Federal Reserve (Fed) Chair Alan Greenspan lobbied hard to defeat her and legitimized the use of unregulated derivatives by lending legal and regulatory support to it and actively promoting them as a way to improve "risk management." Stiglitz says that many people understood the risk, "but the deregulators were in charge at the Fed, at the Securities and Exchange Commission, and elsewhere, and they decided to do nothing."[35]

The essential issue is that buying derivatives is not investing, for they don't create anything.[36] Ellen Brown, author of *Web of Debt*, says they not only create nothing, but they serve to enrich non-producers at the expense of the people who *do* create real goods and services. According to the Comptroller of the Currency, the books of U.S. banks now carry more than $180 trillion in derivatives. This amount is three times the annual gross domestic product (GDP) of all the countries in the world combined.[37] When this banking scheme began to collapse, the government provided bailout money directly from the taxpayers.

Alan Greenspan said regulation of participants was unnecessary because it was in their self-interest to be prudent. They would "regulate themselves." Obviously, this was false. The financial meltdown occurred because the dynamics were not well understood, and the incentives of many key participants were not congruent with the larger needs of society, or even of their own firms in many cases.

People got enormous bonuses for how well they did over the past quarter. They figured if it went bad, they could get out before the other guys—but most could not get out in time. The structure of a system defines how people will behave; when the system was designed to limit risks, they were limited. When the limits were removed, people took more and more risks. Individually, they knew they were doing very well—it was the *system* that was flawed. As they kept adding more and more leverage, the system became more and more unstable until it blew up—which was inevitable.

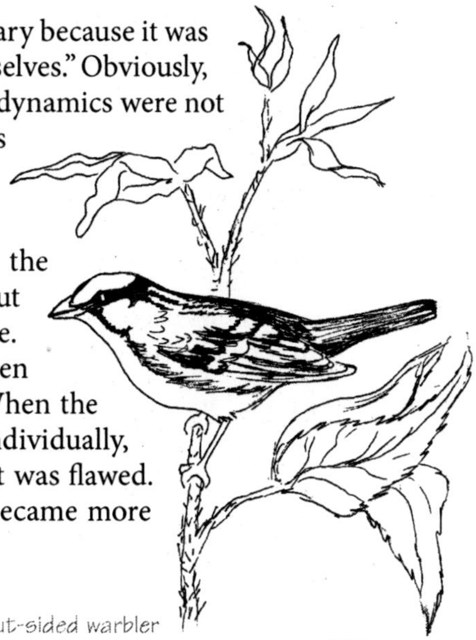

Chestnut-sided warbler

An added factor in the financial meltdown is the large tax cuts passed by Congress in 2001, 2002, and 2003, which cost $1.7 trillion in revenue losses and did little to stimulate the economy.[38] These tax cuts shaped the background conditions of the current crisis. Because they did little to stimulate the economy, the Fed had to create the stimulation with unprecedented low interest rates and liquidity. This flood of liquidity made money available in mortgage markets, even to those who would normally not be able to borrow. This action succeeded in forestalling an economic downturn, which would normally have occurred (as it had in the 1970s) because of the soaring oil prices caused by the Iraq War. Because of our dependence on oil imports, we had to spend several hundred billion more to purchase oil—money that otherwise would have been spent on American goods. America's household saving rate went to zero. Stiglitz says, "It should have been clear that we were living on borrowed money and borrowed time."[39]

The Bush tax cuts included a cut on the capital gains tax rate. Stiglitz says this decision was based on values: "those who speculated (read: gambled) and won were taxed more lightly than wage earners who simply worked hard. But more than that, the decision encouraged leveraging, because interest was tax-deductible."[40] The 50 highest-paid private investment fund managers in 2007 averaged $588 million in compensation, each, for the year.[41] And because of favorable treatment from the Internal Revenue Service (IRS), their income was considered a form of capital gains, so they paid only 15% tax on it.

The cost of the Iraq War has also made us less able to deal with the financial crisis. The total costs are now $2 trillion, but this number does not include many hidden costs, such as the lifetime of disability and health-care benefits for the tens of thousands of wounded veterans; as many as 20% of the wounded have suffered devastating brain and spinal injuries.[42] The full long-term costs of the war will be more than $3 trillion.[43]

The individual investor never sees the whole system because she cannot see what will happen over time, but a complex system can only be understood when you analyze its dynamics over time and look at how everything is connected. Prudent behavior and the incentives are focused on the extremely short term. Payout is not based on how the firm does over the next decade, but over the next quarter. This creates a shell game. There are no incentives in capitalism for the long term. Genuine prudent behavior is the kind demonstrated by the small-town banker with a long-term relationship with the community (their clients). In any case, prudent behavior is local, but what is important in a complex system are the interconnections throughout the entire system. Only regulations can address the whole system. They can impose significant limitations on the kind of states the system is capable of achieving.

But aside from the collapse, the entire economy has become unstable and weak. Only industry, agriculture, and their accompanying trades and services introduce real value into the economy. But the financial sector creates *nothing of real value*. One writer said it is "tantamount to squeezing value out of already created value. . . . After all, why invest in actual economic growth when the real money is to be made buying and selling other people's mortgages?"[44]

The collapse of the financial markets demonstrates the risks of not understanding that the economy is a complex system. Economists are oriented toward equilibrium analysis, which is appropriate for a system that can go to a static state of stable, even growth (that is, reach equilibrium). However, the economy is a complex system; it is therefore dynamic and unstable and will not reach equilibrium. The whole philosophy is a misfit because economists don't take a systems view. The "free market" without regulations is inherently unstable. Economist Hyman Minsky wrote *Stabilizing an Unstable Economy,* which is about why and how an unregulated economy will always be boom or bust, either going up or down. The ecological vulnerabilities of Earth have many parallels with the financial meltdown of 2008. They are both the result of the same thinking—that we can pretend that there are no limits. We need to stop viewing the economy in isolation; we now need to view it as part of a larger whole that includes the consequences it has on Earth and communities. Otherwise, the natural world will collapse, just as the financial markets did.

Our Current Economic System No Longer Works

The financial crisis that began in fall 2008 has finally discredited neo-liberal economics. We can now see the house of cards it created. In addition to forcing a low wage policy and the destruction of our natural systems, it can be seen that it wasn't sound economics, either. As bad as it has been, this crisis may force a fundamental reassessment of how we live our lives together. With this reassessment, we may have a chance to stop global warming and restore Earth.

Economists are now competing to try to make sense of our failed financial system. An article in *Newsweek* said that at the January 2010 annual meeting of the American Economic Association, panel after panel was discussing "how, exactly, they'd gotten things so wrong."[55] Joseph Stiglitz gave the keynote speech, in which he debunked the profession's key tenet—namely, that markets can be trusted.[56] The key topic for the meeting was the crisis of conscience in economics itself. For the first time in decades, the profession is rethinking all the big questions. Since Reagan, the consensus was that you just had to make the pie grow bigger by unshackling markets and investors. Wealth, and thus health, happiness, and all other good things, would eventually trickle down to everyone. Although it was a simple and elegant way of looking at the world, Krugman says that in it, beauty trumped truth.[57] Now that this view has been proved false, many new theories are competing to take its place. All are based, to one extent or another, on the idea that people (individuals, firms, industries) often do not act in their self-interest—they do not act the way Alan Greenspan believed they would—and that markets are not always efficient or resilient. In many cases, the new theories blend economics with entirely different disciplines.[58]

Peach

Chapter 6

Views of Political Power that Contribute to the Crisis

Political power, both nationally and internationally, is dominated by the power of corporations. This power of corporations corrupts our entire democratic political process, for Congress is overwhelmingly beholden to corporate interests and is unable to make decisions that would go against these interests. Democratic rule by the people has become an illusion in the U.S. The first step in getting our democracy back is to see how the power of corporations actually operates. We can never stop global warming if corporations control the U.S. political process, as well as the world economy. The anti-democratic processes we now have are unsustainable, for the predatory power of corporations is destroying Earth.

Fascism in the United States

The enormous power of corporations in neoliberal economics has led to a growing concern about fascism in the U.S. The consequences of global warming will provide an easy excuse for an authoritarian, totalitarian, or fascist government to close down our open society. In fact, global warming makes a much better excuse for that than "fighting terrorism."

> When fascism comes to America it will be wrapped in a flag and carrying a cross.
>
> Sinclair Lewis

The problem is complicated because most Americans don't know what fascism is, even though we fought World War II to stop it. So the more we know about the characteristics of fascism, the more we can understand what kind of government will make change impossible. There is a vast difference between conservatism and fascism; however, the situation is now very serious because during the Bush administration, our country moved toward fascism.[1]

Mussolini claimed to have named fascism yet remarked, "Fascism should more appropriately be called corporatism because it is a merger of state and corporate power."[2] The 1983 edition of the American Heritage Dictionary defines fascism as "a system of government that exercises a dictatorship of the extreme right, typically through the merging of state and business leadership, together with belligerent nationalism."[3]

Dr. Laurence Britt, a political scientist, studied the fascist regimes of Hitler (Germany), Mussolini (Italy), Franco (Spain), Suharto (Indonesia), and Pinochet (Chile). He created a list of 14 points that define fascism:[4]

1. Powerful and continuing expressions of nationalism
2. Disdain for the importance of human rights
3. Identification of enemies/scapegoats as a unifying cause
4. Supremacy of the military/avid militarism
5. Rampant sexism
6. Controlled mass media
7. Obsession with national security
8. Religion and ruling elite tied together (a perception is manufactured that opposing the power elite is tantamount to an attack on religion)
9. Power of corporations protected
10. Power of labor suppressed or eliminated
11. Disdain for and suppression of intellectuals and the arts
12. Obsession with crime and punishment
13. Rampant cronyism and corruption
14. Fraudulent elections

During World War II, many in the U.S. became concerned about fascism developing here. In early April 1944, the *New York Times* asked Vice President Henry Wallace to write an article answering the following questions: What is a fascist? How many fascists have we? How dangerous are they? Wallace saw a tide of fascism rising in America, and in his answer to these questions, he said, "They claim to be super-patriots, but they would destroy every liberty guaranteed by the Constitution. They demand free enterprise, but are the spokesmen for monopoly and vested interest."[5] Finally, Wallace said, "Democracy, to crush fascism internally, must . . . put human beings first and dollars second. It must appeal to reason and decency and not to violence and deceit."[6]

Even the right-wing John Birch Society opposed Bush's warrantless wiretap program, saying, "This is to say that from the administration's perspective, the president is, in effect, our living constitution. This is, in a specific and unmistakable sense, fascist."[7] On the left, Robert F. Kennedy, Jr., specifically indicts the Bush administration for fascism in his book *Crimes against Nature: How George W. Bush and his Corporate Pals Are Plundering the Country and Hijacking Our Democracy.*

How Corporations Gained More Rights than Citizens

The main players in neo-liberalism are corporations. The corporation is a legal form that allows unlimited concentration of economic power, power used only to benefit stockholders, without any accountability for the consequences of their actions on society or on the environment.

Lowland Gorilla

The following history of how corporations gained more rights than individuals is "public" knowledge, but few people know about it. Most people do not know that corporations have more legal rights than they do. The U.S. Constitution gave states the power to grant or deny permission for a business to operate in their state. At the time, people understood that in granting a charter, the state assumed the corporation would serve the public good. After the Civil War, the number of corporations increased, but since states gave them their charters, they had duties to fulfill to keep their charter and were accountable to the people. However, this focus on the public good changed as corporations began to acquire enormous power in the mid-nineteenth century.

> Slavery is the legal fiction that a person is property. Corporate personhood is the legal fiction that property is a person.
>
> Molly Morgan

In 1865, the Thirteenth Amendment freed American slaves. In 1868, three years later, the Fourteenth Amendment gave citizenship rights to all persons born or naturalized in the U.S. This amendment granted rights to former slaves. However, in 1886 (in a case called Santa Clara County *v.* Southern Pacific Railroad Co.), the Supreme Court decided that they could use the Fourteenth Amendment to rule that a corporation was a person and had all the rights and privileges of a person. The court made this ruling without hearing any arguments, without any public debate or any law passed in Congress. At the time that the court extended the rights of personhood to corporations, they still denied the same rights to all women, all Native Americans, and most African-American men.[8]

For the first 20 years following this ruling, the Supreme Court saw more than 300 cases based on the Fourteenth Amendment. Of these, only 19 cases dealt with the rights of African-Americans, while 288 dealt with granting rights to corporations.[9] Corporations then worked to get the entire Bill of Rights applied to corporations. In 1893, they received Fifth Amendment protection of due process. In 1906, the courts granted corporations Fourth Amendment rights, giving them search and seizure protection. Because of this, the Occupational Safety and Health Administration (OSHA) and the EPA have to schedule their inspections at a time convenient to corporate managers. If you think the air, land, or water in your community is being poisoned or the workers mistreated, neither you nor the government can go on corporate property to get information without first getting legal permission from the corporation.[10]

In 1925, corporations acquired freedom of the press and speech. Then in 1976, in the Buckley *v.* Valeo decision, the Court declared that spending money to influence elections was a form of "free speech" protected by the First Amendment and largely beyond democratic control, that is, money is equal to speech.[11] Several other cases since 1976 have expanded the First Amendment rights of for-profit corporations to include protection of advertising, negative or compelled speech, and commercial speech and unlimited expenditure in support of candidates, parties, and ballot issues.[12]

If corporations did not have personhood, we could prohibit all corporate political activity—including contributions to candidates or parties, or lobbying (as was the case, originally). Without their protections under the Fifth and Fourteenth Amendments, the government could prevent corporations from merging or owning stock in other companies. If they were not a "person," we could distribute leaflets against them in shopping malls. These activities are currently illegal because the courts give the property rights of "corporate persons" priority over the First Amendment right to free speech for human persons. This ruling prevents the labeling of genetically modified foods because the First Amendment protects the right of "corporate persons" *not* to speak. The "civil rights" of corporations means that communities cannot officially "discriminate" in favor of local stores. If this "personhood" of corporations were changed, we could stop advertising for tobacco, guns, and other dangerous products.

However, if we did abolish corporate personhood, none of these things would change automatically. Congress would need to write new laws, and we would need to challenge old laws in court. Defining a corporation as a person has far-reaching consequences. It has resulted in granting more "rights" to the rights of property (organized in the corporate form) than rights of individual persons as citizens. Yet a corporation is not a person, it is a contract on a piece of paper.

In addition to being "persons," corporations have "limited liability." David Korten, in *Corporate Pathology and the Suicide Economy*, says this term, the "limited liability" of corporations, means that the stockholders are exempt from any liability for destructive corporate behavior beyond the value of their shares, no matter how destructive or illegal the behavior. All profit belongs only to stockholders, but they have no liability for what the company does. Directors and managers also face almost no risk of personal fine or imprisonment for crimes committed by the corporations in their charge.[13]

The Women's International League for Peace and Freedom has a campaign called Abolish Corporate Personhood. In a speech on this campaign, Molly Morgan said, "When you look at two-plus centuries of US legal history, the pattern is that people acquire rights by amendment to the Constitution—a long, drawn-out, difficult process—and corporations acquire them by Supreme Court decisions. Rights for corporations, because they're about property, is about who is excluded; rights for human beings is about who is included."[14]

How Corporations Gained More Power than Governments

In addition to having more rights than people, corporations now have more power than governments. Of the 100 largest economies in the world, 51 are multinational corporations, and only 49 are nations. Along with tremendous power, multinational corporations are free of any constraints of location or nationality.

Peregrine Falcon

Any real progress on global warming cannot occur within the current global economic structure. The current model of neo-liberal economics is also called free trade, the corporate globalization model, or the "Washington Consensus." Corporations have institutionalized this globalization model in treaties and in the World Trade Organization (WTO). The WTO was formed in 1995 to replace the General Agreement on Tariffs and Trade (GATT). The WTO sets the rules for the global trading system and resolves disputes between its member states.

To facilitate trade, the U.S. has led the creation of trade treaties in the Americas. These are structures similar to those the WTO creates to support worldwide globalization. The first treaty in the Americas, approved in 1994, was the North American Free Trade Agreement (NAFTA) between the U.S., Canada, and Mexico. The rules of the WTO and the agreements in NAFTA focus on three areas: agriculture, services, and non-agricultural market access (or NAMA), that is, goods other than agriculture.

> Voice or no voice, the people can always be brought to the bidding of the leaders. That is easy. All you have to do is to tell them they are being attacked, and denounce the pacifists for lack of patriotism and exposing the country to danger. It works the same in every country.
>
> Hermann Goering
> Hitler's Reich Marshall, 1939

Since 1994, corporations have been working to expand NAFTA to 33 countries in Central America, South America, and the Caribbean (except Cuba). This new treaty, called the Free Trade Area of the Americas (FTAA), was supposed to be completed in 2004 and ratified by January 1, 2005. However, there was massive citizen opposition in Latin America; in addition, Brazil, Argentina, and Venezuela would not agree to Washington's terms of trade, and the talks on FTAA ended.

Opposition to trade treaties grew in part because NAFTA turned out to be a disaster for working people and the environment in all three countries. This free trade between the countries of North American under NAFTA represents an active commitment to a low-wage policy. In turn, lower wages means that returns to those who own capital in all three countries will go up.[15] NAFTA has also done serious damage to Canadian health care, American manufacturing, and Mexican corn farmers. Agriculture, jobs, and services in Mexico, the U.S., and Canada have all suffered. More than 1.5 million Mexican farmers have lost their livelihoods as a result of NAFTA, resulting in skyrocketing immigration to the U.S. Over the past few years, developing countries have begun to stand firm against signing treaties that harm their economies.

There are many ways to create development and regional integration, but multinational corporations choose to focus only on trade and created rules and regulations to maximize their profit at the expense of workers and the environment. Yet, they insist that this approach is inevitable and is universally beneficial, and that in any case, "There Is No Alternative."

There are two huge problems with this model: first, the trade agreements override any national laws enacted to protect the environment, workers, and democracy. Second, the entire structure focuses all economic activity on trade for export. Earth is 25,000 miles around at the equator, and moving goods around the world requires massive amounts of fossil fuels.

Our current form of capitalism has now become as rigid and centrally controlled as communism was—only this time, it is the decisions of global corporations, the World Bank, the WTO, and trade treaties that control the economy of the entire world. Vandana Shiva says this alliance is creating a "centrally controlled, authoritarian rule like communism in its control, but different in the objective of profits as the only end of power." She says these corporations are undemocratic, dictatorial, authoritarian, and centralized; therefore, she says, "Free-trade is about corporate freedom and citizen disenfranchisement."[17]

Proverbs for Paranoids 3: "If they can get you asking the wrong questions, they don't have to worry about answers."

Thomas Pynchon
Gravity's Rainbow

One rule that is included in NAFTA and WTO negotiations is that corporations have the right to sue foreign governments directly, without approval from their own government. This means that corporations, without any leaders elected by the public, can control government policy, including environmental policies. The trade agreements say that a country cannot seize (expropriate) the assets of a company without that country compensating the company. Chapter 11 of the NAFTA agreement extended this protection with the words "tantamount to expropriation," meaning a country does not have to actually seize assets to be in violation. If governments (local or national) pass laws that would limit the future profits a company could have made if the law had not been enacted, the corporations consider these laws to be "tantamount to expropriation." Companies now watch laws proposed by states or local governments, and if a proposed law would limit the money the corporation could make, they notify the government. If the local government enacts this law, the corporation can sue them.

A three-person tribunal at the WTO, convened in secret, hears all these cases. William Greider of *The Nation* says, "I think of it actually as kind of an exclusive court for capital. American citizens not admitted, even American legislators not admitted. And if that doesn't up-end democracy, I don't know what does."[18] So far, in every environmental dispute that has come before the WTO, its three-judge panel has ruled in favor of business and against environmental or other protective policies. NAFTA makes no provision for a full appeal to U.S. courts, and it sets no caps on the amount of damages that can be awarded a corporation. Corporations assume that environmental laws will limit corporate profits (which is not always true). Therefore, environmental laws become, if not illegal, at least very expensive for taxpayers. Greider said, "The boundaries on this are literally the infinity of the legal imagination."[19]

Corporations are also working to privatize water by getting governments to reclassify water from a *right* to a *need*. If water is a right, then national governments must protect its people's access to water.

Trout Lily

However, if it is only a need, then corporations can "own" water and charge whatever they want for it. The WTO could allow the commercial export of water by prohibiting the use of export controls on any "good" for any purpose. This means that corporations could challenge, as a form of protectionism, any quotas or bans a government imposes on the export of water for environmental reasons.

In these treaties, governments have given up their sovereign right to pass laws safeguarding the environment and workers. The treaties are undemocratic in the power they give to corporations, but the negotiations are also undemocratic—there is no transparency, no input from citizens.

How Corporate Power Controls U.S. Politics

Corporations maintain their power because they control politics and the media. Although the Obama campaign changed financing to some degree, it is still true that to change the power of money in elections we need publicly funded elections and free airtime for candidates.

The structure of any system dictates the kinds of behaviors and outcomes it creates; therefore, to change any system, we have to change its structure. The current structure of U.S. electoral politics is simple.

On the one hand, we have corporations that

- can benefit greatly from laws and regulations made in their favor and
- have a great deal of money.

On the other hand, we have legislators who

- have the power to make laws and regulations and
- need a great deal of money to be re-elected.

Of the four elements, we can only change one—the need legislators have for money. Our entire political system has been paralyzed because it has been bought and paid for by corporate money. Our election system ought to be bought and paid for by American citizens. This requires publicly funded elections and making corporate donations illegal, as they once were. The candidates' need for money has corrupted politics. Currently, Noreen Hertz, in *The Silent Takeover*, says, "In the absence of clear ideological distinctions, parties can most effectively differentiate themselves in terms of marketing strategy and spending."[20] In essence, she says, people vote for the best ad campaign. This simply means that the better-funded candidate usually wins.

This issue was made much worse when, on January 21, 2010, in a 5-4 ruling, the U.S. Supreme Court held as unconstitutional provisions of the 2002 Bipartisan Campaign Reform Act, known as the McCain-Feingold Act, that prohibited corporations from spending general treasury funds on "electioneering communications" and independent campaign advertisements.[21] Sen. Russ Feingold called this ruling "a terrible mistake."[22]

Corporations and labor unions were already free to say whatever they wanted about political candidates, but they could not use unlimited general treasury funds; they were required to use money raised by their political action committees (PACs) from employees and members.[23] This allowed for some transparency, for anyone could go through a PAC's federal disclosure filing and see the name of every individual who put money into it. And every contributor must be a citizen of the U.S. But under this Supreme Court ruling, corporations can support candidates without limit.[24] Tara Malloy, attorney with the Campaign Legal Center of Washington, D.C., said that under the new law, human-people (as opposed to corporate-people) may only give $2,300 to a presidential campaign. But corporate interests like hedge fund billionaires, who typically operate through dozens of corporate entities, may now give unlimited sums through each of them.[25]

Greed is not a virtue and sharing is not a sin.

David Korten

Republican presidents have appointed 12 of the 15 new justices over the past four decades. Explaining the case, the Republican National Committee's political director said that to have a chance of matching Obama's small donations, "we need to be on an equal footing, and we think that law [the McCain-Feingold Act] keeps us from doing that."[26]

The *New York Times* said the ruling represented a sharp doctrinal shift, and it will have major political and practical consequences. Specialists in campaign finance law said they expected the decision to reshape the way elections were conducted. President Obama called it "a major victory for big oil, Wall Street banks, health insurance companies and the other powerful interests that marshal their power every day in Washington to drown out the voices of everyday Americans."[27]

How Corporate Power Controls Public Information

Corporations also have the power to control information. Conservatives charge that the media has a liberal bias, which may be true in terms of abortion and gay marriage, but economically, the media is very conservative. Democracy requires the free flow of information, yet corporations control the information that the public receives. Corporate control of the media is possible because only a very few huge corporations own all newspapers and TV stations all over the country. Paul Krugman says the media is not controlled by big business, it *is* big business, so that now, he says, we get our news from AOLTimeWarnerGeneralElectricDisney WestinghouseNewsCorp.[28] But their greatest power comes from controlling what is *not* covered—such as the enormous threats of global warming.

In 2002, on his Public Broadcast System (PBS) program *NOW,* Bill Moyers said, "The world of the mass media is shrinking. How a handful of companies came to exercise such control over the media is one of the astonishing stories of our time. But there are real consequences to what's happening that affect a democracy."[29] As of 2002, six companies controlled 90% of the programming.

Senegal Hoopoe

There are 500 channels, but only a few corporations control them all. Moyers began with the total number of media:

Approx. # of daily newspapers in North America	1,800
Approx. # of magazines in North America	11,000
Approx. # of radio stations in North America	11,000
Approx. # of television stations in North America	2,000
Approx. # of book publishers in North America	3,000

Recently, there has been a great deal of concentration of media ownership. The number of companies owning a controlling interest in the media listed above has changed over time. In 1984, 50 companies held the controlling interest in these media, but:

In 1987, 26 companies held controlling interest

In 1996, 10 companies held controlling interest

In 2002, only *6 companies* held controlling interest in all media

President Clinton signed the Telecommunications Act of 1996, which lifted the 40-station limit that any corporation could own.[30] This caused the radio industry to consolidate. Clear Channel Communications now owns 1,200 stations in all 50 states, reaching, according to their Web site, more than 110 million listeners every week. They were involved in highly controversial programming during the 2004 presidential election. Right before the election, they forced all their stations to pre-empt normal programming and air as *news* a lengthy program widely considered highly partisan and pro-Bush.

Viacom's Infinity radio network holds more than 180 radio stations in 41 markets, concentrated in the 50 largest radio markets in the U.S. In 1999, Infinity owned and operated six of the nation's top 10 radio stations.[31] Comcast gained control of 70% of the cable subscriber base in the nation's top 20 media markets because of a 2002 Federal Communications Commission (FCC) decision, but FCC chair Michael Powell said he saw no need to investigate its actions.[32] This concentrated control limits diversity, creates programming unconcerned with local issues, and creates commercial and homogenized news and entertainment.[33] By law, the airwaves belong to the public and are supposed to operate in the public interest. The FCC, established in 1934, is supposed to see that the nation's broadcasters serve the public interest, not just commercial interests. Originally, to get a license, the FCC required stations to function as a "public trustee." This included making a reasonable attempt to cover opposing points of views. The FCC also required that stations perform public service in reporting on crucial issues in their communities, requirements known as the "Fairness Doctrine." Soon after becoming FCC chair under President Reagan, Michael Fowler worked to do away with the Fairness Doctrine. A 1987 court case ruled that because Congress had not mandated the doctrine, the FCC no longer had to enforce it.[34]

It is not an accident that few people know that we need to cut our emissions by 80% by

2020. So much about the climate crisis has been kept from the public. The media makes the reality of global warming seem uncertain, but even when presented as true, the focus is on voluntary measures. Ross Gelbspan says, "To me as a journalist, this whole campaign [of disinformation about global warming] goes way beyond the traditional reach of public relations spin. To me, this effort basically amounts to the privatization of truth."[35]

People understand the world based primarily on what the mass media tells them, yet what they hear involves a very narrow range of opinions. The media rarely deals with anything important. Instead, they focus on stories that are diverting: disaster and conflict, sports, scandals, personalities, or celebrities. Bill Moyers says, "Corporate media consolidation is a corrosive social force. It robs people of their voice in public affairs and pollutes the political culture." In fact, he says, "As journalism goes, so goes democracy."[36] Frank Blethen, Publisher and Chief Executive Officer (CEO) of the *Seattle Times*, said, "If we go out 20 years from now with the same pace of concentration of media ownership we've had for the last 20 we will not have a democracy. There's simply no way."

In Moyers's PBS documentary titled "Buying the War," he says the Bush administration deceived Americans into supporting an unprovoked war on another country. "The dominant media remains in denial about their role in passing on the government's unverified claims as facts. That's the great danger. It's not simply that they dominate the story we tell ourselves publicly every day. It's that they don't allow other alternative competing narratives to emerge, against which the people could measure the veracity of all the claims." The media has colluded with "the powers that be in a culture of deception that subverts the thing most necessary to freedom, and that is the truth."[37]

For instance, in the last week of January 2003, Comcast, the nation's No. 1 owner of cable television companies, rejected the Peace Action Education Fund's request to purchase airtime for ads opposing our attacking Iraq. The 30-second ads featured calm restatements of mainstream concerns by a diverse group of Americans. Yet Comcast would not air them on CNN in Washington during the week of the President's State of the Union address. In denying airtime, they said, "We must decline to run any spot that fails to substantiate certain claims or charges." At around the same time, CNN, Fox, and NBC would not sell airtime on their national networks for an ad from the Win Without War coalition, which includes the National Council of Churches and other groups. CNN spokesperson Megan Mahoney said, "We do not accept international advocacy ads on regions in conflict." Therefore, it was easy to shut down any real discussion of the Iraq invasion.

The corporate-owned media's greatest power comes from their ability to control what is *not* said. In reporting the arguments over President Obama's March 5, 2009, healthcare summit, major newspaper, broadcast, and cable reporting rarely mentioned the idea of a single-payer national health insurance program, even though a majority of Americans favor a single-payer system. Instead, there were often references to "socialized medicine" and "government-run" healthcare.[38]

Pine Cone

Large, corporate-owned media organizations are also free-trade proponents. They rarely present any dissenting views or criticism of corporate economics, resulting in a very simplistic, superficial view of the issues. They may mention an individual corporation negatively, but any corporate problems are always considered an exception. They dismiss or ignore any economic or political ideas that go against the interests of corporate America. In *The Media Monopoly,* Ben Bagdikian described the context of reporting: "Items are more likely to be pursued in depth if they portray flaws in the *public, tax-supported* sector of American life, and less likely to be pursued if they portray flaws in the *private corporate* sector. . . . Over time, this results in the public impression that public-sector activities are essentially flawed and should be limited while private enterprises are essentially sound and have no need to change."[39] In addition, no amount of corporate scandals persuades people that the way our economy is structured is fundamentally unsound.

The media conveys information that tells us about who we are, what we believe, and who we want to be. *All* communication constructs how we see reality and produce meaning. All communication conveys ideas, information, or news from some "point of view." All communication projects a bias. It is essential that we become very conscious of how this is done so that we can evaluate information rather than accept it as fact.

Newspapers and TV stations compete to attract advertisers; therefore, they need to offer coverage sympathetic to big-business interests to attract big-business advertising dollars. Media corporations are selling a product, and that product is their audience, whose attention is sold to advertisers, who pay for advertising time to influence consumers. The TV shows or news content is just the medium used to attract viewers. Advertising promotes consumerism and tries to replace civic values with commercial ones. Bagdikian says, "An article on genuine social suffering might interrupt the 'buying' mood on which most ads for luxuries depend. Therefore, programming is carefully non-controversial and light; it focuses on entertainment, celebrities, and fashion."[40] In "The Propaganda Model: An Overview," David Cromwell says, "Each big story is approached from the same 'hard-hitting' journalistic angle which is marked by asking tough questions about peripheral issues, but leaving the structure of society unprobed."[41] We debate small questions in order to ignore the big questions; this perpetuates the myth of individual responsibility and ignores systemic causes.

> What suffers in the atmosphere of immediacy is analysis. What suffers in this search for speed is depth. The media in the wealthy world are becoming increasingly simplistic, superficial, and celebrity-focused.
>
> Laurie Garrett
> AIDS Politics and the Journalists' Role

Corporations want us to believe that what is in their interest is also in ours. They can use their media power to marginalize any other interest and label them a "special-interest group." By limiting reporting to isolated issues, activists are required to respond issue by issue, which steers debate toward short-term compromises and makes rethinking the entire system impossible. The corporations that own the media can determine the boundaries of acceptable public debate and the ideas that have airtime or space in a newspaper. They decide which issues to ignore and present their messages as self-evident. Simplistic labels, such as the "evil empire" or "axis of evil," are as memorable as advertising slogans. Environmentalists are labeled "eco-terrorists," or as in the case of protests over the WTO, as "anarchists." Concern over climate change is a "liberal" plot to bring in regulations and taxes. The loss of species has been depreciated as an ill-advised fight to "save the spotted owl," rather than the threat of mass extinctions.

Corporations, by having unrestricted and exclusive control of our media, are able to develop and ingrain in people the worldview they want them to adopt, particularly regarding questions of corporate power and threats to it, like global warming. During the Cold War, I saw a discussion on TV between an American and a Russian. The American commented that we did not have censorship in this country. The Russian replied, "No, you don't have censorship, because you don't need it." Our media is self-censored, and we are socialized into what is acceptable opinion; the State doesn't have to do it for us. Today, corporations control what is acceptable opinion and create a de facto censorship. There is no way for the general population to know the effects of U.S. foreign policy on other countries, unless they read foreign newspapers on the Internet.

[Genuine freedom of the press] would allow honest commentators to generate intense public debate around the question of how a corporate media can possibly be relied upon to report the truth on corporate abuses of people and the planet.

Mia Jarlov
What Fourth Estate?

The information about global warming is on the Internet, not in the media. Now, the attempt to control information is moving to the Internet. The Internet has created a new kind of public and democratic cultural space and has added a new location for civic society. Now, corporations are trying to privatize the Internet, charge for its use, and limit access to sites they choose. Cable broadband companies want to build proprietary "walled gardens" and introduce different tiers of Internet service. Companies also want to gain the power to control how information is created and consumed. This would simply shut down all access to alternative opinions.

The reshaping of the Internet began in 2005, when the FCC decided to reclassify broadband transmission facilities as Title I "information services" rather than Title II "telecommunications services." This had real consequences, because the telecommunications traffic (phones) has nondiscrimination obligations and consumer protections. This means that the phone company cannot control what comes over their lines, no matter what company it is coming from. By reclassifying the Internet as "information services," those safeguards no longer apply to Internet content.[42] Krugman says, "Broadband providers that face neither effective competition nor regulation may well make it difficult for their customers to get access to sites outside their proprietary domain—ending the Internet as we know it. And there's a political dimension too. What happens when a few media conglomerates control not only

Green Frog

what you can watch, but what you can download?"[43] There are many groups fighting for "net neutrality" to keep cable companies that control the distribution of the technology from also controlling the content that people can see on the Internet.

How Corporations Use Developing Countries' Debt to Control Their Economies

Corporations control not only the U.S., but also developing countries. This control is the result of the creation of Third World debt. Individuals and corporations can declare bankruptcy, but under international law, countries cannot. These debts of developing countries must be forgiven if the world is to have a chance to address global warming.

Jubilee USA is an organization working for debt cancellation; they take their name from the biblical concept of Jubilee, in which all debts are forgiven every 50 years. Jubilee USA gives a history of the debt crisis in developing countries. Beginning in the 1960s, the U.S. government spent more than it earned and so printed more dollars, which made the dollar fall in value.[44] Oil was priced in dollars, so in 1973, the major oil-producing countries raised prices, and the profits from the price hike were deposited in Western banks. Interest rates plummeted, and the banks, faced with an international financial crisis, loaned out the money fast to stop the drop in the value of the dollar.

> Every child in Africa is born with a financial burden, which a lifetime of work cannot repay. The debt is a new form of slavery, as vicious as the slave trade.
>
> All African Conference of Churches

At the time, the economies of developing countries were doing fairly well; however, they wanted money to maintain development and meet the rising costs of oil. Most countries had never borrowed from commercial banks before. However, developing countries, having been encouraged by the WTO to grow cash crops, found that prices fell for the materials they sold, like copper, coffee, tea, cotton, and cocoa. The WTO and the World Bank had advised too many countries to produce the same crops, so prices for these crops fell, and interest rates and the cost of oil rose further. Developing countries were trapped. They were earning less for their exports and paying more on what they needed to import. In time, they had to borrow more money just to pay the interest.

In the early 1980s, Mexico, Argentina, and Brazil all came close to defaulting on their loans, threatening the entire international credit system. Therefore, for the first time, the International Monetary Fund (IMF) and the World Bank joined to offer new loans to help pay the interest. The Bretton Woods Agreement created the IMF and the World Bank in 1944 to prevent future conflicts by lending money for reconstruction and development. The original purpose of the IMF and the World Bank was to finance reconstruction of Western Europe after World War II and to rebuild and stabilize the international payments system. Originally, they could not control the economic decisions of governments and they could not intervene in the national policy of foreign governments. But these institutions have now become something much different. The IMF shifted from short-term, balance-of-payment loans, mainly to industrialized countries, to medium-term loans for developing nations. The World Bank added policy-linked loans to its development projects for building infrastructure.[45] They claim this process is designed to restore economic health to indebted countries.

Rick Rowden, in *A World of Debt,* says that it was Ronald Reagan's administration and that of Britain's Margaret Thatcher that shifted these two institutions to a neo-liberal economics ideology, and the WTO and the World Bank began to micromanage more than 90 economies of the developing world. The focus was on reducing the role of the state and increasing the role of the private international corporations and investors in these countries' economic development.[46] Today, the IMF and the World Bank can intervene anywhere and force governments to participate in a world economy, even if it is harmful to their people.

All countries are developing countries. It just depends on what we are developing towards.

Willis Harmon
Creative Work

The IMF and the World Bank promote "structural adjustment programs" (which is another way of saying neo-liberal economics). This shift is also called getting countries to adopt "sound economic policies" (that is, neo-liberal economics). Soren Ambrose, Senior Policy Analyst with the 50 Years Is Enough Network, says that these structural adjustment programs simply "reorient economies to benefit corporate interests while reducing spending on social programs and locally oriented production."[47] What actually happens is that countries are required—in order to get a loan—to open their markets to globalization, privatize their utilities and other state-run services, and ignore the needs of their own populations or the environment. John Cavanagh and Jerry Mander, in *Alternatives to Economic Globalization: Another World Is Possible*, report that the structural adjustment programs require governments to

- Cut government spending on education, health care, and the environment and cut price subsides for basic necessities, such as food grains and cooking oils.
- Devalue the national currency and increase exports by accelerating the plunder of natural resources, reducing real wages, and subsidizing export-oriented foreign investments.
- Liberalize (open) financial markets to attract speculative, short-term portfolio investments that create enormous financial instability and foreign liabilities while serving little, if any, useful purpose.
- Increase interest rates to attract foreign capital that has fled its home country, thereby increasing bankruptcies of domestic businesses and imposing new hardships on indebted individuals.
- Eliminate tariffs and other controls on imports, thereby increasing the import of consumer goods purchased with borrowed foreign exchange, undermining local industry and agricultural producers unable to compete with cheap imports, increasing the strain on foreign exchange accounts, and deepening external indebtedness.[48]

Red-kneed Tarantula

Our social assumption is that only capitalism can reduce poverty, when in reality it has created much more poverty on a global basis. Because the developing countries are required to cut public spending on basic social services, both poverty and inequality have increased. People in developing countries are very aware that the economic policies that the U.S. has imposed on them are responsible for greatly increasing poverty. The debt of developing countries has become a modern form of colonialism.

Following World War II, when the classic forms of colonial rule became unacceptable, international debt became a favored instrument for gaining leverage over local economies. Subsequently, economies were forced open to foreign corporate ownership and control through debt restructuring and trade agreements.

David Korten
Agenda for a New Economy

Rick Rowden shows that over the past 150 years, industrialization in Europe, the U.S., Japan, and the Four Tigers of Asia (Singapore, South Korea, Hong Kong, and Taiwan) has relied on an extensive partnership between industry and the state. The industrialization of these countries involved several decades or more of government help. This help included providing "protective trade barriers, large subsidies to domestic industry, support for public utilities and state-owned industries, tax breaks and other incentives for research and development to diversify the economy, and controls on currency and capital."[49] Yet the WTO programs require a reduction in the state's role in the economic development process. This means the state cannot help domestic industry or provide needed public services. Rowden says, "*No country in history has ever industrialized under such a process.*"[50]

Many organizations are working to get developing countries' debt cancelled, including Oxfam International, Bread for the World, 50 Years Is Enough Network, Drop the Debt, and Jubilee Plus, which has ties to 68 countries and was formerly called the Jubilee 2000 Network. There is also a Jubilee USA Network of religious organizations and development organizations. In addition to these groups, world leaders are calling for total debt cancellation, including the pope, former U.N. Secretary General Kofi Annan, the U.N. Conference on Trade and Development, the Meltzer Commission, and others. Delegates from more than 50 developing countries established the Jubilee Movement International at a conference held in Mali in April 2001. They released a "Statement of the Jubilee Movement International for Economic and Social Justice," which reads:

> We assert that the debt burden borne by our countries is the inevitable consequence of the world's unequal political and economic structural relationships. Over the last 30 years, the period of the so-called "Washington Consensus," the basic values and rights that some of us enjoyed, and that should be common to all humanity, have steadily been eroded, *and subordinated to the rights of money*. Today money rights are more powerful than human or environmental rights. . . . We are committed to a transformation of the current globalization model.[51]

Ambrose explains that even when these loans are made responsibly, the money usually ends up back in the wealthier countries. The terms of World Bank programs implicitly require that Western consultants, engineers, and construction companies be hired. In fact, most World Bank projects (dams, roads, power plants, etc.) are not in and of themselves direct generators of income for the developing countries, but are infrastructure projects that allow Western private companies to do business.[52]

Opposing this View of Development

A major political movement in Latin America with coalitions of many civic groups has united into the Hemispheric Social Alliance (HSA), the first organization of its kind in the world. They call the trade treaties "the purchase of democracy for corporate interests." The HSA was created to oppose the trade treaties. It coordinates regular actions, campaigns, and strategy meetings, with hundreds of organizations, social movements, and regional networks participating.[53] The coalition mobilized a Continental Campaign against the FTAA treaty.

> The annual income per person in Zaire is $110. To pay off the country's debt, each resident of Zaire, would have to raise $236. In 1985, Live Aid raised $422 billion to combat famine in Ethiopia. All African countries pay back $422 billion on foreign debts every week.
>
> *Worldwatch* Magazine
> Matters of Scale
> A Fate Worse than Debt

This group is working to promote their own vision of an alternative to FTAA and has developed the statement "Alternatives for the Americas."[54] The alternative concept starts with the premise that the people should have a voice in determining the future of regional integration. Deborah James, the Global Economy Director of Global Exchange, describes what the HSA advocates:

> Enforcement of core labor and environmental rights and standards should be at the center of any new agreements. Governments should be encouraged to adopt new local, national and global rules to discourage harmful speculative activity, and to encourage lasting investments in productive and sustainable local economic activities. Access for foreign products and investments should be negotiated with adequate concern for national development plans and priorities. Protection of critical sectors, such as food production, must be the right of each country so as to ensure the rights and well being of all people.[55]

Hundreds of social movements and organizations in the HSA collectively produced and endorsed this document.[56]

Ecological Debt

Ecological destruction is also creating a new argument about debt. Organizations fighting for debt cancellation are insisting on a redefinition and reassessment of "who owes whom." Jubilee South is a coalition of debt cancellation campaigns from Latin America, the Caribbean, Asia-Pacific, and Africa. They argue that the developing countries' governments should repudiate their debts because of the exploitation of slavery, colonialism, and the current economic system. Then, in reassessing "who owes whom," they argue that developed countries owe developing countries an ecological debt. Developed countries have created almost all the global warming, yet the effects will be hardest—by far—on developing countries. This

Common Puffin

ecological debt

disparity is a debt owed *by* developed countries *to* developing countries.

Friends of the Earth International defines ecological debt as "the obligation and responsibility that industrialized countries of the North have with the Third World, for the exploitation of their natural goods such as oil, minerals, forests, biodiversity, indigenous and peasant knowledge, marine goods, and for the disproportionate and illegitimate use of the atmosphere and of the oceans to dump waste, including the greenhouse gases."[57] They argue that ecological debt is created when the rich use more than their fair share of the finite environmental space. Therefore, a realistic global deal on debt would acknowledge the right of all people to share equally in the global commons of the atmosphere. The financial services initiative of UNEP estimates that the extra economic costs of disasters attributable to global warming are running at more than $300 billion annually. The best guess of development groups is that climate change–related disasters could cost developing countries up to $9.3 trillion over the next 20 years.[58]

Economic Democracy versus Economic Aristocracy

The power of corporations has taken over our democracy. Our culture tries to separate politics and economics in order to hide the inherent conflicts between democracy and neo-liberal capitalism. We will never be able to make the changes we need unless we have a democracy, but capitalism always rewards those in power—people who then work to maintain their privileges. Dealing with global warming requires democratic debate and choices, especially about economic choices.

> The vast inequality of this new Gilded Age didn't just happen. ... What happened is the rich declared class war and spent what it took to win.
>
> Bill Moyers

Oppression and domination take many forms, but the forms that exist in the economy are generally invisible. True democracy means that people can participate in making the decisions that affect their lives, but the decisions that have the most impact on people are economic. What really affects people's lives are factories closing, jobs moving overseas, and their community's health being harmed by poisoned water, air, and soil. People cannot control or protect their health or livelihood without economic democracy, and political democracy becomes irrelevant if economic interests control elections. Therefore, one primary role of democracy is the social control of economic forces.

However, making the economy democratically accountable is not part of democratic theory. Again, the theory divides society into public and private spheres and explicitly says that the power of the state can only operate in the public sphere. However, there are two different divisions of public and private spheres. In one, the state is public and families and communities are private; in the other, the state is public and the economy is private. Since the power of the state can only operate in the public realm, the power of money is declared private and nonpolitical and is hidden from any political democratic accountability. The result of these artificial partitions is that an economic aristocracy has come to control our politics and our culture. Thom Hartmann, in *Unequal Protection: The Rise of Corporate Dominance and the Theft of Human Rights,* says, "What you've seen is the subsuming of the political process to the corporate agenda."[59] This conflict has always existed but has become extreme over the past 30 years.

The totalitarian power of money severely limits democracy by lobbying, controlling the media, and investing in candidates. Margorie Kelly, editor of *Business Ethics* and author of *The Divine Right of Capital: Dethroning the Corporate Aristocracy,* says, "Free market mythology is a smokescreen that disguises the real nature of elite power—much like [the concept of] the divine right of kings. It allows elites to run our economy to suit them, without interference and with a veneer of legitimacy."[60]

The world might be laid waste in the interest of not wasting investor money.

Margorie Kelly

People only questioned the rule of the king after the phrase "divine right of kings" was created, and from then on, it was no longer a given. Kelly invented the phrase "the divine right of capital" to try to make our unconscious assumptions visible. She says that in our country, control through wealth is the root of many problems. Capitalism does not have countless social problems, such as corporate pollution, low wages, or corporate welfare, but instead has only one problem: the power of wealth, a problem she calls "wealth discrimination," which is the concept that more wealth for the wealthy is the single greatest need.[61]

Kelly says, "At root, what really governs corporations is an idea that is the intellectual descendent of the *great chain of being*: the notion that only those who possess wealth matter. Implicitly, they are of a higher class of persons who alone are considered real members of corporate society; hence only they have a vote."[62] Therefore, our government is becoming a plutocracy (government by the wealthy), or what she calls an "economic aristocracy." The government's primary function is no longer to protect human welfare, human rights, and public resources; it is to support the ability of corporations to make money. Cavanagh and Mander, in *Alternatives to Economic Globalization*, say the same, that now the function of the government is to "serve corporate interests, protect corporate property, guarantee corporate profits, break up unions, sell off public assets at give-away prices, stifle dissent, and make sure that people fulfill their roles as obedient workers and compliant consumers."[63]

Corporate control is a result of the *design* of the system itself, in particular, the design of capitalism based on the enormous power of multinational corporations. This design is a form of aristocracy that Kelly calls "wealth privilege," meaning that the power, control, and choices all belong to the wealthy: "If equality under the law is the hallmark of democracy, privilege sanctioned by law is the hallmark of aristocracy."[64] Therefore, in our economic aristocracy, we talk of employee productivity but not stockholder productivity; we assume company assets have

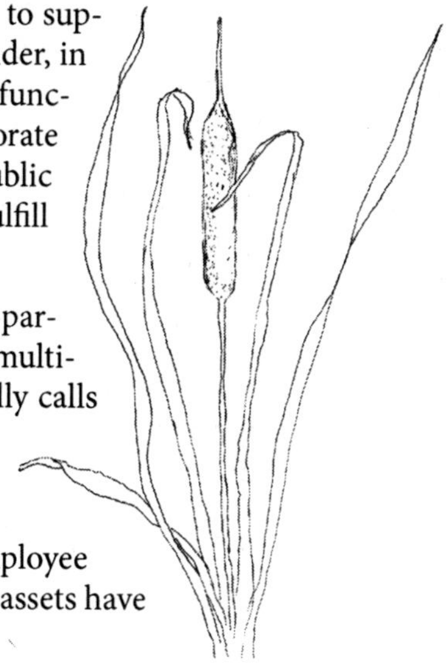

Cattails

value, but community assets have no value: "In terms of accounting, employees have no value. Money has value, objects have value, ideas (intellectual property) have value, and even some airy thing called goodwill has value. Employees, by contrast, have a negative value: They appear on the income statement as an expense—and expenses are aimed at a singular goal: to be reduced."[65] Yet a corporation is its employees. Particularly when knowledge is a corporation's primary work, the primary capital is the intellectual capital of employee knowledge—their expertise, experience, and connections. So Kelly asks, "Why are the rights of working persons inferior to those of wealthy persons? Why does the corporate board have a fiduciary duty to wealth holders but not to employees?"[66]

We assume that stockholders have an exclusive right to all the company's profits because we say they "own" the company, making the employees relatively unimportant. Therefore, all decisions must be made to further the stockholders' short-term profit. The concept of stockholder ownership developed when most corporations were capital intensive. (In 1900, three-fourths of the stocks traded were railroads.) Today, the stockholder's contribution of capital is mostly an illusion. Only when a company issues new common stock do corporations actually see the money represented by the stock. For the most part, stockholders are not providing money to a corporation; they are exchanging bets with other stockholders on the presumed future value of the company. According to the figures from the Federal Reserve, in recent years, about $1 in every $100 traded on public markets has been reaching corporations.[67]

> Today we chase after corporate pollution as one problem, low wages as another problem, and corporate welfare as still a third problem. They're all manifestations of wealth discrimination – the insistence that more wealth for the wealthy is the single greatest need. When we recognize this core issue, our separate battles will become one. And that battle will gain momentum.
>
> Margorie Kelly

Therefore, 99% of the money in the stock market never goes to corporations; speculators merely trade it back and forth. Some economists have suggested that stock ownership is being democratized, but Kelly says that the wealthiest 1% have doubled their share of national household wealth from 20% to close to 40%. This massive concentration "gives the wealthy virtual sovereignty over both our economic and our political systems."[68]

However, rather than individuals owning stock, institutional investors such as insurance funds, pension funds, and mutual funds hold most stock, and the job performance of the fund managers is based on short-term profits. Business consultant Peter Drucker calls this "pension-fund capitalism" and says it will probably become the universal form of ownership in developed countries because of their aging population.[69] The money invested is not entrepreneurial capital, but rather is workers' deferred wages invested for their pensions. Other workers (the staff of the pension fund), not entrepreneurs, do the investing. Drucker says that this shift is as different from the original capitalism as socialism is.[70]

The social mantra—just increase economic growth and everyone will eventually be better off (also known as "trickle-down economics") maintains the power of the wealthy. Krugman, in a *New York Times Magazine* article titled "For Richer," said that due to the recent "tectonic shifts" in the distribution of income and wealth, we are living in another Gilded Age like the 1920s. Over the past 30 years, most corporate employees have received only modest salary increases, from an average of $32,522 in 1970 to $35,964 in 1999, or about a 20% increase in 29 years. However, in that same 30-year span, the average pay of the top 100 CEOs went from $1.3 million (39 times the pay of an average worker) to $37.5 million (more than 1,000 times the pay of the average worker, an increase of 2,500%).[71]

The New Deal and World War II created the American middle class, which continued into the 1970s. However, over the past 30 years, a transformation has occurred that is eliminating the middle class. Krugman says, "Much more than economists and free-market advocates like to imagine, wages—particularly at the top—are determined by social norms. During the 1940's and 1950's new social norms of equality were created, discrediting the extremes of the Gilded Age. What happened in the 1980's and 1990's was that those norms unraveled, replaced by an ethos of 'anything goes.' And as result, there was another explosion of income at the top of the scale."[72] He also says the total income that goes to the rich is not small. The top 1% now gets 16% of total pretax income (double the 8% of 30 years ago). This also means that the top 1% gets as large a share as the bottom 40%.[73] In fact, Krugman says the great economic growth of the nation has had little impact on ordinary families.[74]

Politics based on race [and] gender … are tolerable to Empire, because they emphasize and perpetuate division. Discussion of class, however, is forbidden, because it exposes common interests and unifying structural issues around which a powerful resistance movement might be built.

David Korten
Agenda for a New Economy

Before 1980, economic growth and job growth went up together. But now, the two operate independent of one another. In fact, increased economic growth in profits has often resulted in the loss of jobs. The average worker now realizes that jobs, health, and retirement benefits are no longer secure. Wives, independent of what they would choose, can no longer afford not to work. A two-income family has become a necessity—some call it the 80-hour workweek—and single parents end up among the working poor. Before the 1980s, the difference in wealth from the poor to the rich was a gradual increase, and the middle class could be anywhere along a long continuum. The gap between rich and poor is now so huge that even the middle class is relatively poor.

The real threat to capitalism is not socialism; the real threat is genuine democracy. This conflict has been a part of America from the beginning. It is a conflict between which takes precedence—property rights (especially property in the corporate form) or personal rights. Robert Dahl, in *A Preface to Economic Democracy*, says:

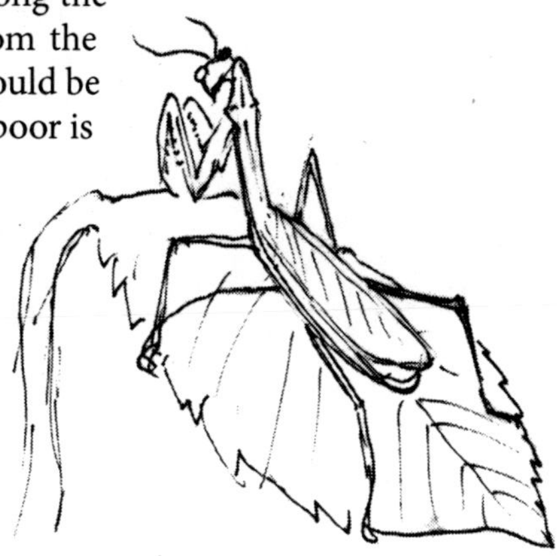

Praying Mantis

> Americans have always been torn between two conflicting visions of what American society is and ought to be. To summarize them over simply, one is a vision of the world's first and grandest attempt to realize democracy, political equality, and political liberty on a continental scale. The other is a vision of a country where unrestricted liberty to acquire unlimited wealth would produce the world's most prosperous society. . . . In the first view, the right to self-government is among the most fundamental of all human rights, and, should they conflict, is superior to the right to property. In the second, property is the superior, self-government the subordinate right.[75]

Dahl cautioned that we need choose between these conflicting ideals and assert the priority of democracy, political equality, and the political rights necessary for self-government over established property rights, economic inequality, and undemocratic authority within corporate enterprises.[76]

Running corporations to serve stockholders means running the economy to benefit the wealthy.

Margorie Kelly

The political power of corporations and the wealthy who control them is making it all but impossible to make the changes needed to halt climate change and our other ecological problems. Corporations now have the power and means to control the content of political discussions and keep the focus away from corporate practices. In addition, they operate within a system of accounting that does not even try to measure the external damage they cause. As Supreme Court Justice Louis Brandeis said, "You can either have great concentrations of wealth or you can have democracy. You cannot have both."[77]

Within our current globalization model, it is not possible to create democratic, socially responsible, and environmentally sustainable trade policies. Now that corporations have more power than nations, no one can stop them. They cannot even stop themselves. We have structured the corporate system so that acting in the interests of environment, human safety, or health is not possible because by law, corporations must maximize shareholder financial interests. Under our current values and beliefs, the individuals creating these agreements are just good business people. Because we ignore what our theories ignore, each step that built globalization seemed logical, fair, and appropriate.

Now, with the general acceptance of the reality of global warming, even corporations are beginning to change. In 2005, in the article "The Business of Saving the World: Can Capitalism Become a Force for a New Global Consciousness?" Helen-Jane Nelson, Director of Cecara Consulting Limited, describes where many companies are now: "We are at a point … where business leaders are beginning to recognize that their businesses are not sustainable, that the whole way that they have been doing business—the pursuit of continual growth—is not sustainable. The planet cannot tolerate it. And they're scared."[78]

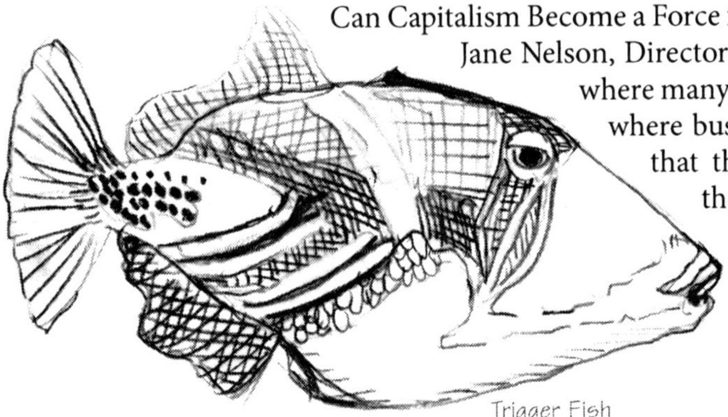

Trigger Fish

Part III

Creating Solutions to Global Warming

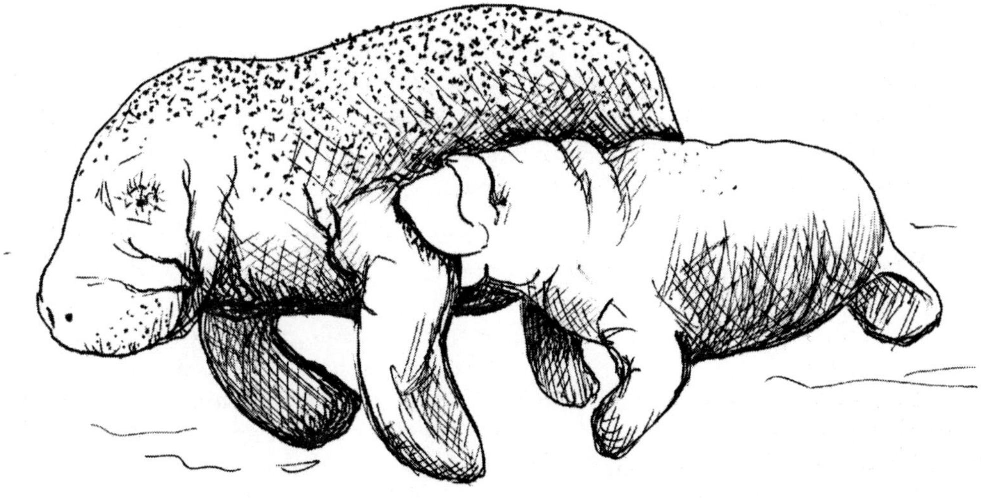

West Indian Manatee and nursing calf

Chapter 7

Creating a Cooperative World Based on Partnership

Addressing global warming is not primarily an issue of technology or economics or energy. Fundamentally, global warming is a question of ethics and morality. Politics is about moral values, and global warming is the defining moral issue of our time. If an evaluation of what is moral addresses only the actions of individuals, we lose sight of the unintended systemic consequences of individual choices. This allows people to ignore their responsibility for our destruction of Earth. Ultimately, there are no isolated, autonomous individuals; therefore, morality and moral consequences have to be viewed in terms of the whole, including all effects.

The environment is, after all, the human habitat, and time after time extinction has followed on loss of habitat when the species at risk was not able to adapt in time. Despite our large numbers, we are an endangered species.

Jack Miles
Global Requiem: The Apocalyptic Moment In Religion, Science, and Art

In the decade from 2000 to 2010, greenhouse gas emissions grew much faster—3 or 4 times faster—than in the 1990s, and almost all of the increase came from the developing world. Addressing global warming is now dependent on changing the direction of developing countries. We need a global crash project to rewire the world with renewable energy, which will also greatly improve the lives of people in developing countries. We need to forgive the debt of developing countries and eradicate poverty. These efforts could shift the world toward efforts at cooperation and mutual care as well as toward stopping global warming.

Ecology is also a major issue of justice: the developed world created the warming we now have, but the terrible consequences will impact the developing world the most. The people living on island nations are arguing that our economic ideology is destroying their countries. In spring 2003, Ambassador Lionel Hurst of the island nation Antigua and Barbuda said, "The most populous and wealthiest of the world face a moral challenge greater than colonialism or slavery. They are failing in that challenge. Men have lost reason in the fossil fuel economy. . . . Inhabitants of small islands have not agreed [to be] sacrificial lambs on the altar of the wealth of the rich."[1]

Global warming requires that progressive values of community and mutual responsibility be extended to cover international issues, as well as the natural world and all of its creatures.

Differing Views of Causation

Morality is based on one's understanding of causation—specifically whether causation is individual or systemic. Morality addresses what people think of as the cause of a problem, and because our two different worldviews have very different understandings of causation, they have very different understandings of morality. This is why the left and the right cannot agree on the definition of a problem and therefore cannot agree on its solutions. The two social worldviews discussed in Chapter 3 also have two very different understandings of morality.

Conservatives understand causation mechanistically: there is one cause for one outcome, causation is always direct, and it focuses on the choices and actions of individuals and their direct and immediate results. George Lakoff says that in this understanding, moral action is learned and understood in the conservative worldview to be "single actions by a single individual that directly result in a clearly defined outcome." Conservative, or what Lakoff called strict father, morality makes people focus on direct causation over and over. In this morality, the response to any problem is internal or external discipline. This linear cause-and-effect thinking leads to finding someone or something to blame.[2]

We assume people are free to act as long as they do not cause harm to others. But "causing harm to others" depends on how you understand causation. In the conservative worldview only direct, person-to-person causation counts. Conservatives do not see small individual actions within massive complex systems as immoral.

In contrast, progressives focus on the whole system and think of causation in systems terms. Therefore, in an interconnected system, an outcome has multiple inputs, most of them indirect, and change occurs by adding up inputs over time and from many places. Lakoff says in the empathy-based (progressive) model, people learn to look for systemic causation or systemic reasons for people being harmed or helped.[3] This is the complex system non-proportional causation described in Chapter 2.

The two worldviews of conservatives and progressives differ fundamentally, so much so that Lakoff says that these two perspectives are virtually opposites. Because the two models have such different understandings of causation, they have very different understandings of what is real, which affects both religion and politics. Lakoff says, "Political decisions affect reality. What is disturbing is that political ideology can so deeply affect the understanding of what is real and so thoroughly hide the real consequences of so many political decisions."[4]

Abdim's Stork

Conservative Morality: The Sufficiency of Private Virtue

Conservatives value a moral life and strive to be good people. They are willing to put energy and commitment into doing what is right and moral. But their understanding of morality is narrow, and their arguments for morality end up being about authority, power, and obedience. In particular, they focus only on the individual and the actions of an individual in isolation from the rest of the world. Their concept of morality, often called "family values," refers to the morality of obedience to a strict father authority in the traditional family. Because their morality focuses on individual action, conservative morality has been called "the sufficiency of private virtue." When people focus on individual behavior, they automatically ignore the indirect consequences their choices have on the whole system.

> We as a nation must undergo a radical revolution of values. ... A nation that continues year after year to spend more money on military defense that on programs of social uplift is approaching spiritual death.
>
> Martin Luther King, Jr.

Conservatives win, Lakoff says, because they generate policy from the very core of their frame: the need to restore the "natural" order of things and the right way to function in the world.[5] In the 1960s, economic conservatives, social conservatives, and religious conservatives all opposed each other. But then they began to focus on what they all had in common.[6] Then, Lakoff says, to forward their values, conservatives set up institutes within and outside universities. They do research, write books, and endow professorships to teach young people the right way to think. In the authority-based model, the top value is preserving and defending the moral system itself; therefore, they built infrastructure, bought up media in advance, and planned ahead.[7] They set up the Heritage Foundation, Olin professorships, the Olin Institute at Harvard, and others. They support their intellectuals and create media opportunities for them. Lakoff says they "have media studios down the hall in institutions so that getting on television is easy. Eighty percent of the talking heads on television are from conservative think tanks. Eighty percent."[8] These think tanks understand the importance of framing ideas and carefully work out how to frame each issue with stories and language. Progressives do very little of this.

Another fundamental difference between these two models also contributes to our ecological crises. When we base choices on rules, authority, and mechanistic cause and effect, the evaluation of acceptable choices is done up front. This eliminates any consideration of feedback from the unintended consequences of decisions. People using the authority-based model would consider re-evaluating choices based on new information to be "flip-flopping" or "waffling." They consider making a decision once and then holding to it to be "steady leadership." But the world is so complex and changing so fast that we need to be able to respond creatively and find novel solutions as we go. When we base choices on mutual care and community or on the systems concept of interdependence, the evaluation of the correct choice is an ongoing dynamic process in which we continually measure and evaluate choices against new information, particularly information about unintended consequences, and change accordingly.

When conservatives address poverty, they define charity as individuals helping individuals but ignore the systems that create poverty or that can give or deny people a fair chance. Some people still think that people are poor because they are lazy or irresponsible. It would be very threatening if people began to suspect that our economy systematically privileges some and excludes others. It would mean that we would have to have collective solutions to poverty, which can only be coordinated on a sufficient scale by the government.

Progressive Morality: We're all in this Together

Progressive morality is defined by empathy, respect, and partnership, together with the responsibility and strength to act on these values. Progressives do not think solely in terms of individuals but think in terms of the whole, with the implicit guiding assumption that "we're all in this together." Therefore, they believe in mutual responsibility in addition to individual responsibility. They think in terms of systemic causation in which the whole is tightly interconnected and no one operates alone. The basic conflict between the two worldviews is between patriarchal family values and cooperative community values.

Progressives focus on the common good, human dignity, fairness, and equality. Their focus on morality is not just on individual behavior but also on structural injustice and systemic moral issues. Therefore, they focus on global warming, poverty, and population growth, rather than on individual behavior issues like abortion.

Because of this ethic of empathy and mutual care, Lakoff says progressives view the role of government as providing *protection* and *empowerment.*[9] Protection would include the army, police, and fire department, but also social security, public health, disaster relief, health care, consumer and worker protection, and environmental protection. Empowerment by the government includes highways and bridges, the Internet and satellite communications, public education, the banking system, and the court system to enforce contracts and protect patents (90% of court cases are for business). So, Lakoff says, "Nobody makes a dime in this country without being empowered by our government."[10]

Systemic morality addresses systems dynamics, not the individual's right to be free of any obligations to others. This morality would go beyond the progressive interest in providing services for people, to instead restructuring the system dynamics that create the problems in the first place. This

Sea Lion

contrast is reflected in Brazilian Archbishop Dom Helder Camara's comment, "When I give food to the poor, they call me a saint. When I ask why the poor have no food, they call me a communist." The first is an expression of direct, single causation; the second expresses complex, indirect causation. This means that, in addition to providing soup kitchens, we need to explore why we need soup kitchens in the first place.

The difference in morality can also be seen in the conservative emphasis on the Ten Commandments. Each of the commandments addresses actions by individuals, as a demand. But Jesus was asked which of the commandments is the most important. He replied that the most important are to love God and to love your neighbor as yourself. He added that there are no commandments greater than these. "To love your neighbor as yourself" addresses our interconnectedness and is a systems perspective.

For I was hungry and you gave me food, I was thirsty and you gave me drink, I was a stranger and you welcomed me, I was naked and you clothed me, I was sick and you visited me, I was in prison and you came to me.

Matthew 25:35-36 ESV

Since individual behavior cannot control systemic causation, government system-wide action may be necessary to create collective solutions. Otherwise, we are limited to only individual solutions, which in terms of global warming are catastrophically inadequate. This emphasis on the individual is so strong in our culture, many people think our "every man for himself" ethics is in the Bible.

In fact, Bill McKibben says 75% of Americans believe that the phrase "God helps those who help themselves" comes from the Bible. But this phrase was Benjamin Franklin's and is a good description of our intense individualism. It is also the direct opposite of the Biblical emphasis on "Love your neighbor as yourself."[11] The Bible also says that when we help those who are hungry, thirsty, a stranger, needing clothes, sick or in prison, it is the same as doing these things for Jesus or God. Love your neighbor as yourself means that we should take care of each other. The United States is the only industrialized nation in the world that does not offer single-payer health care. We need to ask, what kind of a society do we want: one in which no one owes anyone else anything, or one in which we take care of each other?[12] Not having a single-payer health care system is incredibly immoral. But we don't see that because conservative religion has been subsumed under conservative economics, two ideas that used to be in conflict.

For instance, in an example of protection, progressives believe that one form of protection and community care is providing everyone with health care. For conservatives, it is a commodity that people should buy. One-third of the cost of private health care insurance goes for profit and administration (mostly for denying care), while Medicare only spends 3% on administration and none on profit. Insurance companies make money by *denying* care. Health insurance will always work this way. This is not the same as care, and it is the opposite of the way in which markets work. In a typical market, companies that provide more of their product make more money.

Health insurance is thus an anti-market phenomenon. Progressives want universal health care because it is a matter of collective protection: you don't need to buy private insurance for police or fire protection. Health care is just as much a matter of fundamental protection, so it shouldn't be sold as insurance.[13]

Ecological Morality: Budget for Reducing Population and Poverty

And the King will answer them, "Truly, I say to you, as you did it to one of the least of these my brothers, you did it to me."

Matthew 25:40

No amount of empowering or protecting individuals in community will adequately address global warming. For that, we need to create social and economic transformation. The morality needed to address the causes of global warming includes empowering and protecting the needs of the global community. Morality is no longer a matter of opinion. Now we can look at Earth and define the concrete requirements to restore Earth and ourselves to health. Meeting these requirements then becomes a moral imperative. We must solve global warming together. However, the solutions we need go far beyond what is normally seen as "progressive." We need a new language in order to change the debate. The only response that will be adequate must put Earth at the center and define all other issues in the context of preserving Earth, that is, we need an ecological morality.

Lester Brown gives four steps to address global warming, or as he puts it, "to save civilization:"

- eradicate poverty,
- stabilize population,
- restore the economy's natural support systems, and
- restructure the economy (addressed in later chapters of this book).

Reducing population requires eliminating poverty, and eliminating poverty requires reducing population. Our economy must change in order to stabilize climate and global warming, but we must also restore nature's natural support systems, which have been so critically compromised. The essential task now is to create a new ecological political theory based on the new moral vision required to preserve Earth, create equity, and support democracy.

The morality that matters now requires that we move to what Jeremy Rifkin calls "a global empathic consciousness." It is no longer sufficient to just address fossil fuel use in the U.S. China, India, and other developing countries must be helped to make more constructive choices on energy. That means that the effort to address global warming must be global and it must be led by the United States.

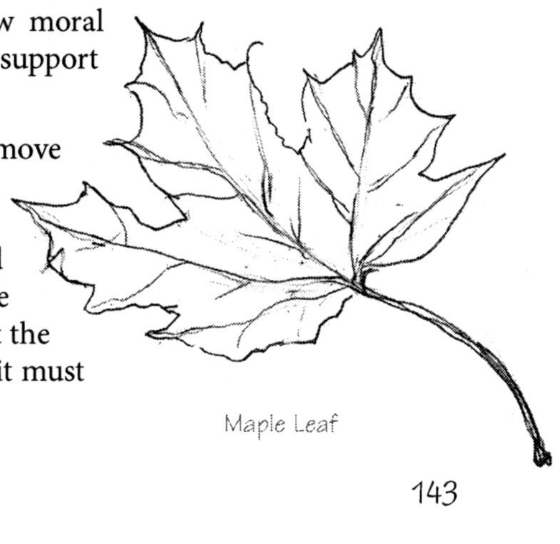

Maple Leaf

In *Plan B 4.0*, Lester Brown has developed two budgets to address the requirements of both people and nature. The first budget is for reaching basic social goals for stabilizing population and eradicating poverty, which are both necessary if we are to save Earth and restore hope to developing countries. The second budget is for restoring Earth's natural systems.

Plan B Budget

Additional Annual Funding Needed to Reach Basic Social Goals

GOAL	FUNDING
	(billion dollars)
Universal primary education	10
Eradication of adult illiteracy (mostly by volunteers)	4
School lunch programs for 44 poorest countries	6
Assistance to preschool children and pregnant women in 44 poorest countries	4
Reproductive health and family planning	17
Universal basic health care	33
Closing the condom gap	3
TOTAL	*77*

To end poverty and slow population growth, all women who want to plan their families need access to family planning services. Poverty is inherited, and the main thing that will raise people out of poverty is education, particularly the education of girls. Fertility falls as female education levels rise.[14] As women and girls become more educated, they are better able to find contraception and use it correctly. Their children are healthier and as infant and childhood death rates fall, they do not have more children with the assumption that some will die. In addition, once a woman is educated, she wants to educate her own children. But education is expensive, so in order to educate her children, she will choose to have fewer of them. Brown's budget includes offering school lunches because they are a significant incentive for children to go to school.

> Whatever your cause, it's a lost cause without population control.
>
> www.overpopulation.org

Funding these social goals becomes the major moral issue of our time. The rough annual budget for interconnected social goals of eradicating poverty and reducing population is $77 billion.[15] These costs are in addition to what is already being done.

If the U.S. funded these additional costs, it would change the dynamics of the world. For one thing, it would be the best way to fight terrorism. Achieving basic social goals would do more to help fight terrorism than the military could ever do. If all these services were to go to developing countries, with the label, "Funded by the USA," the effect could be profound. This effort would change world perceptions. Terrorists are angry with the

U.S. government and our economic policies, which have greatly reduced the services their governments are able to provide their people.

Arnold Toynbee, who died in 1975, predicted, "The 20th century will be chiefly remembered by future generations not as an era of political conflicts or technical inventions, but as an age in which human society dared to think of the welfare of the whole human race as a practical objective." We have had to wait for the twenty-first century, but for us to solve our ecological crises we will need to finally redefine morality as the effort to secure the welfare of the whole human race, with everyone having enough and none too much.

Global Warming and Issues of Family Planning

> World population is expected to hit 7 billion in late 2011.
>
> Census Bureau

Because of the imperative of reducing both population and poverty, the major focus of conservative morality on opposing contraception and abortion creates enormous problems. Our political arguments over contraception and abortion have far-reaching implications for our survival. The increasing world population drives all other problems, making all of them worse. Reducing the world's population is now a matter of planetary life and death. Lester Brown says, "If we cannot stabilize climate and we cannot stabilize population, *there is not an ecosystem on Earth that we can save.*"[16]

Even without the coming increase in population, scientists expect global warming to result in the deaths of tens of millions of people due to increased storms, rising sea levels, floods, severe water shortages, food shortages, and new epidemics. In addition to deaths, there will be hundreds of millions of environmental refugees. We need an understanding of morality that addresses these issues without defining morality only in terms of the unborn.

The world's population is growing at a rate of 80 million a year. As of August 2010, the world population is about 6,864,000,000.[17] World population is expected to hit 7 billion in 2011, 8 billion by 2020, and at least 9 billion by 2050, which represents an almost 50% increase over the current population. Nine billion would be like adding two more Chinas to the world. World population reached:

1 billion in 1804,
2 billion in 1927 (123 years later),
3 billion in 1960 (33 years later),
4 billion in 1974 (14 years later),
5 billion in 1987 (13 years later),
6 billion in 1999 (12 years later).[18]

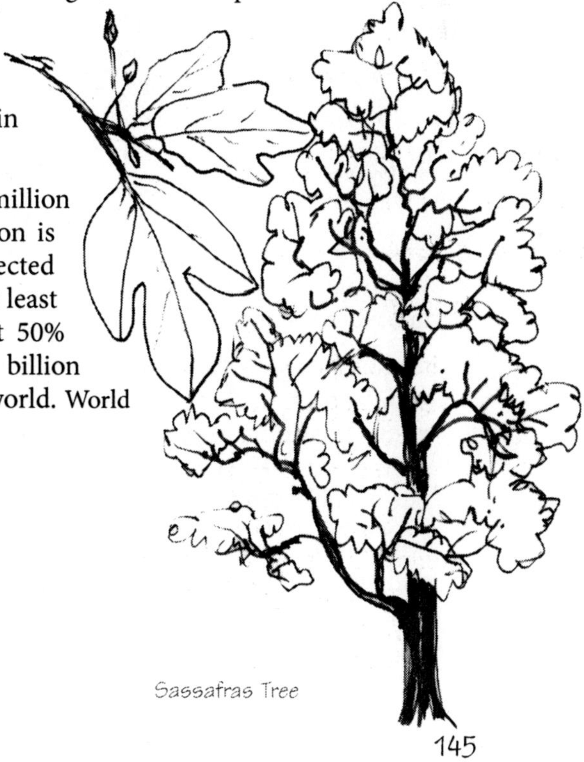

Sassafras Tree

World population has more than doubled since 1960. By one estimate, if current birthrates continued until the end of this century, there would be 25 billion people. Obviously, if we are running out of water, land, and food, and warming the atmosphere now, with only 6.8 billion people, this increase is not possible. Human population *will* be limited. The only question is whether it will be limited by family planning or by disease, famine, food and water shortages, social chaos, and wars.

Almost all future population growth will be in the developing world, where many countries are already overpopulated and already have stressed ecosystems.[19] Reducing population will be difficult because developing countries have the largest generation of youth in history. In less developed areas, 29% of the population is under 15, and 19% is 15–24, so that almost 50% of the population is under 24. In the least developed countries, it is 40% under 15 and 20% 15–24, so that 60% of the population is under 24.[20] These young people will create an explosion of births because they are just entering their reproductive years. If current trends continue, the combined populations of the 48 countries classified as least developed will almost triple by mid-century—from 658 million to 1.8 billion. All developing countries are trying to reduce their birth rates. If this fight is not won in India within two decades, India's population could nearly double by 2050, and India already has severe water shortages.[21] Unless there are reductions in fertility, the world population could increase by nearly twice as much as currently projected.

> Half the world's people live in countries where water tables are falling as aquifers are being depleted. In Sana'a, the capital of Yemen—home to 2 million people—water tables are falling fast. Tap water is available only once every 4 days; in Taiz, a smaller city to the south, it is once every 20 days.
>
> *World on Edge*, Quick Facts

Lester Brown says rapid population growth in recent decades is causing many countries to show signs of "demographic fatigue." Governments are already stretched to the limit and now must simultaneously try to educate increasing numbers of children, create jobs for increasing numbers of young job seekers, and deal with the environmental effects of population growth. The rapid increase in young people guarantees that there will not be enough jobs for the number of educated youth, so it creates a large number of young men that are underemployed, frustrated, and resentful. When a young man is unemployed, he also cannot marry. These problems fuel recruitment into extremist groups.[22]

Almost all countries are working to move through the "demographic transition," from a population with short lives and large families to one with long lives and small families. In this transition, women are in the workforce, infant and maternal mortality is low, and contraception is widely available. Drops in the birth rate eventually reduce the growth rate of adolescents looking for jobs and increase the population's average age.[23]

Issues of the availability of contraception and especially abortion are complex and difficult, but the context of the argument changes when we look at them in terms of global systemic results, rather than in terms of the rights of the individual, whether of the woman or the fetus. We live on a finite planet that can only hold a finite number of people—a number that has already been far exceeded.

The cultural understanding of contraception and abortion in the U.S. directly affects the availability of contraception and abortion in developing countries. Therefore, how we deal with these issues will directly affect world population growth.

Reducing the Number of Abortions: By State Control of Women or by Preventing Pregnancy

Defining the abortion debate in terms of "pro-life" and "pro-choice" distorts and confuses the issue. We need to start by making explicit what the main issue is: *everyone* wants to reduce the number of abortions. *No one* is "pro-abortion." But progressives fight the issue on pro-life terms. They need to redefine the argument as a conflict over *how* to reduce the number of abortions, either by state control of women or by preventing pregnancy by increasing the education about and availability of contraception, and as a last resort, the availability of safe abortions.

For progressives, the pro-life focus on criminalizing abortion is a denial of the full personhood of women. In the pro-life world, any woman from 13 to 50 who chances to get pregnant forfeits all personhood on conception, and she becomes a machine, an incubator, controlled by the state. Pro-choice advocates are reacting to this control of women and the loss of the ability of families to plan and control when they want a child. Pro-choice advocates want women and couples to be able to determine when they will bear a child, yet they still want to reduce the number of abortions.

Throughout history, the only way not to get pregnant was abstinence. Communities have always had a deep, vested interest in being sure that any child born had a family ready to take care of it. But for the past 50 years, that moral imperative can now be met by using contraception. We now need to emphasize responsible sex in terms of both sexual partners taking full responsibility for ensuring that the woman will not become pregnant, which requires wide access to affordable contraception.

In addition to information about and access to contraception, the second focus of progressives is on supporting parents and empowering women. The choice to have an abortion never happens in a vacuum. The structure of the system can limit individual choices, and pro-choice means maximizing the ability of a woman to make her own choice about bearing a child, including being able to choose *not* to have an abortion. To reduce the number of abortions, we need to provide a comprehensive approach including, reducing the economic pressures that can force a woman to have an abortion. We can focus on programs to support new parents, such as the Maternal and

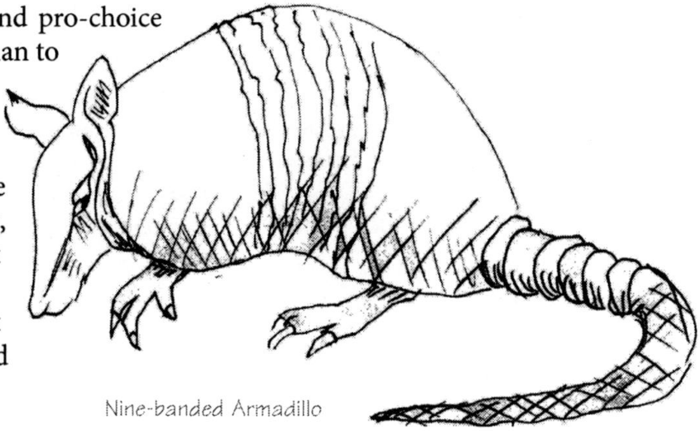

Nine-banded Armadillo

Child Health Block Grant and the Healthy Start program, which provide health care and other services for pregnant women and young children who lack other access to care. In addition, we need childcare school programs, domestic violence prevention programs, and the variety of family supports financed through the Community Services Block Grant.

Another focus of progressives is to expand adoption awareness programs and increase and expand the adoption tax credit.

Pro-Life is also about Denying Access to Contraception

There is an assault on contraception that is well organized and well funded. To prevent pregnancy, it is essential that people have information about and access to contraception. In the U.S., only 11% of sexually active women do not use contraception, but they make up 50% of those having abortions.[24] And 80% of self-described pro-life voters strongly support contraception. Yet not a single pro-life organization in the U.S. supports contraception, and they have given funding only to abstinence programs that do not allow any discussion of contraception (which obviously greatly increases the number of pregnancies and therefore abortions).[25]

> The world is now adding close to 80 million people per year. Even worse, the overwhelming majority of these people are being added in countries where cropland is scarce, soils are eroding, and irrigation wells are going dry.
>
> Lester Brown
> *World on the Edge*

Many pro-life individuals are unaware of this broader agenda of national pro-life organizations.[26] For instance, in October 2007, President Bush appointed Dr. Susan Orr to head the government's family planning program responsible for providing birth control and other family planning services to the poor. She had called birth control the "culture of death."[27] And the *Pro-life Activists' Encyclopedia* by the American Life League says, "Contraception cannot be separated from abortion. . . . [T]hey are virtually indistinguishable in a psychological, physical, and legal sense. . . .Those individuals who use artificial contraception take the critical step of separating sex from procreation. Contraception not abortion was the first step down the slippery slope."[28] The American Life League also says the condom's "biggest flaw" is that "those using it to prevent conception of another human being are offending God. . . . Any action, including condom use, which proposes to render procreation impossible, is intrinsically evil."[29]

Abstinence-only programs teach teenagers that safe sex is a dangerous myth and that birth control is unsafe (including the condom). Yet one-half of all teens 15 to 19 are sexually active. These abstinence-only programs lead to riskier sex among teens, more pregnancies, and therefore more abortions. We need sex education programs that present the full range of options, which includes abstinence as one option. Promoting abstinence as a legitimate choice could be very empowering to young women who are being pressured into sex.

One approach used to stop the availability of contraception is to redefine pregnancy.

Pro-life organizations argue that pregnancy begins at conception, which politically has a far-reaching affect. In the past, laws have always used the medical definition agreed to by the American Medical Association, the British Medical Association, and the American College of Obstetricians and Gynecologists, who all define pregnancy as starting at implantation of the fertilized egg on the wall of the uterus.[30] But pro-life organizations want to define pregnancy as starting at conception, which is a biologically unknowable moment—there is no test to determine whether a woman's egg has been fertilized. The effect of this redefinition is to redefine most forms of birth control as abortion, including the pill, the shot, the patch, the ring, and the IUD, because they can work after fertilization and before implantation. With this redefinition, Cristina Page, who writes on the pro-choice movement, says for pro-life organizations, something that prevents pregnancy is the same as ending a pregnancy.[31] Now there is also an effort to have pharmacists refuse to fill prescriptions for contraception based on their religious objection.[32]

The Real Issue: The Traditional Family versus the Modern Family

The real debate between pro-life (national organizations) and pro-choice advocates is about two visions of the American family. Page says the Christian right is trying to reclaim the traditional family, in which the man is the head of the family, the wife has only those choices he gives her, the wife stays at home and cares for the children, and the husband earns the family income. In the modern family, the husband and wife see themselves as equal partners, the wife earns part of the family income, the father is nurturing and participates in the care of the children, and they jointly make decisions about whether and when to have a baby.

Birth control greatly increases *women's* power and autonomy and threatens the traditional family in which men are in control and are to be obeyed. Much of the opposition to abortion is a response to this loss of male control. Roe *v.* Wade began a radical shift in American society, in the American family, and in the personal power available to women, which is a fundamental threat to male authority. The simple reality is that women have an ability men lack: only women can bear a child. This disparity has great political significance. For a man to give birth to an heir, he must have exclusive sexual access to and therefore control over a woman. The authority of men has always been closely related to men's control of women's bodies.

We now need to ask, in a world of disastrous problems directly related to increases in population—poverty, famine, environmental decline, global warming, disease, food and water shortages, land shortages, and large-scale refugee problems—what does "pro-life" really mean?

Sun Bear

Morality only makes sense if we measure it in terms of the full consequences of our choices. We need to change the scale at which we measure morality, or measure "pro-life." What if what looks like pro-life on an individual level results in tens of millions of deaths on the global level?

We need to creatively address the needs, rights, and responsibilities of men, women, the unborn, and the planet in a critically overpopulated world. We need to begin to find ways to get a community-formed consensus over the relationship between the individual and community, on limits imposed on both, as well as on mutual responsibilities. All life is sacred: children, adults, and the elderly. It does not make sense to say that the life of the unborn is the only life that counts—unless, that is, the real goal is to control women and restore the hierarchical power of men.

Gay Marriage and the Traditional Family

The debates over gay marriage are related to the issue of abortion in that they are also concerned with preserving the traditional family. Conservatives condemn gay marriage because it is a fundamental challenge to their understanding of hierarchical authority and ranked differences. Hierarchical authority rests on the assumption that there are "natural" hierarchies of command and obedience, of superior and subordinate. Marriage is the basic hierarchical relationship in society. By their definition, a marriage must include two ranked categories of individuals, a husband who is always male, and a wife who is always female. The husband is always the superior, and his authority is central. All other hierarchical relationships flow from this basic relationship.

If gay marriage becomes legal, it would mean we accept as a legitimate "marriage" a relationship in which there is no "naturally" designated subordinate and that is therefore a relationship of equals. This is a threat to the entire hierarchical worldview, so it undermines conservative morality and the traditional family. Feminism and equality in marriage present the same problem, but they are not as visible as legalizing gay marriage would be. Supporting legal, committed marriage for gays has nothing to do with heterosexual marriage, unless the point is maintaining the power of the husband in the traditional family.

For liberals the basic issue is whether being a homosexual is something people do or something they *are*. Science now says homosexuals are definitely born being homosexual. When conservatives speak of homosexuality as a "lifestyle," they mean to imply that it is a choice. For progressives, living authentically as gay or lesbian is a question of personal freedom. Jesus said that there is neither Jew nor Greek, bond nor free, male nor female; certainly today he would add that there is neither gay nor straight.

The real issue is what kind of society do we want? We need to encourage and support communities of people in committed relationships of care and compassion so that everyone has a shared stake in building caring, stable communities. Supporting committed marriages of all types

Magpie

will make our communities stronger. We live in a world torn apart by hatred. In such a world it is a waste of time and energy to argue over which groups of people should be allowed to love each other.

One related issue is the "don't ask, don't tell" rule in the military. It is telling that the military is so concerned that heterosexual men might receive unwanted sexual advances from homosexual men. But they have little concern about the unwanted sexual advances women in the military receive from heterosexual men, including assault and rape. An unwanted sexual advance is sexual harassment and should not be tolerated, no matter who is doing it to whom.

Co-opting the Religious Voice in Politics

The primary political function of the emphasis on issues like abortion and gay marriage is to drain off enormous amounts of political energy, allowing the real conservative focus of supporting the wealthy to go unchallenged. Whenever there are threats to economic interests, politicians can just yell, abortion! or gay marriage! and support is ensured.

However, more and more conservative Christians are declaring that the dominance of the fundamentalist identity of the religious right has gone too far. Evangelicals, such as the evangelical Rev. Gregory A. Boyd, are challenging fundamentalist voices. He criticized the "hypocrisy and pettiness" of Christians who focus on "sexual issues" like homosexuality and abortion. He said Christians these days are constantly outraged about sex and perceived violations of their rights to display their faith in public. "Those are the two buttons to push if you want to get Christians to act," he says. "And those are the two buttons Jesus never pushed."[33]

But an even greater problem is that the abortion debate has distorted and restricted the use of religion in political debates. Because pro-life policies are so intrusive and based on limiting freedoms, the culture has begun to assume that any use of religious language in politics will do the same. But that was not always the case. The civil rights movement was all about "imposing" their religious beliefs on the nation, but it was about expanding freedoms rather than restricting them. We need to reframe the abortion debate in order to restore the legitimacy of using religious values and language in public political debates.

Orca and calf (Killer Whale)

Ecological Morality: Budget for Restoring Earth

Historically the U.S. has contributed three times as much CO_2 to the atmosphere as any other country; therefore, we have a major responsibility for addressing global warming. Solving this crisis requires the moral leadership of the United States. Our survival requires us to cooperate with the rest of the world and make it clear that we are willing to share Earth's resources with other countries. It will require an unprecedented global cooperation, but it must start with us.

Plan B Budget

Additional Annual Funding Needed to Restore the Earth

GOAL	FUNDING
	(billion dollars)
Planting trees to reduce flooding and conserve soil	6
Planting trees to sequester carbon	17
Protecting topsoil on cropland	24
Restoring rangelands	9
Restoring fisheries	13
Protecting biological diversity	31
Stabilizing water tables	10
TOTAL	*110*

Brown's second budget is for restoring the natural systems of Earth. In *Plan B 4.0*, (above) Brown gives estimates of the costs for undoing the damage already done (these costs again are in addition to what is already being done). He gives the annual budget for restoring Earth's systems:[34]

This is what global moral leadership would look like. The combined total for both sets of goals is $187 billion annually. For this amount of money, the world could be transformed, Earth renewed, and basic care provided for the poor of the world, which in turn would reduce population. But for this to happen, the U.S. would need to see itself as one country among other sovereign countries. That would require a major change in the way we see ourselves and the world. We have a choice: we can continue to try to dominate the world militarily and economically, or we can stop trying to control the world for our benefit and begin to build equal partnerships to address our common crises.

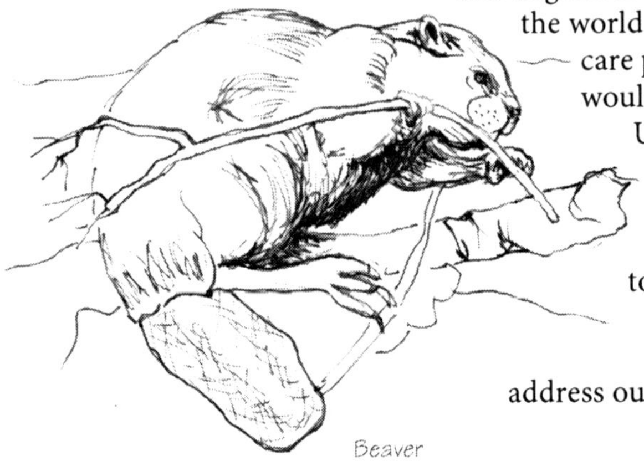

Beaver

Redefining National Security and Cutting the Defense Budget

The $187 billion for ending poverty and restoring Earth should come from our military budget because global warming is also an issue of national security.

Instead, our military budget is bankrupting our nation, yet no one will object to any proposed military budget. Legitimate defense would be a fraction of what we now have. The total U.S. budget for the 2011 fiscal year is about $3 trillion, not counting the funds collected for Social Security. The budget for the military is about $1.6 trillion. That includes the Pentagon budget request of $708 billion, plus an estimated $200 billion in supplemental funding, called, "overseas contingency funding" to fund the wars in Afghanistan and Iraq, $40 billion or more in "black box" intelligence agency funding, $94 billion in non–Department of Defense military spending, $100 billion in veterans benefits and health care spending, and $400 billion in interest on debt raised to pay for prior wars and the standing military.[35]

The nationalist not only does not disapprove of atrocities committed by his own side, but he has a remarkable capacity for not even hearing about them.

George Orwell

The 2011 military budget is the largest in history, in inflation-adjusted dollars, even greater than the spending in World War II. Military spending in all its forms represents 53.3% of the total U.S. federal spending.[36] The enormous increases in military and defense spending began in 1998, well before 9/11. Between 1998 and the current proposed 2011 budget, the defense budget has *doubled*, yet only half of this increase is attributable to our two wars.[37]

U.S. military spending constitutes half of the entire global spending on war and weaponry, meaning that our military spending is as much as that of the rest of the world combined. The military spending of our allies—countries like France, Britain, Germany, Italy, and Japan—accounts for 21% of the world's military spending. The 12 top spenders among big military-spending nations are either allies of the U.S. or are friendly countries like Brazil and India. Therefore, America and its friends and allies account for more than 70% of all military spending worldwide. By comparison, China spends only about $130 billion on its military. Russia spends less than $80 billion a year, and of our "enemies," Iran spends $4.8 billion and North Korea spends an estimated $5 billion.[38] Our military is still organized to fight large-scale wars with opponents like the Soviet Union—which no longer exists.[39] This defense budget is unsustainable and ineffective for the real threats we face.

This total military budget includes maintaining our 761 military bases in foreign countries and an additional 104 bases in U.S. territories.[40] This does not include the estimated bases in Iraq (300) or Afghanistan (400).[41] These bases are in

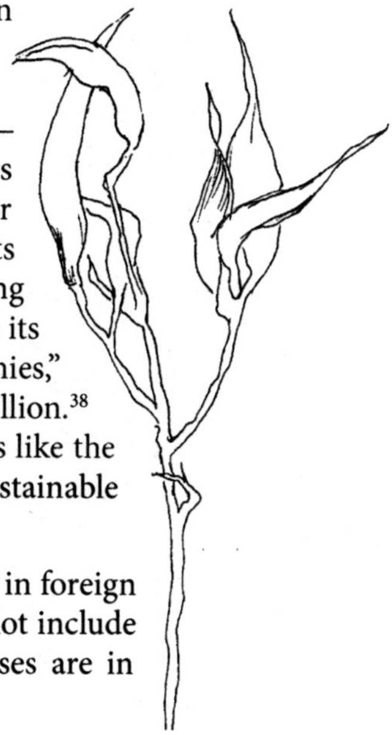

Whorled Milkweed

about 60 countries, although that number may be low. Issues of jurisdiction with host countries are defined in "Status of Forces Agreements." During the Cold War, these were normally public documents, but they are now often classified as secret. According to Pentagon records, the Pentagon has formal agreements with 93 countries.[42]

Americans are ignorant of the impacts of these bases, for they get little or no media coverage. U.S. overseas military bases frequently give rise to major social protests in the host countries. The Prime Minister of Japan was forced to resign in May 2010 because he broke his campaign promise to expel the U.S. base on Okinawa. Service personnel who commit crimes are exempt from local laws. Bases also create a very toxic environment, but the Agreements exempt the bases from the host country's environmental laws, and they also create many social problems. The perception of U.S. military bases as intrusions on national sovereignty is widespread in host countries. Russia and China see the build-up of bases in Afghanistan, Pakistan, and three of the former Soviet republics of Central Asia as posing threats to their security. We have begun to surround China by our bases at Manas in Kyrgyzstan, 250 miles from the western Chinese border, where U.S. planes are landing daily, as well as U.S. bases to the east in Japan, to the south in South Korea, and Washington's military support for Taiwan.[43]

> Every gun that is made, every warship launched, every rocket fired, signifies in the final sense a theft from those who hunger and are not fed, those who are cold and are not clothed.
>
> President Dwight D. Eisenhower
> April 16, 1953

The establishment of U.S. bases in Saudi Arabia during the Gulf War was regarded by some Saudis as "an occupation of the holiest land of Islam, to be repelled at virtually any cost."[44] The September 11, 2001, attacks were in part a response to the American bases in Saudi Arabia. The presence of these bases is the reason for Osama Bin Laden's first public "fatwa" (Declaration of War) against the U.S. in 1996. The fatwa is titled "Declaration of War against the Americans Occupying the Land of the Two Holy Places" (Mecca and Medina). Fifteen of the 19 hijackers on 9/11 were from Saudi Arabia.[45]

New bases are now being acquired in South Asia, the Middle East and Africa, and Latin America and the Caribbean, and in regions where U.S. bases have not existed previously, such as the Balkans and Central Asia. The U.S. is presently on a course of intense imperial expansion, in part as a means of promoting its political and economic interests. The present war on terrorism is in many ways a consequence of this projection of U.S. power, yet that is now being used to justify the further projection of that power.[46]

The Military Is Expected to Prevent the Emergence of Threats

The enormous military budgets began with the end of the Cold War, when even though the threats decreased, our leadership kept increasing their security goals. The Project on Defense Alternatives said that these goals soon went beyond "the traditional goals of defense and deterrence, aiming to use military power to actually *prevent the emergence* of threats and to 'shape' the strategic environment. US defense planners also elevated the importance of lesser and hypothetical threats, thus requiring the military to prepare for many more, lower-probability contingencies."[47]

Preventing the emergence of threats requires treating other nations as if they are considering aggression even if they are not. This increases tensions and creates the outcome that the "prevention" is meant to stop. Carl Conetta, Co-director of the Project on Defense Alternatives, says, "Thus, in addressing the nuclear programs of both North Korea and Iran, our coercive efforts spurred, rather than retarded, the behavior we had hoped to stop." Our attempts at prevention became a provocation. Yet politicians like this approach because it gives the illusion that we can control events.

This approach emphasizes uncertainties, which heightens fears and suspicion. Conetta says they lowered the bar on plausibility scenarios and focused on "worse case" eventualities.[48] Using force gives everyone the illusion that we can make events turn out the way we want them to, but in fact it creates profound and unpredictable chaotic events.[49]

Chalmers Johnson, in *Nemesis: The Last Days of the American Republic*, says, "The real reason for constructing this new ring of American bases along the equator is to expand our empire and reinforce our military domination of the world."[50] Johnson states, "By the end of the 1990s, the neoconservatives were developing their grandiose theories to promote overt imperialism by the 'lone superpower'—including preventive and preemptive unilateral military action, spreading democracy abroad at the point of a gun, obstructing the rise of any 'near-peer' country or bloc of countries that might challenge U.S. military supremacy, and a vision of a 'democratic' Middle East that would supply us with all the oil we wanted."[51]

Yet our primary military threats are from a handful of terrorists armed with little more than a willingness to die for their cause. The only beneficiaries of this stupid security policy are the Wall Street corporations that profit from defense expenditures.

David Korten
Agenda for a New Economy

It used to be assumed that force would be used only as a last resort, mostly as a defense against aggression, and used only when forced upon us. But the U.S. has lost the ability to think of alternatives to a military response to events. This emphasis on a military response has militarized diplomatic functions. This security policy is counterproductive and unsustainable. Any real progress requires that we reassess the role of force and the armed forces in U.S. policy.[52]

Military Operations against Terrorists Are Not the Solution

Much of U.S. aggression is defended on the basis of stopping terrorism. But terrorism is misunderstood, and using the military to fight it is counterproductive. Governments are said to have a monopoly on violence because only the police and the military can legitimately use violence. If anyone else uses violence, it's a crime. Terrorists are non-state actors by definition; therefore, terrorism is a crime. Using the military is like bombing New York City to get rid of organized crime.

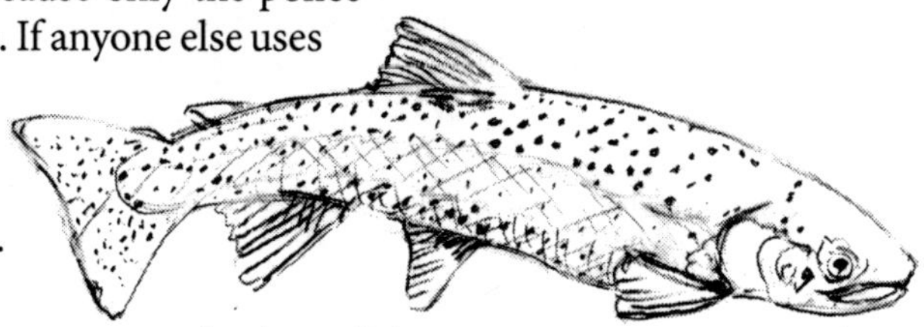

Cutthroat Salmon

The consequences are obvious—in Iraq and Afghanistan and now Pakistan, we are killing large numbers of civilians, even by the Pentagon's own reports. It's amazing how routine the killing of civilians has become. These deaths fuel recruitment of new terrorists, and so our way of fighting terrorism just creates more terrorists. We need to demilitarize our Afghanistan policy. Terrorism should be fought as international crimes of mass murder, with the coordinated efforts of the CIA, FBI, Interpol, MI6, and others. Carl Conetta says:

> From the start, the global war on terrorism drew criticism. The most trenchant critics questioned the logic of proposing to wage war on something that was not a political entity, but a tactic or means of warfare. Not only was the putative target unbounded, but extinguishing the sources of terrorism in general might be a task that no amount of military power could accomplish. This made a mash of strategy, which at minimum requires that the target and goals of a war be well defined."[53]

This approach also creates perpetual war; there is no way to sign a peace treaty with a method of fighting. Conetta explains that "in many conflicts with an Islamic element—for instance, in Chechnya, the Philippines, Israel-Palestine, Indonesia, Xinjiang, and Kashmir—local conditions and real grievances play a major or even principal part in driving violence. In such cases, Islam may serve only as the idiom of militancy, not its source. . . .The 'global Islamic insurgency' does not exist except as a construct in the minds of jihadi fanatics, of neoconservative thinkers, and of the authors of Pentagon Reports. What does exist are a number of separate local insurgencies with a strong Islamic element."[54]

> But today it is failing states that provide the greatest threat to global order and stability. As Foreign Policy notes, "World leaders once worried about who was amassing power; now they worry about the absence of it."
>
> Lester R. Brown
> *World on the Edge*

President George W. Bush changed the name from the "War on Terrorism" to "The Long War against Islamic Extremism." This concept, Carl Conetta says, "seems tailor-made to tempt war with the Muslim world."[55] This is no way to ensure our national security. Radical Islamic parties have grown substantially in more than a dozen countries by a call to fight the "war against Islam." The U.S. is responding as though bin Laden and al-Zawahiri were Hitler or Stalin, directing hundreds of divisions to our gates. This view utterly misconstrues the actual dynamics of their influence and distracts from the real dangers they pose.[56]

The basic issue is the morality of dominance and Empire. Chalmers Johnson says that for our Empire, "America's version of the colony is the military base."[57] Today, he says, "the 'war on terrorism' is at best only a small part of the reason for all our military strategizing.[58] This creates a world in which everyone feels threatened by us, for many countries deeply resent the U.S. military bases and our use of force. Defense Secretary Robert Gates has repeatedly said that, in terms of Afghanistan, a military victory is impossible. The only way to a peaceful resolution is by winning the "hearts and minds" of the Afghan people. Bombing them and killing many civilians will never do that.

Terrorism will never go away; it is now a fact of modern life. But responding with violence simply escalates the violence. It is also alienating our allies, who now believe the U.S. is having a negative influence on world affairs. Voters in Britain, Italy, and Spain have voted against their governments for their pro-American stance.[59]

Military Budgets Are Used to Maintain Our Access to Foreign Oil

There are two issues that drive the use of the military. One is militarizing our response to terrorism, which Obama has renewed. The other is our need for access to and control of foreign oil. Our war in Iraq (the world's second-largest oil producer) was largely about oil. President Bush wanted to invade Iraq for many reasons, but both Bush and Cheney are oil men. They wanted to create a "regime change" in which we could have open access to Iraq's oil. *New York Times* columnist Thomas Freidman says we are funding both sides of the war—our petro-dollars fund the governments that support terrorists. The military's role is to protect our access to foreign oil, including providing every oil tanker that leaves the Middle East with a military escort.

> In the councils of government, we must guard against the acquisition of unwarranted influence, whether sought or unsought, by the military-industrial complex. The potential for the disastrous rise of misplaced power exists and will persist. We must never let the weight of this combination endanger our liberties or democratic processes.
>
> President Dwight D. Eisenhower
> Farewell address to the nation, 1961

In Afghanistan, part of the reason for the war is again access to oil and natural gas. Corporations, especially the Unocal Corporation, and our government have wanted to build a secure corridor for U.S. controlled oil and natural gas pipelines from the Caspian Sea in Central Asia through Afghanistan and Pakistan to the Arabian Sea. The war in Afghanistan and the creation of U.S. bases in Central Asia are seen as an opportunity to make this possible. [60]

Perhaps now, with the oil spill in the Gulf of Mexico, we will finally get serious about renewable energy and plug-in hybrid cars (see Chapter 12). Stopping our addiction to oil would have many benefits:

- It would address global warming.
- It would protect the oceans and the environment.
- It would cut our trade deficit in half.
- The hundreds of billions of dollars we send to the Middle East for oil would instead be available for our own economy.
- Our oil money would stop funding networks of Islamic militants, as well as the money for Iran to develop weapons of mass destruction, or Palestine to get the money for nuclear weapon technology.[61]
- If we did not need to protect our access to oil in other people's countries, we would have much less need for a massive military presence all over the world.

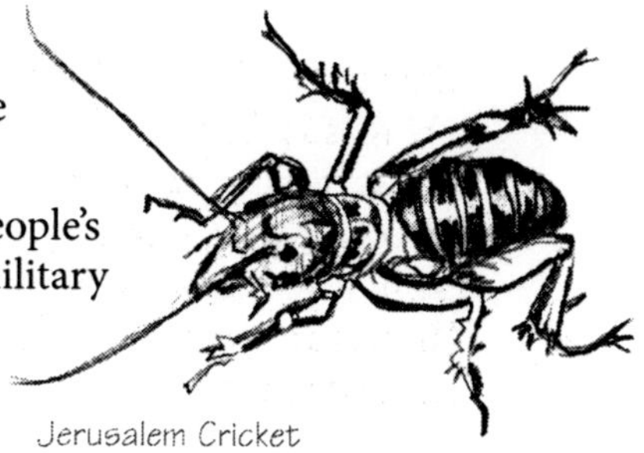

Jerusalem Cricket

The Real Threat Today Is Failed States

In any case, Lester Brown says we have entered a new era in geopolitics. In the twentieth century, the main threat to international security was superpower conflict; today, it is failing states. It is not the concentration of power but its absence that puts us at risk.[62] States fail when national governments can no longer provide personal security, food security, and basic social services such as education and health care. In a failed state, the government loses control of part or all of their territory. When governments lose their monopoly on power, law and order begin to disintegrate. After a point, countries can become so dangerous that food relief workers are no longer safe and their programs are halted. In Somalia and Afghanistan, deteriorating conditions have already put such programs in jeopardy.[63] Failing states are of international concern because they are a source of terrorists, drugs, weapons, and refugees, and they threaten political stability everywhere.[64]

> The means by which we live have outdistanced the ends for which we live. Our scientific power has outrun our spiritual power. We have guided missiles and misguided men.
>
> Martin Luther King, Jr.

Our global civilization depends on a functioning network of politically healthy nation-states to control the spread of infectious disease, manage the international monetary system, and reach scores of other common goals. It will be a major problem if the system for controlling infectious diseases—such as polio, SARS, or avian flu—breaks down.[65]

The poverty budget of $77 billion would also help in avoiding failed states.

Dominance or Partnership

If we treated terrorism as international crime and converted to renewable energy and plug-in hybrid cars, we could cut military expenditures by at least 70% (remember, defense expenditures have doubled since 1998). We could use the money saved to fund renewable energy, education, health, to get new industries started, and to eradicate poverty worldwide.

A fundamental issue is whether the United States can foresee accepting China as an equal—first regionally, then globally. Conetta says the Pentagon wants to "integrate China as a 'responsible stakeholder' in an international order *led* by the United States. This subordinate relationship is unacceptable to Beijing and will become more so as China grows more powerful."[66] In the U.S. there is a tendency to see any threat to our status as "sole superpower" as a threat to national security. Therefore, without regard to the cost, risks, or unintended consequences, the U.S. has an abiding faith in the effectiveness of war and our armed force, so much so that the U.S. is in a state of permanent war.

In the future, 90% of the growth in greenhouse gas emissions will come from developing countries. To address global warming, we have to help the developing world, particularly China and India, to move rapidly to renewable energy and hybrid cars and local food. Developing countries are trapped because of debt and an economy based on export, so that they are very dependent on the world economy. It is essential that Third World debt be forgiven and that sustainable technology be given to them and their own people trained to use it, rather than having multinational companies make money doing it for them. Then we need to make a determined effort to eradicate poverty and restore Earth's natural systems. These efforts will create more peace and genuine security than the military could ever do.

The new reality is that the world is only one poor harvest away from chaos. It is time to redefine security. The principal threats to our future are no longer armed aggression but instead climate change, population growth, water shortages, spreading hunger, and failing states.

Lester Brown
World on the Edge

The presence of the U.S. in the world has been driven by the neoconservative dream of endless U.S. global domination. Imperialism is a policy of extending our rule over foreign countries. Our culture has had the sense that the United States ought to and has the right to control the world both militarily and economically. Because it is the U.S. that is dominating the world, we are the ones that need to choose and model partnership. Why do we consider bombing people "realistic" while partnership and diplomacy is "naive?" Why can't we see that our behavior creates the response we are trying to stop? Why do we choose a hierarchical, coercive, command and obedience view of power?

The only vision of power we have is authoritarian. We talk about democracy, but in reality, we don't want consensus, we want control. Can we develop the ability to relate rather than dominate? Lester Brown's budget for eradicating poverty would change world dynamics. In addition to not creating additional harm, it could change much of the world dynamics that create terrorism in the first place.

Remember that nature has the ultimate weapons of mass destruction and the military is irrelevant in protecting us from them. The world now has a common crisis and only an approach of partnership and cooperation will adequately address it. Gaining the cooperation of developing countries is going to be very limited if we have colonized the whole world for our benefit. We control the world militarily because we believe in dominance and Empire. Is it possible for us to see ourselves as one country among equals and to help everyone achieve their rightful share as well as their own autonomy? We need to agree that the age of Empire is over. Instead of one country above all others, we need to move on to a world united as one, with a shared destiny and a common enemy: global warming.

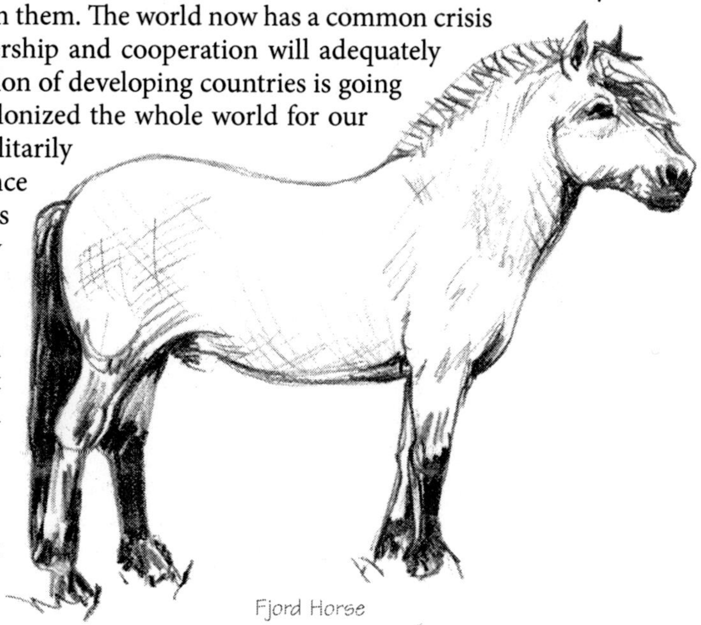

Fjord Horse

Case Study: A Contrast in Values and Rethinking Justice

It is difficult for most people to visualize the primary values of empathy and respect in operation. One way to see what it would look like is to review ideas about justice in the two worldviews.

A new understanding of justice is developing that is inherently a part of the new worldview, and this new concept is a good illustration of how differently the two models understand the world. The manner in which we understand and enforce justice is a fundamental characteristic of a society's worldview. Conservatives see justice in terms of crime and punishment. The new concept assumes there are many factors creating crime and that to think only in terms of judgment and blame of individuals ignores many relevant factors.

Justice as Retribution

Justice in our current worldview emphasizes authority, obedience, and punishment. It sees any crime as an affront to the authority of the state, and the response is retribution, that is, punishment. This approach emphasizes punishment by the state and separation of the offenders from society, and it assumes that longer sentences and harsher conditions will help "fight crime." In this understanding of justice, offenders do not learn the personal and social harm they have done, nor do they accept responsibility for their actions. In addition, victims have no role in this process or any opportunity to find healing and forgiveness. In *Kaleidoscope of Justice*, Harry Mika and Howard Zehr list the assumptions of retributive justice:

- Guilt must be determined
- The guilty must get their "just deserts"
- Just deserts require the infliction of pain
- Justice is measured by this process
- The breaking of the law defines the offence.[67]

This emphasis on punishment has now gotten to the point that we punish people convicted of a crime even after they get out of prison. When someone gets out of prison with a felony conviction, he or she is permanently unemployable. Virtually all employers carefully screen out anyone with this background. In recent years, states have also passed legislation that lengthens the list of jobs that people with a criminal conviction cannot hold. In New York, there are more than 100 prohibited job categories, including plumbing, real estate, barbering, education, health care, and private security. [68]

Justice as Restoration

The new worldview can shape a new understanding of justice. Our criminal justice system is in a state of crisis, with both courts and prisons overloaded. This has caused some people to rethink the whole justice system. The shift is part of the larger shift from power-based organizations and practices to those that are relationship based.[69]

"Restorative justice" is the name for a new understanding of crime. It is a value-based approach. Rather than defining crime as an affront to the authority, power, and laws of the state, it sees crime as an injury to the victim and the community. In restorative justice, community values guide decisions on policy, programs, and practice. The first response to crime should be to repair, to the greatest degree possible, the harm done by the crime. This model emphasizes restitution rather than punishment: if you do something wrong, do something right to make up for it. Restorative justice assumes that the major origin of crime is in social conditions and broken relationships in the community. Therefore, crime prevention is dependent on communities' taking responsibility for remedying these conditions and healing broken relationships. Many countries are developing these practices of restorative justice.[70]

The Center for Restorative Justice at Suffolk University says that "restorative justice is a broad term which encompasses a growing social movement to institutionalize peaceful approaches to harm, problem-solving and violations of legal and human rights. Restorative justice is a value-based approach to conflict and harm. These values are often identified as inclusion, democracy, responsibility, reparation, safety, healing and reintegration. But one value is more essential than any other—Respect."[71] One emphasis is on creating a "culture of care" in our schools to stop a culture of bullying, which they define as the assertion of power through aggression and domination. When working at schools, the emphasis is placed on healing in community by building personal relationships.

The Center for Restorative Justice and Peacemaking at the University of Minnesota says that restorative justice places victims, communities, and offenders in active roles in order to

- Empower victims in their search for closure,
- Impress upon offenders the real human impact of their behavior, and
- Promote restitution to victims and communities.

Restorative justice is a set of principles based on problem solving and can involve various practices. Generally, the victim helps define the harm done and offers ideas about how to repair the harm. Facilitators encourage the offender to accept personal accountability for his/her harmful actions and to take action to repair it. The community's role is to help the victim to heal and to reintegrate offenders into the community, while monitoring and enforcing community standards of behavior. All parties affected by a crime collectively seek ways to recover from the resulting trauma. Although the process requires voluntary cooperation from everyone, many people choose to participate in the process, even for serious crimes, although, for serious crimes, the process occurs alongside traditional criminal justice, rather than in place of it.

Philippine Eagle Endangered

There are three goals in restorative justice. The offender should develop social competency. In particular, offenders who enter the juvenile system should leave more capable than when they entered. The victim should feel accountability was achieved, for an offense creates an obligation to the victim and, often, to the community. The community should have community safety, for the justice system has a responsibility to protect the public.[72]

This problem-solving systems approach to crime involves the affected parties and the community in an active relationship with criminal justice agencies. The restorative justice process requires that we

- Focus on the harms of wrongdoing, more than on the rules that have been broken.
- Show equal concern and commitment to victims and offenders, involving both in the process of justice.
- Work toward the restoration of victims, empowering them and responding to their needs as they see them.
- Support offenders while encouraging them to accept and carry out their obligations.
- Recognize that while obligations may be difficult for offenders, they should not be intended as harms, and they must be achievable.
- Provide opportunities for appropriate, direct or indirect, voluntary dialogue between victims and offenders.
- Involve and empower the affected community through the justice process, and increase the community's capacity to recognize and respond to the community's part in the problem.
- Encourage collaboration and reintegration rather than coercion and isolation.
- Pay attention to the unintended consequences of our actions and programs.
- Show respect to all parties, including victims, offenders, and justice colleagues.[73]

The chart on the next page compares the two approaches to justice.[74]

Restorative justice is a systems view. The contrast between retributive and restorative justice clearly mirrors the contrasts between an authority-based morality and an empathy-based morality.[75] Restorative justice demonstrates the kind of community involvement and commitment that are required to create a culture that can preserve Earth. Because of its sense of seeking forgiveness and restitution, many churches have been involved in these efforts, as has the Boston Theological Institute.[75]

Justice and Drug Policy

In a similar process, David Kennedy, professor at New York's John Jay College of Criminal Justice, invented a new way for the police to deal with drug dealers.[76] Currently, the method police use includes being very aggressive and knocking down doors. This made the police "the enemy" in inner city neighborhoods.

In 2004, the High Point, N.C., police tried Kennedy's approach. The police would videotape the drug dealers selling drugs and prepare a case that would send them to prison. They brought the dealers in, showed them the evidence against them, and then said they'd give them a second chance and let them go. Working with their families, the police helped the drug dealers find job training and mentors. By giving the kids a second chance, they won the trust of the community. After four years, police in High Point had taken the drug dealers off the street corner and created a 57% drop in violent crime in the targeted area since the program began.

When law enforcement pushes harder and harder, it drives the community farther and farther away. Kennedy's program demonstrated that this spiral can be consciously stopped and even reversed. Kennedy's research shows that a very small number of people—dozens, not hundreds—cause the violence in the cities' worst areas. Most people just want a safe place to live but feel anger toward heavy-handed police and their every-day intrusion into these communities.[78]This new approach changes the way in which law enforcement work is viewed by the community.[79]

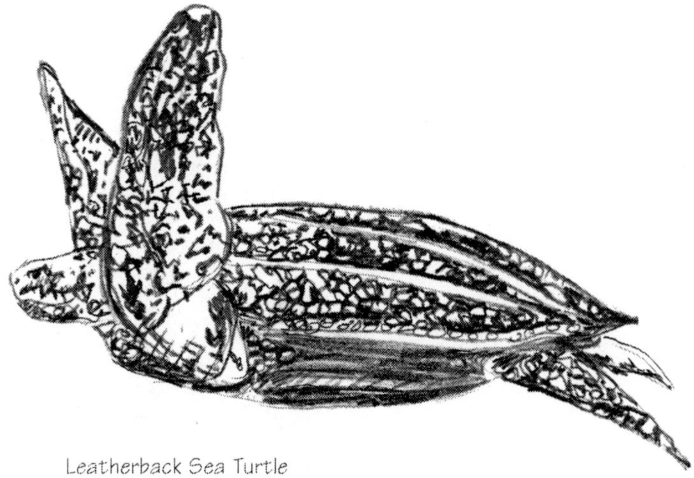

Leatherback Sea Turtle
On ten most endangered list (ICF)

These charts contrast concepts on justice and in politics as they are understood in the two different worldviews. Our current worldview is in the left column and the new ecological worldview is in the right column. Although these charts are quite linear, they help clarify the differences between mechanistic thinking and systems thinking,

Two Worldviews of Justice

RETRIBUTIVE JUSTICE	RESTORATIVE JUSTICE
Victim's needs ignored.	Victim's needs central
Wrongs create guilt.	Wrongs create liabilities and obligations
Crime defined by violation of rules and relationships	Crime defined by harm to people
Crime seen as categorically different from other harms and conflicts	Crime recognized as related to other harms and conflict
State as victim	People and relationships as victims
State and offender are primary parties	Victim and offender are primary parties
Interpersonal dimensions irrelevant	Interpersonal dimensions central
Offense defined in technical legal terms	Offense understood in full context: moral, social, economic, and political
Guilt is absolute, either/or	There are degrees of responsibility
Guilt is indelible	Guilt is removable through repentance and reparation
Debt is abstract	Debt is concrete
Debt paid by taking punishment	Debt paid by making right
Accountability = taking one's "medicine"	Accountability = taking responsibility
Blame-fixing central	Problem-solving central
Focus on past	Focus on future
Battle, adversarial model normative	Dialogue normative
Imposition of pain normative	Restoration/reparation normative
Harm by offender balanced by harm to offender	Harm by offender balanced by making right
One social injury added to another	Emphasis on repair of social injuries

Javan Rhinoceros
On ten most endangered list (ICF)

Two Worldviews in Politics

CONSERVATIVE POLITICAL WORLDVIEW	PROGRESSIVE POLITICAL WORLDVIEW
Autonomous individuals are central: Fragmented	Relationships in community are central: Whole
Patriarchy, dominance, and hierarchy	Partnership and equality
Patriarchal family values	Cooperative community values
Privilege	Equality
Self-interest	Common good, reciprocity, mutuality, and empathy
Central authority	Mutual trust
Information/knowledge/logic	Imagination/intuition/wisdom
Observing consciousness	Participating consciousness
Voting, creating winners and losers	Deliberative democracy and consensus building
Objectivity and observation	Creativity and participation
Responsibility to oneself	Responsibility for the whole
You're on your own	We're all in this together
The world is almost static, change is gradual and local	The world is dynamic and can undergo abrupt, discontinuous transitions from one state to another
Autonomy	Community
Alienation	Connectedness
Empire	Earth Community
Authority	Empathy
Human nature is greedy and violent	Human nature can be nurturing and cooperative
Power-based organizations	Relationship-based organizations
Masculine dominated	Gender balanced

Chapter 8

Ecological Economics: Putting Earth First

Redesigning our economic systems is basic to addressing global warming, but most people are unaware of this reality. Conservatives tend to see global warming as a "liberal" issue because the solutions involve government restrictions on economic behavior. However, even most environmentalists and progressives only propose responses to global warming that leave our current economic system unchanged. Our current form of extreme neoliberal economics resulted in the financial collapse of 2008 and is causing a collapse of Earth's natural systems. This economic system is unsustainable and needs to be replaced before we experience another sudden collapse. There is an alternative to the economic worldview described in Chapters 5 and 6, but few people know about it. This chapter and the next explore how profoundly differently our economy could be structured. Creating this new ecological economy will require a radical change in consciousness.

Wall Street Capitalism versus Main Street Market

In discussing how to create a new ecological economy, ecologists differentiate between capitalism and markets. Ecologists still want markets. Markets are a mechanism for determining the prices of things for the purpose of exchange, but capitalism is a much larger system of regulations, institutions, practices, and ethics within which markets operate. We can separate the two and can use markets under an entirely different set of regulations, institutional practices, and ethics. David Korten's name for the alternative to our current capitalism is mindful markets or main street markets. In distinguishing main street markets from Wall Street capitalism, Korten says, "A capitalist system, which is about concentrating control over productive assets in the hands of the few to their exclusive benefit, is very different from a market system, which is about people creating their means of livelihood through the production and exchange of goods and services."[1]

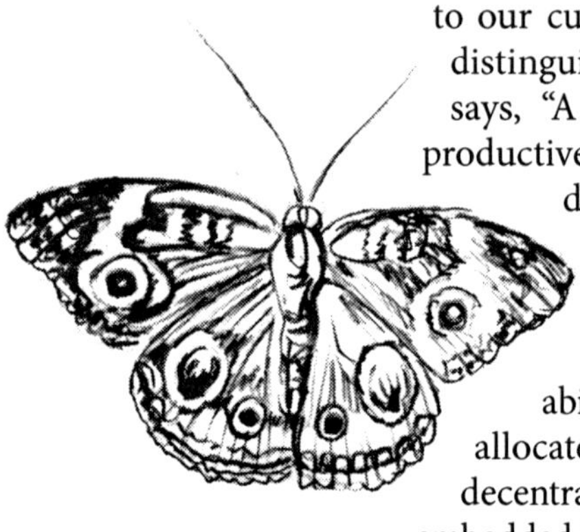
Buckeye Butterfly

Ecologists want to retain the market because of the market's ability to use scattered, fragmentary information to efficiently allocate resources by allowing decentralized decision-making. This decentralized decision-making is based on information (the feedback embedded in prices) cascading throughout the market, allowing the economy to self-organize. However, the market does not need to be free sense of having no limits; it only needs to be fluid, within socially determined limits, so that it can "self-organize."

Every person must learn to think like the Earth, to act like Earth, to be Earth.

Gerald Barney

Our culture teaches that free market exchanges that allow for the maximum profit for stockholders are a reasonable goal of society. But the only thing the market can choose is efficiency, yet no one asks, Efficiency for whom? or Over what time frame? The market is only a mechanism—it is not a moral imperative. Society needs to define the context and set the limits within which the market operates. David Korten says there are two ways to make products cheaper for the consumer. The first is to increase efficiency, which is useful. But the second is to reduce environmental and employment standards. Reducing the wages paid for a given amount of productive work is not an increase in efficiency, but a lowering of standards.[2]

In *Agenda for a New Economy: From Phantom Wealth to Real Wealth*, David Korten says that we need to base economics on the value of life rather than on the value of money. He insists that the only legitimate function of an economic system is to serve life. Korten compares the two understandings of economics in this chart.[3] Markets need to exist in the context of community values that focus on the common good because markets also create a way of life, and they affect what kind of people we become.

Wall Street Capitalism vs. Main Street Markets

	WALL STREET CAPITALISM	MAIN STREET MARKETS
Dominant Driver	Making money	Creating livelihoods
Defining Activity	Using money to make money for those who have money	Employing available resources to meet basic needs of the community
Firm Size	Very large	Small and medium
Costs	Externalized to the public	Internalized by the user
Ownership	Impersonal and absentee	Personal and rooted
Financial Capital	Global with no borders	Local/national with clear borders
Purpose of Investment	Maximize private profit	Increase beneficial output
The Role of Profit	An end to be maximized	A means to sustain viability
Efficiency Measure	Returns to financial capital	Returns to living capital
Coordinating Mechanisms	Central planning by mega-corporations	Self-organizing markets and networks
Cooperation	Can occur among competitors to escape the discipline of competition	Can occur among people and communities to advance the common good
Purpose of Competition	Eliminates the unfit	Stimulates efficiency and innovation
Government Role	Protects the interests of property	Advance the human interest
Trade	Free and unregulated	Fair and balanced
Political Orientation	A democracy of dollars	A democracy of persons

Currently, our economics encourages a way of life in which our entire society is focused on competition, greed, self-interest, and consumption. These values make it difficult to maintain a fair, equitable society or any type of moral economy. When the economy makes choices only on the basis of profit, it sends jobs overseas, creates more poverty, and destroys the life-support systems of nature. Yet the market cannot measure or even acknowledge these effects.

Our economic system, like our political system, focuses on the individual and ignores the effects on the whole system. If I want to buy a TV, I can choose between dozens of models. But there is no market in which an individual can buy clean air, a safe community, or higher literacy rates. Clean air is not available on the "open market." Its availability requires democratically chosen limits.

The fundamental fallacy of our free-market system is the assumption that community, nature, or third-party interests are unaffected by the exchanges between buyer and seller. That is, the market cannot recognize that private choices have public consequences.

Markets Need Resilience as Well as Efficiency

Bernard Lietaer is one of the leading innovators on economic systems. He gives a complex systems view of neo-liberal economics and argues that this will help us understand the monetary and economic dynamics as they actually manifest in the real world. He says we need to use systems theory to identify the underlying cause of the financial meltdown crisis; however, our current view of economics is based on a mechanistic view of the world, which sees market efficiency as the primary goal. This has created an economics with "views on the mechanics of 'rational actors' and the reliable self-restraint of General Equilibrium Theory."[4] This approach dominates mainstream academic economic literature, as well as boardrooms and politics.[5]

> [In] the current dominant system, markets respond to the wants of those with money and disregard even the most basic needs of those who do not have the means to pay.
>
> John Cavanagh and Jerry Mander
> *Alternatives to Economic Globalization*

Complex systems follow certain universal principles and patterns that apply to economic systems as well as natural ones. Decades of studying natural ecosystems has created sophisticated mathematical understandings of how a network structure affects an ecosystem's long-term viability. Lietaer says that they have found that this viability is the result of a balance between efficiency and resilience.[6] Efficiency measures the ability of a system to "perform in a sufficiently organized and efficient manner as to maintain its integrity over time." Resilience measures the ability of a system to recover from a disturbance by having flexible fallback positions and a diversity of actions that can be used to meet changes and crises.[7]

To achieve both efficiency and resilience, two key structure-related variables are required: diversity and interconnectivity. Diversity is found in the existence of different types of agents acting as "nodes" in the network, and interconnectivity is found in the number of pathways between agents. They each play a role in both efficiency and resilience, but they are opposites. Lietaer says, "A system's resilience is enhanced by more diversity and more connections, because there are more channels to fall back on in times of trouble or change. Efficiency, on the other hand, increases through streamlining, which usually means reducing diversity and connectivity."[8]

Diversity and connectivity can also interact in unexpected ways—for example, when diversity is extremely low, more connectivity is a source of instability in the system, not resilience. This is what happened in our financial crisis—since almost all big banks took on the same overleveraged exposure, they would all have to react the same way to a tightening of credit (there was no diversity), but their extensive interconnectivity meant credit problems would get widely transmitted, triggering an exponential increase in credit problems in the system. But when diversity is high, more connectivity will lead to more stability.

Market fundamentalists are right in their recognition of the creative potential of self-organization. They are dead wrong in their assumption that this creative potential is optimized by the narrow pursuit of individual financial gain. They miss the essential moral dimension of community life.

David Korten

Lietaer's main point is that nature does not select for maximum efficiency but rather for an optimal balance between efficiency and resilience. If either dominates, the system is unstable. "Too much efficiency (from too little diversity and connectivity) leads to brittleness and too much resilience (from too much diversity and connectivity) leads to stagnation."[9] This concept is another way of expressing the "edge of chaos" idea discussed in Chapter 2 on complex systems science; where complex systems are free to adapt, they will seek out a balance between too much chaos and too much fixed structure. There is now a way of quantitatively measuring all the relevant components separately: total throughput, efficiency, and resilience. Lietaer says, "The underlying mathematics are well-behaved enough so that there exists only one single maximum for a given network system."[10] Researchers have found that the optimum state has roughly twice as much resilience as efficiency.

Until recently, total throughput and efficiency have been the only means for us to identify the relative success of a system. GDP measures size, and many economists promote endless growth in terms of size because they assume growth is a sufficient measure of health. Lietaer says this is a poor measure of sustainable viability because it ignores network structure. This measure cannot "distinguish between a resilient economy and a bubble that is doomed to burst; or between healthy development, … or explosive growth in monetary exchanges simply due to runaway speculation."[11] Economics, like ecosystems, has its most critical parameters within a very specific and narrow range that can be computed empirically with precision. Lietaer calls these parameters the "Window of Viability."[12]

These findings arise from the very structure of a complex network system, and therefore they remain valid for any complex network with a similar structure, regardless of what is being processed in the system: "It can be biomass

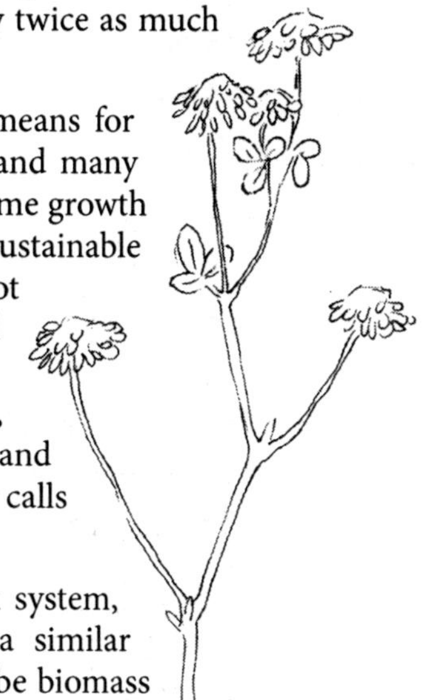

Clover Pods

in an ecosystem, information in a biological system, electrons in an electrical power distribution network, or money in an economic system. This is precisely one of the strong points of using a web-like network approach instead of machine-like metaphor."[13] The fields of engineering, business, and economics are all focused almost exclusively on efficiency, which results in an overly efficient and dangerously brittle or unstable system. Placing too much emphasis on efficiency automatically tends to increase size and consolidation at the expense of diversity, connectivity, and resilience until the entire system becomes unstable and collapses.

Markets Must Have Accurate Prices: Internalizing Externalities

> Full-cost pricing of energy, materials, and land use could expose the real inefficiencies of factory farming, conventional construction, and urban sprawl.
>
> David Korten

In a market economy, goods compete for buyers, who make choices based on price, but these prices are often distorted or inaccurate. Only when prices are accurate are we actually using a market that satisfies even a narrow definition of "market efficiency." To create accurate prices, ecologists highlight the difference between *price* and *cost*. Paul Hawken says, "We know the price of everything, but the cost of nothing. Price is what the buyer pays. Cost is what society pays."[14] Companies try to "externalize" as many costs as possible so that they can lower the price of goods and therefore be more competitive. These externalized costs are costs the corporation can get society to pay. The economic theory of a free market is based on the absence of externalities. If these externalities are minor, it does not have a great impact, but now what is being externalized and ignored is the capacity of Earth to support life. The price of goods must include all the costs of mining, manufacturing, transporting, and selling goods, as well as the costs of breaking down the waste of discarded goods.

The most effective way to create true pricing is with a tax on goods that are harmful and inaccurately priced. Using taxes is a systemic response that uses the feedback of the higher prices to affect what people buy. We need to understand taxes as information deliberately given to the market. Currently, the only information taxes give is that they were politically possible. c are particularly effective in supporting accurate costs because they

- have built-in enforcement (people pay the tax when they buy the goods);
- are systemic (they shift the entire economy);
- use market mechanisms (price) so that the effects permeate the economy;
- are efficient; and
- encourage innovative responses, because companies will want to create goods that are not destructive or toxic and therefore not taxed.

Ecologists want the consumer to pay the full cost of the goods they buy, rather than the poor?

have the taxpayer subsidize the price. The price of goods will be higher, but we could lower employment taxes and lower income taxes to offset the higher prices. This can also help create a tax code that makes sense. People keep more of their income, but they have to pay the full costs of their choices as consumers. This response is called tax shifting because ecologists want to shift taxes from income to goods.

We need to tax what we want less of (pollution) and not tax what we want more of (jobs). These taxes are called "revenue neutral" because ecologists want the government to collect the same total amount of money but change what is and is not taxed. In particular, ecologists want to reduce the most regressive tax—employment taxes (Social Security, Medicare, and unemployment insurance)—and substitute a tax on the use of scarce resources and on the creation of wastes.

So far, most tax shifting has occurred in Europe. Europeans like it because it creates jobs, since environmentally sustainable choices are generally more labor-intensive. They also like that it means shifting the tax burden from income tax to a tax on energy and raw materials. The Netherlands uses taxes to reduce the release of heavy metals, including cadmium, copper, lead, mercury, and zinc. Between 1976 and the mid-1990s, these taxes reduced the industrial discharge of these various elements by 86% to 97% each. In addition, Dutch firms that developed the pollution control equipment used to achieve these reductions gained an advantage in the industry.[15]

The way out is literally unthinkable within the iron reasoning of the corporate machine. The first step is a new way of thinking—a new consciousness or worldview that enables us to recognize how everything is interdependent and how connection to a larger purpose is critical for personal and professional success.

Elizabeth Debold
The Business of Saving the World

Taxes encourage innovation and new approaches. With taxes, there is a built-in incentive to change behavior because as an industry reduces the pollution it causes, the tax on its products will decrease.

The new taxes in Europe also include taxes on sulfur emissions, coal mining, land filling, electricity sales, and car ownership. Other possibilities include taxes on excessive water use, the conversion of cropland to nonfarm uses, tree cutting, pesticide use, and the use of cyanide in gold mining, that is, any taxes on pollution, resource depletion, or degrading ecosystems. Over time, taxes on environmentally destructive activities could increase substantially. In fact, Lester Brown says, they might one day account for most of our tax collection, replacing much of our current income and employment taxes.[16]

Chinstrap Penguin Family

Using the Market to Reduce CO_2 Emissions

Reducing greenhouse gas emissions is a special case for markets. With the catastrophic oil spill in the Gulf of Mexico, we now need to acknowledge that our addiction to oil is immoral. The alternatives are discussed in Chapter 12.

Global warming is caused by a market failure, and to solve it we need to create a genuine market for energy. Øystein Dahle, former Vice President of Exxon for Norway and the North Sea, said, "Socialism collapsed because it did not allow the market to tell the economic truth. Capitalism may collapse because it does not allow the market to tell the ecological truth."[17]

Robert Kennedy Jr. also says the climate crisis is the result of "a distorted market run amok."[18] He says it is a myth that oil and coal are cheaper and more efficient than alternative fuels. The oil industry receives $17 billion a year in direct subsidies, but much greater subsidies are hidden. He says:

> According to Terry Tamminen, former director of the California EPA, the true costs of our oil dependence run as high as $807 billion a year. If all the hidden costs that Americans currently pay for oil were reflected in the price at the pump, gasoline would cost more than $13 a gallon. In short, taxpayers and consumers are essentially giving the oil industry a subsidy of $10 for every gallon of gas sold in America. If we simply eliminated those subsidies and created a truly free market, renewable sources of energy would beat oil—as well as nuclear power and coal, which receive equally grotesque subsidies.[19]

Kennedy says that even without the oil spill in the Gulf of Mexico, oil pollution causes at least $4.6 billion in damages each year to crops, forests, rivers, buildings, and monuments. It costs $54.7 billion for medical treatment for the illnesses caused by oil pollution. We also spend $100 billion each year for the military to defend the oil industry's infrastructure around the world, by maintaining bases in the Middle East and providing military escorts for oil tankers travelling to America. In addition, the $3 trillion war in Iraq is energy related.[20] Alan Greenspan, former head of the Federal Reserve, says in his memoir, *The Age of Turbulence: Adventures in a New World*, "I am saddened that it is politically inconvenient to acknowledge what everyone knows: the Iraq war is largely about oil."[21]

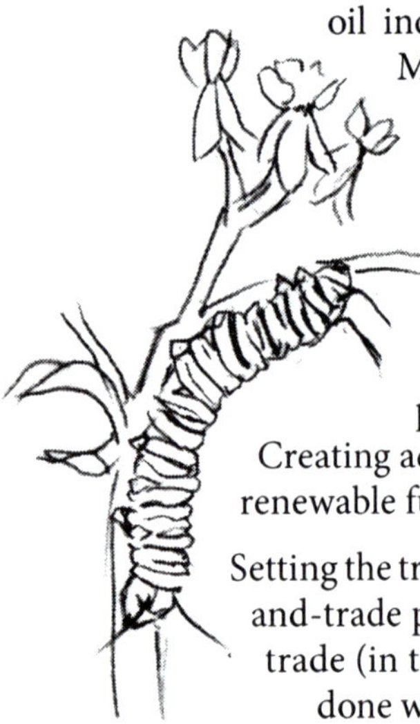

Monarch Caterpillar

The price of coal, oil, and gas must change in order to eliminate these price distortions and support the development of renewable energy. Creating accurate prices could create a flood of investments in energy efficiency and renewable fuels like wind and solar.

Setting the true cost of oil can be done by 1) increasing the price of emissions or 2) a cap-and-trade program for emissions. The method promoted in Washington is cap-and-trade (in two bills: Waxman-Markey and Boxer-Kerry). Currently, nothing is being done with either method. But how the methods work is important to understand because instead of phasing in these costs, we could be blindsided by peak oil, which would have the same effect on prices but would devastate the economy.

Cap-and-trade involves setting an overall cap on total emissions and selling trading permits to pollute up to the amount of the cap. Over time, the emissions cap is lowered in order to continually reduce our emissions. This increases the cost of energy so that people use less. The increased cost of permits that companies must buy encourages them to move investments to renewables. In the Waxman-Markey bill, 85% of the permits would be given away for free rather than auctioned off to raise money that could be rebated to families to compensate for higher energy costs. The media rarely mentions a possible rebate to families because companies want the permits to be given to them for free.

The biggest problem with this bill is that the cap is too high. But if we are serious, the cap could be reduced over time, forcing companies to find ways to reduce their pollution and encouraging innovation. The cap has mandatory targets (the amount by which emissions must be reduced) and timetables. Once the framework is put in place, the targets and timetables can become increasingly restrictive. The second problem with the bill is the offsets allowed, or rip-offsets, as Joseph Romm calls them.

> The global economy is now so large that society can no longer safely pretend it operates within a limitless ecosystem. Developing an economy that can be sustained within the finite biosphere requires new ways of thinking.
>
> Herman Daly
> Economics In a Full World
> *Scientific American*

Instead of cap-and-trade, James Hansen wants a solution he calls "fee-and-dividend," which would apply a carbon tax at the source (the mine, wellhead, or port of entry) with an amount that rises continually. This would make anything made with petroleum more expensive than goods made without it. Most ecologists want the money raised to reduce payroll taxes. However, Hansen does not want to do this because about one-half of the population is not on payrolls, either because of retirement or involuntary unemployment. He suggests that at most 50% of the collected carbon fee should be used for payroll tax deduction.[22]

Hansen calls his idea a fee instead of a tax because he wants all the revenues to be returned to the public on an equal, per capita basis, so those with below-average carbon footprints come out ahead, since the increased cost of electricity would affect them less. In this fee-and-dividend approach, people maintaining a carbon footprint smaller than average will receive more in the dividend than they pay for increased energy costs.[23]

Hansen says, "By the time the fee reached the equivalent of \$1/gallon of gasoline (\$115/ton of CO_2) the rebate in the United States would be \$2000-3000 per adult or \$6000-9000 for a family with two children."[24]

In December 2009, Sen. Susan Collins (R-Maine) and

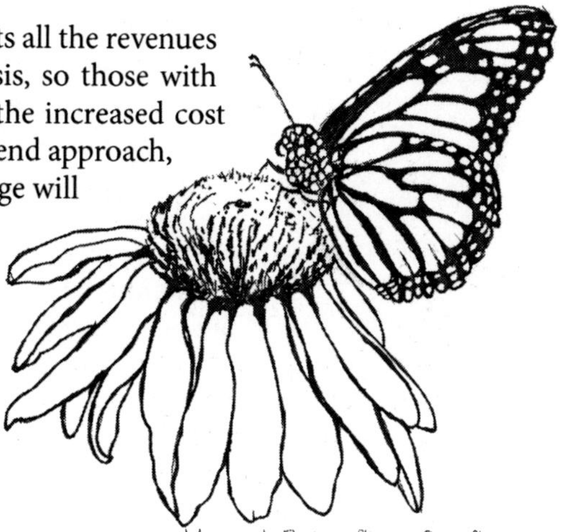

Monarch Butterfly on Coneflower

Sen. Maria Cantwell (D-Wash.) introduced a bill that would replace cap-and-trade with a system that would set a price on CO_2 emissions and return all the revenue to consumers instead of industry.[25]

These fees would affect all activities that use fossil fuels, directly or indirectly. It would also make petroleum-based chemicals more expensive. These are the chemicals that are polluting our land, air, and water and are a major source of illness and birth defects. Any chemical that is made from petroleum can also be made from plants, and these would not be toxic (see Chapter 14). The tax would also increase the price of pesticides and fertilizers, allowing organic farming to compete equally with industrial farming.

> The global economy is now so large that society can no longer safely pretend it operates within a limitless ecosystem. Developing an economy that can be sustained within the finite biosphere requires new ways of thinking.
>
> Herman Daly
> Economics In a Full World
> *Scientific American*

Markets and Peak Oil

The world economy could soon hit a brick wall called peak oil. Peak oil means that demand has outrun supply and that global production will begin to decline.[26] Jeff Rubin is an energy expert and was the chief economist at CIBC World Markets for 20 years. In his book, *Why Your World is about to Get a Whole Lot Smaller: Oil and the End of Globalization,* he explains the impact of peak oil. Its impact will depend on whether we act before it hits, or whether, like with the financial crisis, we will be blindsided.

The problem is that markets are unable to react in the present to a future event. In addition, markets can react to relative scarcity, but not to absolute scarcity. Economic theory assumes that demand will match supply, for as demand increases, prices go up and supply increases. The theory assumes that natural resources are limitless, but now everyone is pumping as much oil as they can and supply is decreasing, even as prices are increasing. Rubin says this problem of supply and demand occurs because

- Although the energy input per unit of GDP has been cut in half, total energy consumption is up by more than 40%.[27]
- Rubin says, "Depletion from existing fields has accelerated to a rate that now takes roughly 4 million barrels per day from world production every year. . . . It's become virtually impossible to both replace and at the same time add to new global production." Therefore, "over the next five years we will need to replace nearly 20 million barrels per day of cheaply produced conventional oil with extremely problematic and high-cost unconventional supply—because that is the only new supply that higher oil prices can raise any more." Only the hard-to-retrieve oil is left.[28]
- One source that is now being exploited is Canadian oil sands, where the oil is very difficult to extract. Each barrel of oil is soaked in 2 tons of sand.[29] Every single barrel of oil from the oil sands pollutes 220 gallons of fresh water and emits more than 220 pounds of CO_2 to the atmosphere. To get this oil means cutting down the boreal

forests and adding toxins to the rivers.[30] A single barrel of oil requires moving four tons of earth.[31] Retrieving it creates a devastated landscape.[32]

- The real measure of available oil will be energy, not finances. Rubin says, "The energy rate of return measures the energy you get back from the energy you expend." It is the easy-to-retrieve oil that is gone. In a conventional oil well for every 1 unit of energy used to extract the oil, 100 units of energy are gained. The Canadian oil sand returns only 3 units of energy for every 1 unit of energy expended for recovery.[33]
- With the oil spill in the Gulf of Mexico and the impact of stronger hurricanes, supply there will decline. Underwater fields, especially deep water fields, deplete twice as fast as conventional oil fields on land.[34]
- The more that technical advances increase the rate at which we can retrieve oil, the faster depletion occurs.
- Rubin says Saudi Arabia is "cannibalizing" their oil. Gasoline there costs 45 to 50 cents. In Venezuela it is 25 cents. They each feel that their citizens have a right to consume as much of this massively subsidized oil as they want, and it is politically impossible to change it. In both China and India domestic oil prices are also held below the world price. Rubin says that as much as 25% of the world's oil supply is sold to over half of the world's population at well below market prices.[35]

We can either spend enormous amounts of money to subsidize finding new oil or we can put that money into developing alternatives to oil. In addition to the problems of finding oil, there are the problems created when we use it. We have to find a way to drastically reduce emissions. The fundamental problem is that we can't use the oil we have left. As James Hansen says, "If we burn even half of Earth's remaining fossil fuels we will destroy the planet as humanity knows it. The added emissions of heat-trapping carbon dioxide will set our Earth irreversibly onto a course toward an ice-free state, a course that will initiate a chain reaction of irreversible and catastrophic climate changes."[30]

Using Markets Instead of the Kyoto Protocol

It seems clear that a political solution to global warming may just be too difficult. Instead, we should use economics. In August 2008 oil prices peaked at an all-time high of $147 per barrel, and in the U.S. people drove 15 billion fewer miles.

We must cut emissions even more and do it immediately, and the best way to do that is to create an economic advantage for the U.S. In doing so, the U.S. could use its economy to help save the world rather than to dominate it. High oil prices will not only reduce our use of oil, it will also end globalization, because distance will cost too much money. The costs of shipping will cancel out the advantages of low wages in China. This shift will radically redefine both economic geography and global trade patterns.

Wood Stork

In the decade from 2000–2010, emissions grew four times faster than in the 1990s, and 90% of that came from the developing world.[37] In China, higher oil prices will increase the demand for coal. China's emissions have doubled since 2000. Two-thirds of China's energy comes from coal, and they plan 500 new coal-fired plants by 2012. Industry in China, uses four times as much energy per unit of GDP than it does in the U.S., yet the average person there uses one-tenth as much energy as the average person in North America.[38] Earth will not survive if the Chinese begin to use energy at the same rate Americans do.

> Higher oil prices are inevitable, super-price spikes are a strong possibility, and even shortages are not out of the question. We must ask ourselves: should we manage the decline in a way that avoids economic catastrophe or do we continue our generally laissez faire attitude toward this major problem?
>
> Tam Hunt
> The Peak Oil Catastrophe-in-Waiting

The challenge now is to reduce emissions in the developing world. But for the Chinese to agree to this, Americans will have to use much less energy per capita. We can't ask China to use less when they already use one-tenth of what Americans do.

The developed world has agreed to give money to developing countries to help them shift to renewable energy. We should also give them related technology for free. But the biggest problems are China and India, and both want to export to the U.S. Rubin says the use of a carbon tax, a cap-and-trade system, along with higher prices, will move our entire economy to increased innovation and new renewable technologies and industries. This will reduce our emissions. He says the next step, both to create a level playing field and also to get the developing world to reduce their emissions, is to create a *carbon tariff*. He says, "What we need to do is to impose a carbon cost on emitters at home, then impose the same standards on imports. . . .The greater the emissions that have gone into producing whatever good is being imported, the higher the tariff on or importing those goods."[39] This, he says, will bring a lot of jobs back home. Therefore:

> Congress needs to realize that the U.S. economy actually has a comparative advantage in carbon emissions and should attempt to galvanize that advantage by putting an economic price on them. ...The higher the price of carbon emissions, the more American-made goods will replace overseas imports that are based on cheap labor and dirty carbon practices. A carbon tariff will insure that China's exports will receive no commercial benefit from emitting greenhouse gases on world export markets, and that together with the impact of soaring oil prices on transatlantic transport costs could soon change the face of the world economy.[40]

This means that the more emissions cost, the greater the cost advantage from emitting less. Oil will be expensive to buy and expensive to burn, so that international competitiveness will now include energy efficiency, carbon efficiency, and transport costs.[41] This will soon bring a lot of long-lost jobs back home. So, Rubin says, "Life as we've known it is up for grabs in a world of expensive fossil fuels."[42]

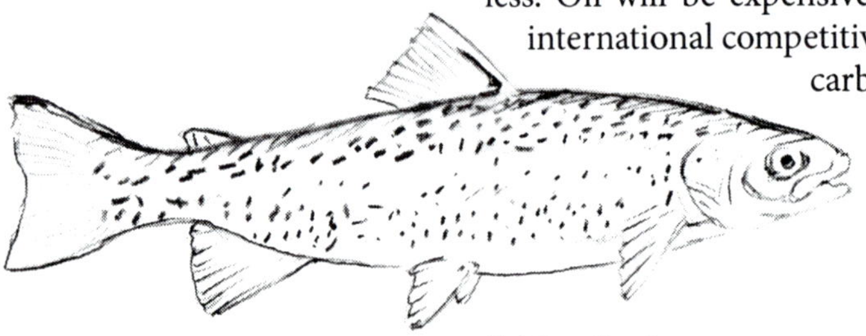

Rainbow Trout

information/abundance

Knowledge as the "Means of Production"

Our entire understanding of economics is changing, and much of this shift is driven by the need to preserve Earth and by the recent financial crisis, which has discredited our current economics and forced us to search for an alternative. But it is also driven by the shift from capital to knowledge as the primary means of production.

Bernard Lietaer, in *The Future of Money*, says the origin of the left-right political debate "had to do with private or public ownership of the 'means of production,' i.e. the factories and machines. However, now the foremost requirement of the 'means of production' is not capital—getting enough money together to produce something. The main thing needed now is knowledge, yet a new political and economic vocabulary to deal with this shift doesn't yet exist."[43]

> Running a for-profit company that's beneficial to humans and the community is not just possible: it's normal.
>
> Thom Hartmann
> *Unequal Protection: The Rise of Corporate Dominance and the Theft of Human Rights*

Herman Daly, a former World Bank economist and the intellectual pioneer of the field of ecological economics, has emphasized that the logical direction for sensible economic development would include *increasing* restrictions on the movement of goods and *decreasing* restrictions on the flow of information. Long-distance transportation and distribution are wasteful and costly and are possible only because subsidies make fossil fuels cheap. Freer movement of information would reduce wasteful duplication of energy and materials. Yet globalization is going in precisely the opposite direction—toward decreasing restrictions on goods (through free trade) and increasing restrictions (through intellectual property rights, etc.) on information. Daly explains that:

> the U.S. imports Danish butter cookies and Denmark imports U.S. butter cookies. Although the gains from trading such similar commodities cannot be great, trade theoriests insist that the welfare of cookie connoisseurs is increased by expanding the range of consumer choice to the limit. Perhaps, but one wonders whether those gains might be realized more cheaply by simply trading recipes? Although one would think so, recipes—trade-related intellectual property rights—are the one thing that free traders really want to protect.[44]

In the past, economic systems have tried to control some key resource that was *scarce*: land, energy, or minerals. Now that *information* is the key resource, this shift will create a very different society, for information follows the rules of abundance. Lietaer summarizes the fundamental differences between information and former key resources.

- *Information is shared, not exchanged.* If I give you information I have, we both have it. What is being bought or sold and then owned is not the information; what is bought is the delivery mechanism. When you buy software, you are not stopping others from using it. He says information is what economists call a "non-rival" product.

Cockroach

- *The most powerful catalyst of transformation is the communications revolution.* In systems theory, increasing the *connections* within a system increases power. Telecommunications has made information transportable through electronic networks, at almost the speed of light, and at a low cost.
- *The basic nature of information is that it tends to leak.* The more it leaks, the more of it we have, and the more of us have it. Government classifications, trade secrecy, intellectual property rights, and confidentiality are all attempts at artificially reducing this natural tendency to leak. Increasingly, these artificial attempts are failing because the actual information cannot be "owned."
- *Information spontaneously tends toward abundance, not scarcity.* Information expands as it is used. (This of course, creates the drawback of information overload.) What remains scarce and competitive is human attention and our ability to understand and use all the information available to us.
- *Information creates conditions for what economists call perfect competition.* Everyone has all information needed, there are zero transaction costs, and there are no barriers to entry for new suppliers. Now the Internet makes transaction costs lower than ever.
- *Information economics turns traditional economic theory upside down.* As means of production, information and knowledge are not subject to the laws of diminishing returns. This creates a law of increasing return. As information becomes more available, it also becomes more valuable. Every new fax machine increases the value of your fax machine, which is the reversal of scarcity determining value.[45]

> Capitalism [is] a word that has gone largely out of fashion. The approved reference now is to the market system. This shift minimizes – indeed deletes – the role of wealth in the economic and social system . . . . It would be hard to think of a change in terminology more in the interest of those to whom money accords power. They have now a functional anonymity.
>
> John Kenneth Galbraith, Economist

The Collaborative Economics of Wikinomics

Because of the accessibility of open information, the classical economic idea that another's gain means your loss is replaced by the idea that enhancing the well-being of others amplifies one's own well-being. The perceived need for secrecy is being replaced with the idea of transparency and openness. Sharing data openly optimizes collaboration and creates additional value for everyone in the network. A venture like Linux, the open-source software, for example, only works when the software, code, and new applications are openly shared among everyone in the network.[46]

Millions of computers are now being connected to collect and analyze data for a range of projects, including the search for new protein structures, finding gravitational waves, studying nanosystems, searching for prime numbers, and developing new drugs. This new distributed computing technology would allow millions of people to aggregate their knowledge and creativity for collaborative learning ventures.[47] These distributed computer projects are rapidly increasing.

In *The Empathic Civilization*, Jeremy Rifkin says "wikinomics" is the new term for the collaborative model of collecting data, sharing knowledge, and problem-solving. Economic activity is no longer an adversarial contest between sellers and buyers, but rather a collaborative enterprise between people with the same interests.[49]

He says, "A 'wiki workplace' refers to a collaborative venture involving scores, hundreds, and even thousands of people—some experts and others amateurs—usually across many different fields, who come together to share their ideas and problem solve. These new flat collaborative learning environments mobilize the collective wisdom of crowds, and their track record is impressive when compared to traditional hierarchically organized corporate learning environments."[50]

Rifkin gives the example of Cisco's CEO, John Chambers, who has transformed one of the world's leading high-tech IT companies into a collaborative work environment. Mike Mitchell, the director of technology communications, says that at Cisco, "We want a culture where it is unacceptable not to share what you know." The company encourages social networking among all of its employees and promotes internal blogs so that everyone can share thoughts and ideas across departments and job descriptions.[51]

In another example, some biotech companies were worried that companies like Monsanto might lock up patents on all of the genes that form the world's store of germ plasm for growing crops. So biologists at GAMBIA, a biotech research institute, have released their own genetic discoveries publicly under an open-source licensing agent called BiOS, which is similar to the kind of licensing agent used by Linux and other open-source software ventures. Although to a more limited degree, some global pharmaceutical companies are beginning put some of their research online in publicly available databases to encourage more collaborative research. Although these efforts are tentative, Rifkin also says this collaboration allows for continuous breakthroughs and new discoveries, making patents increasingly outmoded mechanisms for creating added value.[52]

From Property Rights to Access Rights

This new "distributed capitalism" model has very different ideas about intellectual property. Rather than seeing patents and copyrights as sacrosanct, in the collaborative economy, open-sourcing of critical information becomes essential to collaboration.[53] Product life cycles often run their course in a few years or in a few months, so there is little value in holding a patent on a process or product for 20 years. Copyright laws are also running up against file sharing, blogging, and open-source collaborative ventures, where volumes of information are routinely made free. So, Rifkin says, holding on to copyrights means restricting, not expanding, one's commercial outreach.[54] On the Internet, millions of people can connect, collaborate, and create value together in real time.

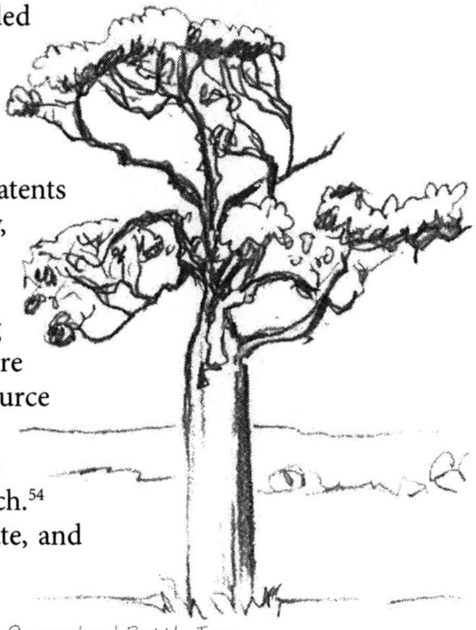

Queensland Bottle Tree

Therefore, in this new collaborative economy, the right of *access* and *inclusion* becomes more important in establishing economic and social relationships than the right of *autonomy* and *exclusion*. This distributed model begins with a very different assumption about human nature; it assumes that it is human nature to want to collaborate with others, for the sheer joy of contributing to the common good.[55] Young people today have a nonhierarchical, networking way of relating to each other and the world, which gives them a greater sensitivity to human diversity. So, Rifkin says. a distributed, collaborative, nonhierarchical society can't help but be a more empathic one.[56]

Ecological Economics Would Limit Globalization

Two more changes define ecological economics: it limits both globalization and economic growth. Our current economics equates mobility with progress—the longer the distance traveled, the more modern the economy. Globalization drains power away from the nation-state and allows international corporations to make the decisions that national governments used to make. Corporate globalization is a contemporary manifestation of colonialism and Empire.

Governments now allow mergers and acquisitions that give enormous power to only a few corporations. These corporations compete with other corporations for ownership of all the world's resources: the land, water, air, people, knowledge, information, and ideas. Ecological economist Herman Daly defines globalization as being "in effect the erasure of national boundaries for economic purposes."[57] He insists that we should not accept globalization as inevitable or celebrate it, because globalization has serious real world consequences. In particular, globalization has become a new form of colonialism.

Countries compete for corporate business by having the cheapest costs, by lowering wages and standards of protection for people's health and the environment. This results in a "race to the bottom," yet countries ignore the damage that "winning" a corporate contract causes.[58] Various discussions of globalization give these as its key characteristics:

- Governments now tolerate mergers and monopolies so that their corporations can become big enough and powerful enough to compete internationally.
- Globalization integrates and converts national economies (including some that had been largely self-reliant) to environmentally and socially harmful export-oriented production. This forces countries to specialize according to the dictates of competitive advantage. This limits the choices local citizens have for earning a living, increases dependence on imports from other countries, and makes all countries less self-sufficient.
- Globalization enforces what Daly calls a worldwide "muddled and self-serving doctrine of trade-related intellectual property rights." This is a direct contradiction of the Jeffersonian belief that "knowledge is the common property of mankind."[59]
- Globalization promotes hypergrowth and the unrestricted exploitation of environmental resources to fuel that growth.

- Globalization dismantles public health, social, and environmental programs that are already in place. Globalization converts public services and the global and community commons into private commodities. Public infrastructure and assets are co-opted for private use by the large, multinational corporations as they bargain with local, state, and national governments.
- Globalization promotes a homogeneous global culture.
- Globalization promotes corporate deregulation and unrestricted movement of capital across borders.
- Globalization replaces the traditional powers of nation-states and local communities with global corporate bureaucracies and extra-national governing bodies with no accountability to any citizenry.[60]

The only way communities can ensure their economic well-being is to stop chasing multinational firms with no community loyalties, and to start investing in community corporations. Prosperity follows when ownership, production, and consumption become intimately connected with place.

David Morris

Not Trade versus Protectionism but Fossil Fuel Use versus No Fossil Fuel Use

The current argument about globalization is defined as a choice between free trade or protectionism, but these choices are now irrelevant. Now, the only choice that matters is, Does it use fossil fuels or not? The central feature of globalization is the focus on export-oriented production, which maximizes the distance goods travel and dramatically increases the use of fossil fuels for transport and shipping activity. Global transport activity today is 25 times what it was 50 years ago.[61] The foundation of globalization is cheap, highly subsidized, polluting energy. Subsidies also artificially lower the costs of transporting food and other goods from the other side of the world.

Peak oil will also end globalization. Peak oil could hit the economy even harder than the financial crisis did, but the effect could be minimized by shifting to renewable energy now.

In addition to massive amounts of fossil fuel, globalization also requires a global infrastructure, and building this infrastructure brings corporations great amounts of profit. These corporations win contracts to create new airports, seaports, oil fields, pipelines for oil, rail lines, and high-speed highways. These are often built in former wilderness areas.[62]

This dramatic increase in the long-distance movement of goods includes trucking, shipping, and air transport. Economic globalization increases shipping by land, which greatly increases the use of large trucks. This requires an enormous use of fossil fuels and requires the building of vast new

Pink Salmon

networks of highways.

Ocean shipping has expanded more than tenfold since the 1950s and is expected to double over the next 10 years. Currently, shipping carries 80% of the world's international trade in goods. The shipping industry consumes more than 140 million tons of fuel annually, polluting oceans, destroying wildlife, and emitting CO_2. The fuel commonly used by ships is a mixture of diesel and low-quality oil known as "Bunker C," which is particularly polluting because of its high levels of carbon and sulfur. If the shipping industry did not use this oil, it would have to be discarded.[63]

> To integrate our lives into Earth's biosystem, we must learn to mimic life's capacity for locally rooted cooperative self-organization.
>
> David Korten

Ocean pollution from shipping has reached crisis levels. In particular, it causes an epidemic of bio-invasions of non-native species, which often out-compete native species, adding to species' extinctions. Ocean shipping also requires increased refrigeration, which contributes to ozone depletion and climate change. The use of packaging and wood pallets for cargo loading is a significant factor in the increased pressure on forests.[64]

One of the biggest causes of climate change is the huge surge in mass air travel. Air transport is even more damaging than shipping and is the fastest-growing transportation sector. Aircraft maximize the impact of emissions because they occur at high altitudes. Boeing forecasts air cargo activity to triple by 2017. Each ton of freight moved by plane uses 49 times more energy per mile than goods moved by ship.

Social Impacts of Globalization

Globalization also destroys any sense of belonging to and being responsible for a community. The mobility of corporations allows them to move factories and jobs to wherever they can make the most money. Corporations pit communities against one another in a contest to lower local taxes to influence the creation of a new factory. Worldwide, 700 million people are unemployed or underemployed, so corporations will continue to move out of the U.S. to find cheap labor. Between the 1970s and the mid-1990s, the U.S. lost more than 75 million jobs; most were replaced, but by jobs that paid much less. Globalization will cause all communities to continue to experience a decline in both the quantity and quality of jobs.

In addition to increasing unemployment, globalization also forces wages to lower levels. Herman Daly says that the trade agreement NAFTA represents an active commitment to a low-wage policy. In turn, lower wages means that returns to those who own capital in all three countries will go up.[65] Herman E. Daly says:

> We have come to speak of global competition as a major value. Are we competing for a good standard of living for most of our people? Eighty percent of the U.S. labor force is classed as non-supervisory employees. What is the value of competing to lower the incomes of eighty percent of U.S. working people? We could do many unwise things to make ourselves more internationally competitive, such as moving back to child labor. That doesn't mean we should. There are two ways to make

> products cheaper for the consumer. One is to increase efficiency. Everyone favors that. The other is to reduce environmental and employment standards. Reducing the wages paid for a given amount of productive work represents a lowering of standards, not an increase in efficiency.[66]

In the U.S., we now have jobless growth (some call it "job-loss" growth), which means that corporations are now creating economic growth in terms of profits without creating new jobs because of outsourcing and other structural changes. Growth in profit has become disconnected from growth in jobs. Eventually, this will mean that fewer and fewer Americans will be able to afford the goods these corporations are producing. However, in the logic of capitalism, that doesn't matter, because corporations will just focus on developing new markets in China and India. Therefore, we will have American corporations hiring foreign workers to create goods for foreign markets—leaving the American worker completely out of the loop, except that his or her taxes will still subsidize these corporations.[67]

> Our economic systems are ours to design.
>
> Zero Waste Alliance

Globalization, the destruction of the environment, and the destruction of communities are interconnected. Globalization creates economic dislocation and social instability, as well as environmental harm. Globalization destroys local economies because it takes local economies out of local control. Therefore, globalization is also inherently undemocratic, for it weakens the ability of local, regional, and national governments to represent the interests of their citizens. Then it substitutes for national governments the power of unelected world-level governing economic bodies (IMF, WTO, World Bank, and trade treaties) and private corporations.

The Alternative to Globalization Is Creating Local and Regional Economies

Globalization harms local economies as well as the environment. The alternative to globalization is local and regional economies, or what Jeremy Rifkin calls "distributed capitalism." Ecologists want locally based, human-scale, community-oriented, fair profit, and nonprofit enterprises. They want local people to be in control of most of the economic activity in their area. This alternative understanding of economics is a more community-centered model in which people have genuine self-determination. This keeps more money local and increases true democracy because the people who feel the impact of decisions are those who make the decisions.

Each town could define the boundaries of its own "local" area because the optimal size will be different, depending on the size of the town, the particular geography, and other factors. In this shift to local economies, the power in financial markets and global corporations shifts to communities, which encourages individual responsibility and allows for individual

Scarlet Macaw

choice and creativity.[68]

> The goal can no longer be individual or corporate wealth; rather it must be community self-reliance, public health, equity, accountability, and democracy.
>
> John Cavanagh and Jerry Mander
> *Alternatives to Economic Globalization*

In *Alternatives to Economic Globalization*, John Cavanagh and Jerry Mander contrast globalization and local economies. They say that when all systems emphasize local production and consumption, it shortens the miles used for economic activity, with fewer miles for food, energy, and work.[69] These local economies would be much more stable than global economies because they are not so complex.

Local economies are something people can create for themselves, rather than globalization, which Cavanagh and Mander say is "something done *to* people rather than *by* them."[70] Local economies would emphasize technology that serves local control, what is often called "appropriate" technologies. Decentralized economies can be much more transparent and democratically accountable. This means emphasizing main street markets, with the economic goal being "community self-reliance, public health, equity, accountability, and democracy," rather than individual or corporate wealth.[71]

A sound local economy is concerned with providing everyone with the necessities of life. It is dangerous to depend on someone outside the community selling something to us, especially if it is necessary for survival.[72] However, this type of local economy is just what the World Bank and the IMF are trying to stop. Mander and Cavanagh say, "Here's the crux of the matter: *systems that emphasize local or regional self-reliance are extremely subversive to free trade, economic globalization, and hyper corporate growth*. . . . Local and regional production created for regional consumption is the archenemy of globalization because it operates on an inherently smaller scale and there are fewer steps in the process."[73] In addition, these economies are based on shared power rather than dominance. The primary way to strengthen local economies is through cooperation rather than competition, which requires communication and cooperative planning.

A major shift in ecological economics is from centralization (in multinational corporations) to decentralization. We can strengthen and stabilize community economies best by

- Locally producing as many types of goods as possible
- Having local ownership rather than the absentee ownership of shareholders
- Creating local economies, which allows information to be shared
- Creating a local energy economy (see Chapter 12)
- Creating a local food economy (see Chapter 13)

Local Production through Import Substitution

We can create local and regional economies with "import-replacing" or "import substitution" industries. In this process, entrepreneurs can study import patterns, identify key dependencies, and then work to reduce them, creating local import-replacing industries. Buying locally greatly reduces the impact economic activity has on the environment.

There are three types of import-substitution programs:

- Information sharing and networking—matching local producers with local suppliers (as Oregon has).
- Buy local programs—encouraging firms and consumers to purchase local products rather than imports.
- Industry targeting—attracting or creating firms and businesses that will engage in or enable import substitution.

For example, the state government of Oregon surveyed all Oregon businesses and created a list of goods that these local businesses imported from out of state. The state then created a list of Oregon businesses that could supply those goods instead. They sent the list to all Oregon businesses and asked them to buy from Oregon firms. For goods not available locally, the state determined what new businesses could be created in the state and then helped the firms get started.[74]

> To put it bluntly, democracy and corporate globalization are mutually exclusive conditions; as are short-term financial values and long-term life values.
>
> David Korten

This process creates an emphasis on small businesses. Small businesses create the most jobs, generate the most patents, and have the most benefits for the economy.

Local Ownership

Ecological thinking also promotes a new kind of ownership. Rather than using the absentee ownership of stockholders, the smaller local businesses can be structured as worker-owned cooperatives or collectives in which employees maintain various degrees of shared ownership and management. This enhances economic democracy, for over time, individuals will have an ownership stake in their homes, in local land and resources, and in the businesses for which they work and from which they buy. These companies are "rooted" in the local community and so they won't just get up and leave.

Michael Schuman, in *Going Local: Creating Self-Reliant Communities in a Global Age,* says ownership can take many forms. Although a company can have employee ownership, at times, employee owners can make short-term choices for profit, just as absentee owners can. He suggests creating *community corporations* that allow only members of a community to own voting shares of stock. Shareholders could exchange or sell the stock freely, but only with other community members. Whenever a shareholder decided to move out of the community, he or she would be obligated to sell the shares, either to other community members or back to the company. Alternatively, a community could create two classes of shareholders: residential

Orangutan

shareholders with voting rights and nonresidential shareholders without them. They could also limit the percentage of the company that any one citizen in the community can own.[75]

> Economies are supposed to serve human ends—not the other way around. We forget at our peril that markets make a good servant, a bad master and a worse religion.
>
> Amory Lovins

Ted Howard, in *Yes! Magazine,* says these "innovations in ownership begin to point the way toward new possibilities for organizing our economy and stabilizing our communities. These new forms of ownership can include community development corporations, consumer and producer cooperatives, employee owned firms, municipal enterprises, for-profit subsidiaries of non-profit organizations, urban land trusts, local currency and barter systems, and community supported agriculture programs."[76]

There are many types of ownership stakes; all help to fairly distribute economic benefits, decision-making responsibility, and accountability. Ecologists are not anti-business; they are pro small and medium-sized businesses but against multinational corporations. Community businesses increase local community self-reliance, stability, and power. Korten describes the type of local economy that ecologists envision:

> Living economies are made up of human-scale enterprises locally owned by people who have a direct stake in the many impacts associated with the enterprise. A firm owned by workers, community members, customers, and/or suppliers who directly bear the consequences of its action is more likely to provide:
>
> - Employees with safe, meaningful, family-wage jobs
> - Customers with useful, safe, high-quality products
> - Suppliers with steady markets and fair dealing
> - Communities with a healthy social and natural environment.[77]

For example, instead of privatizing city services, municipalities have begun to establish city-owned and city-operated profit-making ventures in order to generate additional revenues, create jobs, and increase responsiveness to community needs and interests. Many cities like Seattle own their own utilities. Cities like San Francisco and Austin, along with rural communities like Mecklenburg County, North Carolina, run municipal programs that turn organic waste into marketable fertilizer and soil supplements. In the Washington, D.C., suburbs, Montgomery County's composting facility sold nearly $500,000 worth of its Leafgro soil supplement to nurseries, retailers, and area residents. City-owned enterprises could also create telecommunications services. Already, some community-owned telecom companies have been created.[78]

When employees, customers, local residents, and others own businesses, businesses become increasingly accountable to the needs of the communities they serve. This kind of cross-ownership is common on the Internet. Moreover, spreading ownership broadly generates greater social equity and expands the number of people with an ownership stake in assets of all kinds, including homes, businesses, land, and natural resources. There are many organizations working to develop networks of locally based businesses, such as Business Alliance for Local Living Economies (BALLE).[79]

Local Production for Local Needs through Community Banking

Traditional lending practices do not favor small loans. People without money are the people most in need of affordable credit for housing, automobiles, college tuition, and business ventures. These loans lead to economic diversity and resilient communities. Local lenders or Community Based Financial Institutions invest in the community in which they are owned and operated. These institutions include community development banks, credit unions, loan funds, venture capital funds, and micro-enterprise loan funds. Personal loans to people with low incomes allow them to change their economic standing and participate in the economic revitalization and resilience of a community. Community Based Financial Institutions also provide opportunities for locally and broadly shared ownership of homes and businesses.

Local financial institutions can serve as bridges between socially minded investors and community-based initiatives. In 1977, to encourage local investment, the government passed the Community Reinvestment Act, which mandates that U.S. banks and credit unions reinvest a portion of their deposits locally. In 1994, President Clinton created the Community Development Finance Initiative Act to expand the availability of credit, investment capital, and financial services in distressed urban and rural communities. The Act works to promote economic revitalization and community development by investing in and assisting Community Development Financial Institutions through equity investments, capital grants, loans, and technical assistance support. Community Development Financial Institutions can use this financial assistance to support many community development activities, including housing for people with low incomes, businesses owned by people with low incomes, basic financial services, commercial facilities that promote job creation or retention, and technical assistance.

Sustainability is the nascent doctrine that economic growth and development must take place, and be maintained over time, within the limits set by ecology in the broadest sense—by the interrelations of human beings and their works, the biosphere and the physical and chemical laws that govern it.

William Ruckelshaus

Local Economies Allow Information to Be Freely Shared

When economies are local or regional, there is little or no economic cost to sharing information rather than competing to hoard it. Today, our system rewards people for hoarding information and allows corporations to have a monopoly on it. David Korten says this shift to local economies will facilitate the free sharing of information, eliminating wasteful redundancies: "Under a system of relatively self-reliant local living economies, communities and nations will not find themselves pitted against one another for jobs, markets, and resources. In the absence of such competition, the free sharing of information, knowledge, and technology will become natural, to the mutual benefit of all."[80]

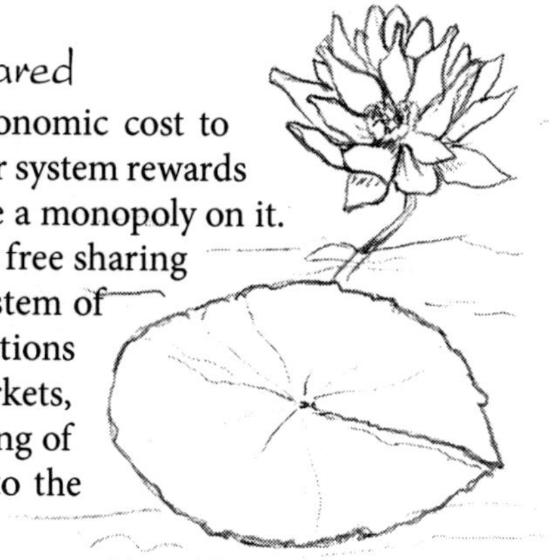

Water Lily

Although the production and sale of goods and services are local or regional, the information used can be global. Distributed capitalism means that there will be a lot of repetition. So people can create networks in which they share information with people in different regions working on the same products or services. They can combine resources, answer each other's questions, and offer support.

In general, a sustainable economy will self-organize within the framework of market rules, and it will be rooted locally and structured to be locally self-reliant; business enterprises will be human-scale and locally owned. People will balance their own needs with their responsibility for the well-being of the whole, thereby building bonds of trust and creating community cohesion and vitality.[81]

Ecological Economics Would Limit Economic Growth

In addition to creating local and regional economies rather than globalization, ecological economics would end economic growth. Yet in our public discussions, no one dares question the logic of continued economic growth. It's an idea most people won't even listen to because it is just too threatening. It is heresy to argue that economic growth must stop and that we need to create a steady-state economy instead.

> At the heart of the current crisis in economic theory and practice is the fact that we are consuming the earth's resources beyond its sustainable capacities of renewal, thus running down that capacity over time; that is, we are consuming natural capital while calling it income.
>
> Herman Daly
> *Beyond Growth*

A fundamental difference between current economics and ecological economics is the denial or acceptance of limits. Particularly in the U.S., with our history of the great frontier, we assume there is always more, that everyone has a right to more, and that any limits are fundamentally wrong and a threat to our freedoms—our right to have everything.

Often, people's first reaction to the idea of stopping economic growth is that it is just not politically possible to question continual growth. Given that problem, Herman Daly says, "Because establishing and maintaining a sustainable economy entails an enormous change of mind and heart by economists, politicians, and voters, one might well be tempted to declare that such a project would be impossible. But the alternative to a sustainable economy, [an] ever growing economy, is biophysically impossible. In choosing between tackling a political impossibility and a biophysical impossibility, I would judge the latter to be the more impossible and take my chances with the former."

We now must create an economy of permanence and accept concepts of sufficiency. However, there need be no limits to creativity, learning, leisure, or nurturing—only limits on stuff.

Herman Daly calls our current blind push for economic growth "growthmania." He then says, "When we add to GNP [Gross National Product] the costs of defending ourselves against the unwanted consequences of growth and happily count that as further growth, we then have hyper-growthmania. When we deplete geological capital and ecological life-support systems and count that depletion as net current income, we arrive at our present state of terminal hyper-growthmania."[83]

Daly has written about the steady-state economy since 1973 and says that if economic growth continues, eventually "we will then have what I call uneconomic growth, producing 'bads' faster than goods—making us poorer, not richer. Once we pass the optimal scale, growth becomes stupid in the short run and impossible to maintain in the long run. Evidence suggests that the U.S. may already have entered the uneconomic growth phase."[84]

> GDP measures the cost not the benefit of economic activity.
>
> David Korten

Ecologists argue that we have been able to ignore the obvious limitations of Earth because we have mislabeled the "whole." Currently, we define the whole world in terms of the global economy and define everything else (environment, community) as a subset of it. However, economies do not operate in a void. Ecologists define the overarching whole as the physical, natural world and define the economy as a subset of the natural world, which supports it. Therefore, once we recognize the limits of the finite biosphere, our economy must operate within these limits.

Economic Growth as a Religion

Our cultural religion is economic growth, which creates a secular salvation (as in the saying "the one who dies with the most toys, wins"). Good and evil are efficiency and inefficiency, or profit and loss; the highest virtue is pursuing self-interest, and greed becomes moral. In this religion, economists are the priests and shopping malls are the cathedrals, where people shop with a religious fervor. Economic growth is a magical belief that allows all social problems to be ignored because it promises that, eventually, economic growth will solve everything.

The economist Robert H. Nelson, in *Economics as Religion,* calls these ideas "economic theology." He says, "Economic efficiency has been the greatest source of social legitimacy in the United States for the past century, and economists have been the priesthood defending this core social value of our era."[85] Our economists emphasize certain values and ignore others, yet they call their profession "value-neutral." Nelson says all economists distinguish between costs that are valid and those that are invalid, so they automatically communicate and endorse a particular set of social values.[86] He continues, "If economists did attempt to measure in a formal way these more subjective cultural elements, many of the principle policy recommendations on which economists rest their professional identity would no longer be defensible—or at least would no longer be defensible on value-neutral grounds, requiring that economists instead explicitly advocate these policies as a matter of making a particular social-value choice."[87]

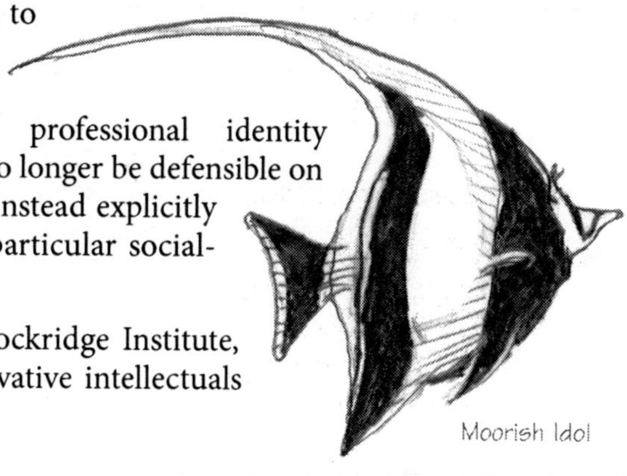

Moorish Idol

In an article in the *Nation*, Fred Block, of the Rockridge Institute, says economics as religion was created by conservative intellectuals in the 1960s and 1970s when they tried to bridge the deep incompatibility between economic conservatives and religious conservatives. They did so by creating a vision that combined the two. They defined Adam Smith's metaphor of the "invisible hand" as a natural force to which people must accommodate and treated "the market" as something like a divine force that always calls forth moral behavior. Conservatives use Smith's invisible hand metaphor to represent markets as natural and government regulations as artificial, creating a quasi-religious belief that unregulated markets will somehow always produce the best possible results. Block says the central flaw is the claim that encouraging acquisitiveness produces good results. This concept, Block says, rests on a fairy tale in which individuals pursuing their own self-interest somehow in the end serve the common good. Making changes is very difficult because the belief in free markets has become a religion and is in fact often called "market fundamentalism."

Growth Is Measured by Inaccurate National Indicators

We cannot define the stable economy we need in terms of constant GDP because GDP does not distinguish between qualitative improvement (development) and quantitative increase (growth) and because it does not distinguish between growth in good things and growth in bad things—they are all lumped together. Daly says that, in general, when we talk of economic growth, we mean growth in the GDP. But he says this is the only level of economics that is expected to grow forever, yet the marginal costs of further growth are never seen to be more than the marginal benefits. Yet in microeconomics, every enterprise has an optimal scale beyond which it should not grow.[88] We are unaware of the problems of economic growth because we don't recognize that our methods for measuring growth are flawed.

> If consumers were to become co-owners of the land and equipment, effectively shareholders, this [ownership] model could be seen as a community corporation.
>
> Michael Shuman
> *Going Local*

We measure economic growth by an increasing GDP. However, the GDP does not measure what we think it measures. It only measures money, and it counts any exchange of money as an increase in the size of the economy and therefore as "good." That is, GDP measures costs but ignores whether these costs are beneficial or destructive. Therefore, we are seeking to maximize costs rather than counting the benefits we want. For example, the attempts to clean up the oil spill in the Gulf of Mexico will add billions to the GDP, but it is hardly a sign of progress.

The government created the then-named Gross National Product (GNP) during World War II to measure the wartime production capacity in the U.S. Now, policymakers, economists, international agencies, and the media use the GDP to measure the nation's economic health and well-being, but this was never the intent. The GDP also ignores everything that happens outside the realm of monetized exchange, regardless of its importance to well-being. It ignores the crucial economic functions performed in the household and volunteer sectors, as well as the contributions of the natural habitat in providing the resources that sustain us. As a result, the GDP not only masks the breakdown of the society and natural habitat, it actually portrays this breakdown as a gain.[89]

Focusing on quantitative wealth means we focus on production for production's sake and on manufacturing and consuming as much "stuff" as possible, all of it maximized by advertising. An added problem is that these economic indicators focus only on the short term and are incapable of giving any indication of what is actually happening over the long term.

Ecologists are using new criteria, rules, and goals to design new indicators (both national and local) as feedback mechanisms. They want to measure concrete social welfare, not the abstract flow of money. In general, ecologists are developing indicators of social welfare in individual and community life, such as literacy, infant mortality, and life expectancy. They treat education as an investment rather than a cost. New indicators also measure purchasing power rather than income. The data to develop these new indicators are already widely available.

> Wall Street turned credit cards into a substitute for paying a decent wage.
>
> David Korten

Ecologists start with consumption, or what actually improves our lives, and then work backward to figure out how to get it, rather than thinking about what they can produce and how to get people to buy it. Amory Lovins, of the Rocky Mountain Institute, says this involves realizing that what we want is not power plants, what we want is "hot showers and cold beer." When you start with what you want, it becomes clear that there are many ways to get it, and the community can make choices. The wealthy can buy bodyguards, live in gated communities, and buy state-of-the-art health care. However, what they really want to purchase is safety and health, not police, prisons, or hospitals. We can measure wealth in terms of health and safety, meaningful work, and supportive relationships, rather than just in terms of money.

Sustainability indicators provide a roadmap for public policy. An effective set of indicators highlights critical social, economic, and environmental problems and provides a way to measure the effectiveness of programs designed to address these problems. Traditional indicators—such as GDP or corporate profits—reflect a faith in growth and efficiency as the primary mechanisms for improving public welfare.

The group Redefining Progress has worked on sustainability indicators for years. They say that governments have been concentrating on the wrong indicators; instead, they believe that the ecological accounting of human demand on nature should be followed as closely as the stock market. The indicators highlight a growing significance of "biocapacity." By measuring our impact on Earth, we will be better able to implement aggressive sustainability policies. The "Ecological Footprint," by Redefining Progress, is the best measure of how much of Earth's limited resources an individual (or business, city, nation, or all of humanity) uses to produce all she consumes and to process or dump all the waste she makes. They provide the scientific data to create the indicators. Today, there is a drastic overshoot in our use of Earth's resources and sinks. They say that in 1961, humans used 62% of Earth's biocapacity. In 2006, it was 144%.[90] If everyone lived in the lifestyle of

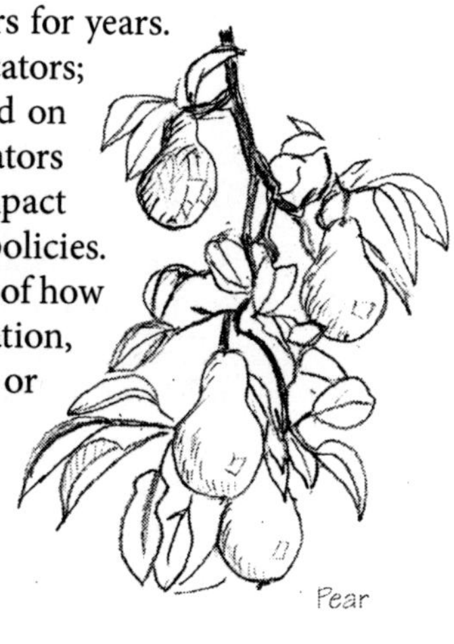

Pear

the average American, we would need five planets.

A second indicator is the Genuine Progress Indicator (GPI), which was created by Redefining Progress in 1995 as an alternative to the GDP. The GPI allows policymakers at the national, state, regional, or local level to measure how well their citizens are doing both economically and socially. It distinguishes between worthwhile growth and non-beneficial economic growth. Their latest update, which plots GPI accounts from 1950 to 2004, shows that by this measure, beneficial economic growth has been stagnant since the 1970s.[91]

The part played by orthodox economists, whose common sense has been insufficient to check their faulty logic, has been disastrous to the latest act.

J. M. Keynes, 1936

The GPI includes measurements of income distribution, so that the GPI rises when the poor receive a larger percentage of national income and falls when their share decreases. The GPI subtracts the costs arising from crime. The GPI includes the value of work done in household and community settings by counting it at the approximate cost of hiring someone to do it. The GPI also takes into account the non-market benefits associated with a more educated population. It measures the depletion of physical resources, considers it to be borrowing from future generations, and counts it as a cost. The GDP ignores any loss of free time, but the GPI treats leisure as something of value. When leisure time increases, the GPI goes up; when Americans have less of it, the GPI goes down. This indicator measures progress by how well we:

- Distribute wealth, income, and access to cultural amenities;
- Diversify and stabilize our economic base;
- Protect and restore native ecosystems; and
- Advance social, economic, and environmental sustainability.

They believe that if policymakers measure what really matters to people—health care, safety, a clean environment, and other indicators of well-being—economic policy would naturally shift toward sustainability. Redefining Progress also hosts and is a member of the Environmental Justice and Climate Change Initiative, which addresses environmental racism. Environmental problems tend to hit low-income families and communities of color first, and hit them hardest. Poorly designed environmental policies impose unnecessary social costs, further burdening these families and communities. They are currently analyzing the impact of climate change on the Latino community.[92]

A third indicator is the Happy Planet Index (HPI), which the New Economics Foundation in London produced in 2006.[93] This combines environmental impact with well-being to measure the environmental efficiency of a country. This is a composite indicator, including three separate components: ecological footprints, life satisfaction, and life expectancy. It uses carbon footprints as a measure of resource use. Their aim is a new economy based on social justice, environmental sustainability, and collective well-being. This measure tries to enhance individual and collective well-being in ways that are environmentally sustainable and socially just. Their goal is to promote the concept of well-being, or a "living better, using less" approach, as a legitimate and useful aim of policy. Like the GPI, in the HPI, the scores of developed nations plummeted from the 1960s to the late 1970s, and despite some gains, HPI scores were higher in 1961 than in 2005.

> Globalization is driven by a set of rules and self-interested institutions that can be changed—if we have a democracy.
>
> John Cavanagh and Jerry Mander
> *Alternatives to Economic Globalization*

David Korten gives his own list of what could be counted. The things we want to increase could include the percentage of food grown locally; attendance at farmers' markets; school attendance and graduation rates; voter participation rates; the number of pedestrian- and bicycle-friendly streets; the acreage of open space near urban villages; youth involvement in community service; the number of neighbors with whom people regularly interact; the percentage of locally owned businesses; or the size of wild salmon runs. On the other hand, we could keep track of those things we want to decrease: divorce rates; the number of single parents; the extent of soil erosion; incarceration rates; infant and child mortality rates; rates of hospitalization for children with asthma; the total area of impervious surfaces; and childhood obesity rates.[94]

But instead of counting what we really want, we count the amount of money that changes hands and assume that if this amount keeps growing, things are getting better. In fact, in order to make this amount grow as fast as possible, we can ignore paying a living wage, or health care, or the destruction of nature. But if instead we begin to count indicators that measure real well-being, we may make less "money" but have more real wealth.

Economic Growth Requires Overconsumption

Limiting economic growth requires a limit to consumption. Yet this transformation to sufficiency will be very difficult to bring about, for we have come to measure status by how much we can afford to consume. Economic growth requires a relentless selling of consumerism, which requires "planned obsolescence" and enormous waste. Advertising doesn't just sell products—it sells a way of life. Our only goal is the satisfaction of wants, with the implicit assumption that one's own wants are more important than someone else's needs because competition has determined who "deserves" to succeed. When we view the world hierarchically and desire status (our position along the hierarchy ladder), we not only desire to accumulate possessions for their own sake, we desire to have more than someone else, particularly if we define our own moral worth by how successful we are.

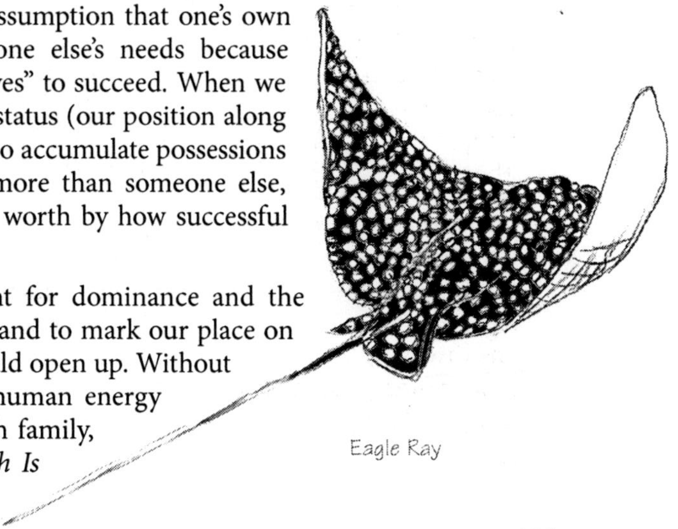
Eagle Ray

If we could let go of the constant fight for dominance and the materialism that forces us to keep score and to mark our place on the ladder, a whole new way of living could open up. Without materialism, tremendous quantities of human energy could be channeled into connecting with family, friends, and communities. In *How Much Is*

Enough? Alan Durning, of the Worldwatch Institute, says that needs are culturally defined and that consumption is addictive. Therefore, creating a sustainable environment will require that we change our values. If we do so, we could be considerably happier, for studies have shown that people are happy not because of consumption, but rather because they are satisfied with family life, especially marriage, with work, and with friendships, and because they have the leisure to develop talents.[95]

> The federal government set the rules that created the U.S. middle class following World War II. When the Wall Street-led political alliance. . . . changed the rules to favor a wealthier class, the middleclass began to disappear. When it changed the rules again in order to make finance rather than industry the dominant economic sector, industrial jobs disappeared.
>
> David Korten

Instead of focusing on more and more stuff, individuals and communities need to create political, physical, and cultural "infrastructures of well-being," but to do so, the choices we need to make must first be available. The *State of the World 2004* says, "Organic produce, reusable beverage bottles, or mass transit obviously cannot be bought if they are not offered for sale. The rules and policies that determine the set of choices available, such as oil subsidies that make fossil energy cheaper than wind power, zoning laws that encourage sprawling development, or building codes that frown on the use of recycled building materials are essentially the 'infrastructure of consumption.'"[96] To overcome these immense structural obstacles to lowering consumption levels, government action is required. This is particularly true in transportation: sprawl creates great distances separating homes, workplaces, schools, and stores, making public transit, biking, and walking difficult or impossible.

Many people are finding that their lifestyle in the U.S. is causing them health problems, enormous stress, and little time for family and friends. Some are "downsizing" or choosing "simple living" in order to improve the quality of life. These people are redefining prosperity to focus on a higher quality of life, rather than just on the accumulation of goods, and have accepted that "enough is plenty." The financial crisis may drastically reduce our consumption rates out of necessity rather than choice.

The Alternative to Growth: A Steady-State Economy

The alternative to continual growth is a no-growth, steady-state economy. Again, the problem is simple: as the economy grows, it gets bigger (uses more resources and creates more waste), but Earth will not get any bigger. Common sense should allow people to realize that continual growth at some point becomes a cancer that will devour its host. But capitalism is incapable of setting an absolute limit on scale (the size of the economy relative to the ecosystem); therefore, there are enormous forces of denial against acknowledging any limits to growth. Some economists use the term "sustainable growth," which is a contradiction in terms: there is no such thing as sustainable growth, any more than there is sustainable cancer. And socialism is not an alternative. Socialism is just as focused on continual economic growth as capitalism is.

Economics in a Full World

Growth means increasing the production and consumption of goods and services. Ecological economics focuses on limiting total consumption by creating a steady-state economy because continual economic growth will drive biological systems to catastrophic collapse. For an excellent summary of the problems with economic growth, see Herman E. Daly's *Scientific American* article "Economics in a Full World," available online.[97] He explains, "As the world becomes full of us and our stuff, it becomes empty of what was here before. To deal with this new pattern of scarcity, scientists need to develop a 'full world' economics to replace our traditional 'empty world' economics."[98] The physical limits of the planet are real and non-negotiable: *the planet is as big as it is going to get.* Because neither the resources nor the "sinks" for wastes are infinite, we have to put limits on the creation of physical goods (that is, on the economy). Paul Hawken puts it succinctly: "Physics trumps economics every time."

The economist Kenneth Boulding has been arguing for a steady-state economics since 1949. In his 1966 essay *The Economics of the Coming Spaceship Earth,* he uses the terms "cowboy economics" and "spaceship economics" instead of Daly's "full" and "empty." His term "cowboy" is symbolic of the open frontier and reckless, exploitative behavior. But he says we should think in terms of a spaceship that would have a closed economy, limited resources, and limited sinks and therefore must have a cyclical ecological system.[99]

The entire concept of a steady-state economy is often misunderstood. *Fortune* magazine equated it to a failed-growth economy.[100] Herman Daly says that nongrowth can come about either because of the failure of a growth economy or because of the success of a steady-state economy (SSE). The failure of a growth economy to grow brings unemployment and suffering. Daly says he advocates an SSE precisely in order to avoid the suffering of a failed-growth economy: "A growth economy and an SSE are as different as an airplane and a helicopter. Growthmania reigns supreme when even the failures of a growth economy become arguments in its defense!"[101] *Fortune* was referring to 1973–1975 recession. Neither economic growth nor economic recession is sustainable.[102]

In summary, Daly gives four policy recommendations for a sustainable economy:

- Stop counting the consumption of natural capital as income.
- Shift the tax base from value added (labor and capital income) to resource throughput (that to which value is added [Such as a carbon tax]).The regressivity of such a consumption tax (the poor would pay a higher percentage of their income than the wealthy would) could be offset by spending the proceeds progressively (that is, focus on aiding the poor), by instituting a tax on luxury items, or by retaining a tax on high incomes.[103]
- Maximize the productivity of natural capital in the short run and invest in increasing its supply in the long run.

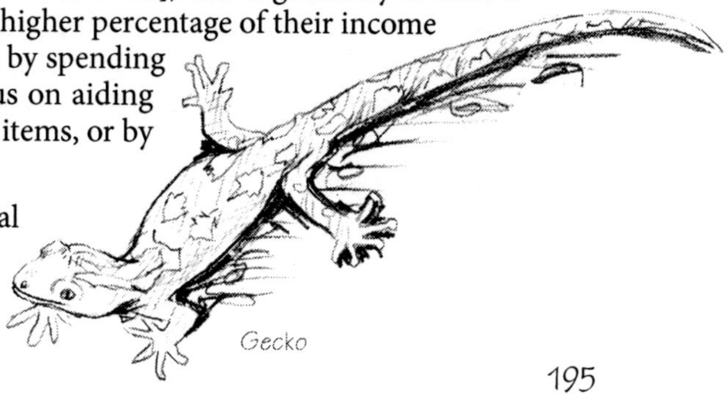
Gecko

- Move away from the ideology of global economic integration by free trade, free capital mobility, and export-led growth—and toward a more nationalist orientation that seeks to develop domestic production for internal markets as the first option, having recourse to international trade only when clearly much more efficient.[104]

Size of the Economy

Moreover, not only does the economy need to stop growing, the optimal size must be smaller than it is now. A sustainable U.S. economy would need to be far smaller than our present one in order to be sustainable indefinitely. We can only achieve a smaller economy by reducing the numbers of people and reducing per capita consumption and pollution. Therefore, a stationary population is a part of a steady-state economy. Population growth, like economic growth, cannot continue forever.

In practice, a policy of maximizing GNP is equivalent to a policy of maximizing depletion and pollution. Herman Daly says that in a stationary state there is a constant stock of physical wealth (capital) and a constant stock of people (population). Therefore, the rate of inflow (birth, production) must equal the rate of outflow (death, consumption). But in addition, the rate of throughput should be as low as possible, that is, the overall size of the economy needs to be smaller than it is now. The scale at which the economy must be maintained must be set by ecology and the ecological thresholds that must not be crossed. Daly says that as these thresholds are approached, 'marginal costs' and 'marginal benefits' become meaningless, particularly since as we reach thresholds, "the 'marginal' cost of one more step may be to fall into the precipice."[105]

> The question is not how much these changes will cost—but how much it will cost if we don't change.
>
> Lester Brown

Low production and low consumption require greater durability of goods and less time sacrificed to production. Chapter 14 addresses the many ways in which goods can be created with fewer materials. However, the limits imposed by increased pollution are now affecting Earth more than the depletion of resources. But this limit is ignored because pollution costs are social, while depletion costs are usually private. Therefore, pollution, like greenhouse gases, exists outside our economic indicators.[106]

Growth Will Be Growth in Leisure

The economy must stop growing, but it does not need to stop developing. In *Beyond Growth: The Economics of Sustainable Development*, Herman Daly says that a sustainable economy must replace "the economic norm of quantitative expansion (growth) with that of qualitative improvements."[107] With total physical goods kept constant, any growth must be in non-physical goods—particularly in leisure and knowledge. In the past, we have taken the benefits of technological progress and increased productivity in the form of more goods rather than more time. We buy more and more stuff and lose the activities that are time-intensive, like friendships, time with family, time to volunteer, and time to meditate and reflect: no one has time to think any more or to join groups.

In the 1930s, Bertrand Russell addressed this choice and proposed a policy of leisure growth rather than commodity growth. He viewed the unemployment question in terms of the distribution of leisure. The following is from his essay "In Praise of Idleness:"

> Suppose that, at a given moment, a certain number of people are engaged in the manufacture of pins. They make as many pins as the world needs, working (say) eight hours a day. Someone makes an invention by which the same number of men can make twice as many pins as before. But the world does not need twice as many pins. Pins are already so cheap that hardly any more will be bought at a lower price. In a sensible world, everybody concerned in the manufacture of pins would take to working four hours instead of eight and everything else would go on as before. But in the actual world this would be thought demoralizing. The men still work eight hours, there are too many pins, some employers go bankrupt, and half the men previously concerned in making pins are thrown out of work. There is in the end just as much leisure as on the other plan, but half the men are totally idle while half are still overworked. In this way it is insured that the unavoidable leisure shall cause misery all round instead of being a universal source of happiness. Can anything more insane be imagined?[108]

The old economy of greed and dominion is dying. A new economy of life and partnership is struggling to be born. The outcome is ours to choose.

David Korten

We have become goods-rich and time-poor.

Neither Technology nor Services Will Allow Continuing Growth

Mainstream economics argues that technology will help reduce the impacts of growth—but it won't if we just use technology to allow for more economic growth. In addition, Daly points out that we assume technology will only solve problems but not create any. We need to distinguish between technology that is part of the solution and technology that is part of the problem, and then we have to figure out how to choose the helpful technology. Currently, profit chooses. In any case, our total dependence on the natural world will never diminish, regardless of technology.[109]

Growth in services is also not a substitute. Advocates of growth frequently say the services can continue to grow indefinitely, since services presumably do not deplete resources or pollute. Daly says there are two fatal flaws in this picture. Some activities use fewer resources than others, but these are not always services, once indirect effects are included. Therefore, when all the indirect as well as the direct aspects of service activities (inputs to service sector, inputs to inputs of service sector, etc.) are added up, it turns out that services do not pollute or deplete significantly less than many industrial activities. Most services also require a substantial physical base, as can be seen, he says, in a university, a hospital, an insurance company, a barbershop, or even a symphony orchestra. Moreover, people in the service sector will spend their incomes, at least in part, on goods.

Jabiru

Our current economics also ignores the needs of future people, who of course have no way to influence today's choices that will impact their lives.[110] The Iroquois say that all decisions must consider the impact on the next seven generations.

Ecological Economics Is an Issue of Distribution and Therefore of Morality and Ethics

The important issue of the stationary state will be distribution, not production. The economic and social implications of the stationary state are enormous and revolutionary. Any questions of distribution rather than production have been avoided because it is argued that the invisible hand of the market makes it unnecessary. This has effectively stopped any discussion of morality in terms of the economy. The collapse of the economy in 2008 resulted because the market is *not* self-regulating. But this belief has allowed corporations and the wealthy to focus on making more money without any questions about who the money should go to. Over the past three decades, productivity has grown by 400%, but all of that increased wealth went to stockholders; none went to workers.

The middle class accepted this view of the market and opposed any question of redistribution because so much of American politics is played out against a backdrop of race. The middle class (who would benefit from any ethical discussion of distribution) object to it because they think it would mean giving wealth to Blacks, who are by definition undeserving, and making sure that Blacks do not gain is more important than the prospect that they themselves would gain.

One of the primary reasons we cannot politically acknowledge limits to growth is because we use growth as a solution to poverty. But we have been growing for some time, and we still have poverty. Daly says it should be obvious that what grows is the reinvested surplus, and the benefits of growth go to the owners of surplus, who are not poor. Not only does growth never overcome subjective poverty, it also increases absolute poverty. Growth requires technical progress, which alters the composition of the labor force, making more places for educated workers and fewer for uneducated. In addition, the system keeps the opportunities to acquire the needed qualifications (except for a few exceptional people) for the families who already have them.[111]

This situation is also maintained by restricting moral issues to demands for obedience in terms of the behavior of individuals, particularly in issues of abortion and homosexuality. This defines all economic (systemic) questions as outside morality. The fundamental issue is to shift from a narrow focus on individual morality to systemic morality ("Love your neighbor as yourself"), which would require that we address economic equality. This becomes revolutionary because capitalism can only address production. It will require politics and ethics to address distribution if the size of the economy cannot grow. If it is only leisure that is growing, it too must be distributed according to some ethical norm of equality.

Ecological economic theory works to develop creative, intelligent, innovative methods to fulfill genuine needs and wants. David Korten says real wealth should also be "a sense of belonging, contribution, joy, relationship, and spiritual connection." He feels the new central focus for society should be learning and human development, in the broadest possible sense. His concept is based on the "learning society" idea of Robert Hutchins. This would create a society "in which the primary goals of the society comprise—for the individual—learning, fulfillment, and fully actualizing one's unique potentialities; and—for the group—ennobling social enterprises."[112]Harmon argues, as does Daly and Cobb, that it is *full participation* that is essential for society, not full employment.[113]

Quantitative wealth is based on production for production's sake, manufacturing and consuming as much "stuff" as possible. Prioritizing consumption instead of production means putting social needs first, something capitalism cannot do. GDP, which is supposed to measure national welfare, measures only production and ignores how it is distributed. Rather than measure wealth *quantitatively* (the amount of money), ecologists want to measure wealth *qualitatively* (genuine well-being for all).

Much of this issue is becoming more aware of what we really want from our economy. We actually could create an economy that fulfills what David Korten calls the "six criteria of economic health." Such an economy would:

1. provide everyone with the opportunity for a healthy, dignified, and fulfilling life
2. bring human consumption into balance with Earth's natural systems
3. nurture relationships within strong, caring communities
4. honor sound, rule-based market principles
5. support an equitable and socially efficient allocation of resources
6. fulfill the democratic ideal of one-person, one-vote citizen sovereignty.[114]

These economic ideas are a good idea in terms of making the economy work for everyone. But they are essential in the context of global warming. I have said that political arguments are over the definition of the problem, not over the solutions. Therefore, the whole issue is a question of whether or not global warming is real, catastrophic or urgent. If it is all three, as I believe (along with 95% of climate scientists), then these economic ideas are necessary for our survival.

The sunset industries of oil, coal, and industrial agriculture have enormous power to control the conversation and make people believe that global warming is not real, or is of little consequence. The thousands of new "green" industries that could revitalize our economy have little power to make things change. Oil companies desperately want to have another 20 years of high profits; if they succeed, the natural world will begin to unravel. Perhaps it is basically a question of human nature. Are enough people capable of shifting their entire worldview and doing so soon enough?

Duck-billed Platypus

Chapter 9

Redesigning Our Money and Banking Systems

Our current money system determines and reinforces established patterns of economic and social exchanges. By design, it requires constant economic growth just to stay even, discounts the future in favor of the short term, creates unemployment, and concentrates money and power in the hands of the already wealthy. It causes ecological breakdowns and undermines our communities. We are waging a war of money against life because we fail to recognize that money is not true wealth. Real wealth includes our own health and the health of our families, communities, and planet.

> The powers of financial capitalism had another far reaching aim, nothing less than to create a world system of financial control in private hands able to dominate the political system of each country and the economy of the world as a whole.
>
> Professor Caroll Quigley
> Georgetown University

Bernard Lietaer, in *The Future of Money,* says our current money system creates and reinforces the core emotions of greed and fear of scarcity but that greed and scarcity do not need to be an indelible part of human nature and behavior. He argues that we can replace this system with a money system that encourages generosity and abundance.[1]

We need to start by deciding where we want society to go and then "design a money system that gets us there."[2] Lietaer says that even if we could pass new laws regulating corporations (as they were regulated 100 years ago) but ignored our current money system, the same old problems would come back: "If we could ban the WTO (World Trade Organization) tomorrow and don't change the money system, the WTO will reincarnate itself sooner or later. If you ignore your money system you are spinning your wheels."[3] He says even most economists only know what money *does*, not what money *is*. Likewise, David Korten says that ultimately, the cause of the economic crisis is an illusion: "the belief that money—a mere number created with a simple accounting entry that has no reality outside the human mind—is wealth."[4]

In taking back control from corporations, we will also need a money system that allows for democratic control, which requires taking back control from the banking industry.

A Short History of Money

Initially, when nomadic tribes settled to form communities, "wealth" was shared food from crops, meat from a hunt, or clothing. People generally distributed wealth according to need, although in some communities, there were hierarchies. In trading with nearby communities, communities established a fair trade exchange through negotiation, perhaps cows for grain, or hides for finished clothing.[5]

Eventually, many people used coins made of valuable metals to trade. The creation of coins was the beginning of the process of replacing the barter system with a currency system. In seventeenth-century Europe, trade was conducted primarily in gold and silver coins. The value of the metal used to make them determined the value of the coins. Because the coins were heavy, goldsmiths began to hold the gold or coins and issue receipts. People traded the paper receipts rather than the actual gold. These receipts were really "I owe you"s (IOUs) from the goldsmiths (who became bankers) and stated that they were holding a person's gold. Therefore, if at any point someone wanted to "cash in" these IOUs, she could get her gold or coins in return.[6]

In the eighteenth century, a Scotsman, John Law, invented a new type of money to replace the use of coins or receipts, and this created the mechanism to finance the industrial revolution. He created paper money (thereby replacing receipts), which, unlike gold coins, could increase indefinitely and was cheaper to make than coins.[7] Then, Law suggested that banks could issue more paper money than they had gold. Banks found that gold could support about 10 times its value in paper money. That is, a bank that held $1 million in gold could safely print and loan out about $10 million in paper money and make money from the interest on the loan.

Following this model, in nineteenth-century America, private banks issued their own banknotes in sums up to 10 times their actual reserves in gold. Until the early 1900s, each bank printed its own money, as measured against its own gold reserves, and printed its name on each note. This is called "fractional reserve" banking, which means that only a fraction of the total money managed by a bank is kept in "reserve" to meet the demands of depositors.

Our Money System Is Based on Debt

"Money" is a national accounting system of who owes what to whom. In the U.S., the private banking industry owns and operates this money system. Banks "create" money from nothing, by "fiat" (that is, it exists because they say it does). All money is "loaned into existence." Although checks and credit cards have largely replaced paper money, the principle remains the same: the banking industry "creates" the money that governments and people then borrow. When you charge a purchase on a credit card, that money did not exist until you created the debt. New money is only created when someone goes into debt. For example, when you borrow $100,000 from a bank to buy a house, the $100,000 you are borrowing does not exist until you borrow it. The banks are not loaning money left by depositors. They "monetize" (turn into cash) your agreement to repay

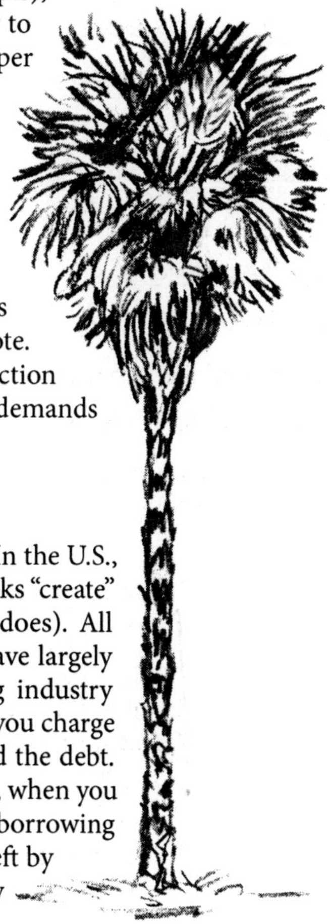

California Fan Palm Tree

the loan. That is, as an accounting entry, they simply enter a debt on your account and an asset on the bank's account. This is how new money is continually put into the economy. Korten says, "Needless to say, granting banks the right to create money with a computer keystroke and then lend it out at interest makes banking very profitable, and Wall Street, which owns the banks, enormously powerful. It also contributes to financial instability and inequality, creates an economic growth imperative, and distorts economic priorities, all costs to society."[8]

I have come to the conclusion that greed and fear of scarcity are in fact being continuously created and amplified as a direct result of the kind of money we are using. ... The scarcity is in our national currencies. In fact, the job of central banks is to create and maintain that currency scarcity. The direct consequence is that we have to fight with each other in order to survive.

Bernard Lietaer
Beyond Greed and Scarcity

Although the bank creates the loan money, they do not create the interest money. In an article titled "A Monetary System for the New Millennium," Canadian money reform advocate Roger Langrick explains the mathematical impossibility inherent in a system of bank-created money lent at interest, for when a bank loans out $100, it wants $110 back (if 10% interest). "The debt in fact is unrepayable. Each time $100 is created for the nation, the nation's overall indebtedness to the system is increased by $110. The only solution at present is increased borrowing to cover the principal plus the interest of what has been borrowed."[9] To put it another way, when the bank "creates" the $100 it loans out, it does not also create the additional amounts to pay the interest—that money does not exist. This money can only be found in new loans. It's like a game of musical chairs. Eventually, someone must default because there is never enough money for everyone to pay back both the principal and the interest. But the process never works forever because it leads to boom-and-bust cycles.

Interest on bank loans causes debt to grow exponentially throughout the world. Virtually the entire money supply is created by banks, with interest owed to the bankers. Ellen Brown, author of *Web of Debt*, explains that, at 5% interest, $1 will become $2 in 14 years, so after 14 years the interest doubles the money owed. Therefore, she says, the U.S. money supply has to double every 14 years just to cover the interest owed on the money that existed at the beginning of this 14-year cycle. "*That means that every 14 years, banks siphon off as much money in interest as there was in the entire economy 14 years earlier*. This tribute is paid for lending something the banks never actually had to lend."[10]

This money system creates a huge Ponzi scheme, which is a pyramid scheme in which earlier investors are paid with the money of later investors rather than from real profits. That is, in this case, earlier borrowers can only pay off their debt with new money created by new borrowers. In this Ponzi scheme, debt is just shuffled around without actually producing anything of real value.[11]

Because the entire system is based on interest, it constantly increases the wealth of the wealthy. Lietaer argues that in our money system, the poor and middle class, who borrow money and pay interest, are continually giving money to the wealthy, who earn interest by investing their money. The wealthy, he says, are "collecting" uninterrupted revenue from every borrower.[12] This is a systematic transfer of wealth from the bottom 80% of the population to the top 10%. Cleverness or industriousness of the players has nothing to do with their success. He says this wealth is due exclusively to our monetary system. In the U.S., the top 1% of Americans now have more personal wealth than the combined wealth of the bottom 92%.[13] Within our current monetary system, money can only trickle *up*.

This banking Ponzi scheme has gone on for hundreds of years. Ellen Brown says, "What distinguishes the legal scheme known as 'fractional reserve' lending from the illegal schemes of Bernie Madoff is that the bankers' scheme is protected by government charter and backstopped with government funds."[14]

There Are Other Ways to Create New Money

There are alternatives. Instead of having banks issue credit, the U.S. government could do it, which it has done before. For a large part of American history, the subject of who should create the money supply was a major debate. A hundred years before the American Revolution, the colonial governments issued their own money, which they both *lent* and *spent* into the economy. When the directors of the Bank of England asked what was responsible for the booming economy of the young colonies, Benjamin Franklin explained,

> In the Colonies, we issue our own paper money. It is called 'Colonial Scrip.' We issue it in proper proportion to make the goods pass easily from the producers to the consumers. In this manner, creating ourselves our own paper money, we control its purchasing power and we have no interest to pay to no one. You see, a legitimate government can both spend and lend money into circulation, while banks can only lend significant amounts of their promissory bank notes, for they can neither give away nor spend but a tiny fraction of the money the people need. Thus, when your bankers here in England place money in circulation, there is always a debt principal to be returned and usury [excessive interest] to be paid. The result is that you have always too little credit in circulation to give the workers full employment. You do not have too many workers, you have too little money in circulation, and that which circulates, all bears the endless burden of unpayable debt and usury.[15]

Koala

After that, King George forbade the colonies from issuing their own currency. The money supply shrank, causing the depression that led to the American Revolution.[16] After the American Revolution, private bankers took control of the money supply.

After Jefferson was no longer president, he realized that the government should have gone back to issuing government credit. He wrote to

John Eppes in 1813, "Although we have so foolishly allowed the field of circulating medium to be filched from us by private individuals, I think we may recover it. . . . The states should be asked to transfer the right of issuing paper money to Congress, in perpetuity."[17] Abraham Lincoln, using the colonial model, created government-issued Greenbacks during the Civil War. This allowed the North to win the war without being plunged into debt to bankers.[18]

In every presidential election between 1872 and 1896, a third national party ran on a platform of banking reform. Ellen Brown says the debate was over "who should create the nation's money and what it should consist of. Should it be created by government, with full accountability to the people? Or should it be created by private banks behind closed doors, for the bank's own private ends?"[19] William Jennings Bryan, when running for president in 1896 and 1900, waged the last serious challenge to this banking system. In the speech that won him the Democratic nomination for president in 1896, Bryan argued for banking reform, said he stood with Jefferson, and said that "*the issue of money is a function of the government and that the banks should go out of the governing business*. . . . [W]hen we have restored the money of the Constitution, all other necessary reforms will be possible, and . . . until that is done there is no reform that can be accomplished."[20]

Today all money except coins is created by banks when they make loans. That includes Federal Reserve Notes (dollar bills), which are issued by the Federal Reserve, a privately-owned banking corporation, and lent to the U.S. government and to commercial banks.

Ellen Brown
Web of Debt

Bryan lost both elections, but he continued to fight for reform in Congress. A major bank panic in 1907 led to a bill called the Aldrich Plan, which would have given control of the banking system to the Wall Street bankers. However, the bill was defeated by opposition led by Bryan. To get their bill past the opposition in Congress, the Wall Street bankers changed the name to the Federal Reserve Act (so that it would sound like it referred to the government) and brought the bill back three days before Christmas. The bill was so obscurely worded, they were able to convince Bryan that the bill did what he wanted. After it passed in 1913, Bryan said, "The right of the government to issue money is not surrendered to the banks; the control over the money so issued is not relinquished by the government."[21]

However, Ellen Brown explains that what happened is that although the national money supply would be *printed* by the U.S. Bureau of Engraving and Printing, it would be issued as an *obligation* or *debt* of the government to a private central bank. Robert Owens, a co-author of the Act, later testified before Congress that the banking industry had conspired to create a series of financial panics in order to rouse the people to demand "reforms" that served the interests of the bankers.[22] Brown says that the Federal Reserve Act was a major coup for the international bankers, who had worked for more than a century to establish a private central bank (the Federal Reserve Bank) in the U.S. with the exclusive right to "monetize" the government's debt. This Act authorized a private central bank to create money and lend it to the government at interest.[23]

This 1913 law set up the Federal Reserve Bank (or Fed) with its 12 branches. In spite of its name, it is not part of the government. It is actually a private corporation, wholly owned by a consortium of private banks, and it protects the banks' interests. In fact, Brown says, "It is so private that its stock is not even traded on the stock exchange. The government doesn't own it. You and I can't own it. It is owned by a consortium of private banks, the biggest of which are Citibank and J.P. Morgan Chase Company."[24] Although the Federal Reserve Bank is a private bank, Congress and the President have some input in appointing members of the Federal Reserve Board.

However, this Board works behind closed doors with the regional banks, without Congressional oversight or control. Bank CEOs sit on the boards of the Fed's 12 branches. In March 2008, the New York Federal Reserve, in private weekend negotiations, advanced $55 billion of the taxpayer's money to J.P. Morgan Chase to buy Bear Stearns (at the bargain basement price of $2 a share, down from a high of $156 a share). Brown says this was a hostile takeover, not approved by the Bear Stearns shareholders or the American voters. The current CEO of J. P. Morgan Chase, Jamie Dimon, sits on the board of the Federal Reserve Bank of New York; therefore, he participated in this decision, an action that normally would represent a conflict of interest that federal statute makes a criminal offense. This happened because of the power of the banking lobby. So, Brown says, "The banking lobby is powerful because private bankers, not the government, create our money and control who gets it."[25]

Senator Ted Stevens asked the Government Accountability Office what authority Congress has to monitor the Fed while they are handing out trillions of dollars to banks. He got back a letter citing an obscure statute that nobody had ever heard of before: the Accounting and Auditing Act of 1950. The relevant section, 31 United States Code 714(b), dictated that Congressional audits of the Federal Reserve may not include "deliberations, decisions and actions on monetary policy matters." This exemption basically includes everything. According to the law, in other words, the Fed simply cannot be audited by anyone, not even Congress.[26]

Today, the debate over who should create the national money supply is rarely heard, mainly because few people (including few in Congress) even realize it's an option, although the issue often arises in times of financial crisis.

Economic Growth Required to Repay Interest

David Korten, like Bernard Lietaer, calls changing the way we create money the "most important reform of all."[27] Economic growth is required when the money system is built on debt. Growth is required in order for people to repay the interest on their loans. Therefore, a stable economy is not possible with the monetary system we have.

One solution is to have Congress return the power to create money to the government, eliminating "fractional reserve" and limiting banks to lending only pre-existing

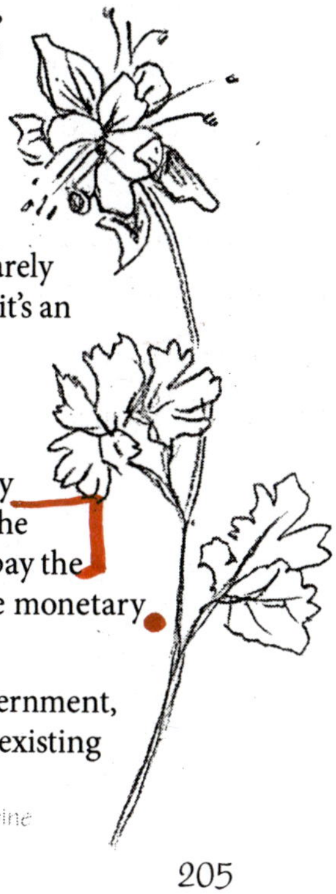

Wild Columbine

funds. If it was the government that created new money, we could then pay off the federal debt, cut taxes, and expand needed government programs. Ellen Brown says paying off the federal debt would not be inflationary because the government would just issue money equal to the total debt, with the dollars just replacing the bonds, leaving the total unchanged.[28] This would save the government about $400 billion in annual interest payments.

> Public banks are able to reduce taxes within their jurisdictions, because their profits are returned to the general fund of the public entity. The costs of public projects undertaken by governmental bodies are also greatly reduced, because public banks do not need to charge interest to themselves. [This reduces] the cost of such projects, on average, by 50%.
>
> Public Banking Institute.org

The fractional reserve Ponzi scheme is bankrupt, yet we continue to bail out the banks. The World Bank counted 96 major banking crises over a recent 25-year period, and taxpayers bailed them out every time.[29] Instead of bailouts, Brown says we need to put banks into a bankruptcy reorganization under the Federal Deposit Insurance Corporation (FDIC), which can wipe their books clean and take the banks' stock in return for getting them up and running again. This would make them truly "national" banks, which could dispense "the full faith and credit of the United States" as a public utility and revive the economy with the sort of money only governments can issue—debt-free legal tender.[30]

David Korten, in *Agenda for a New Economy,* also advocates a new banking system. He says, "The existing system is of the wrong design, and it is corrupt beyond repair. Because it is beyond fixing, we had best devote our attention to replacing it with a new system of a radically different design. . . . Virtually every dollar in the system is borrowed, because we rely on banks to create our money by lending it into existence. No debt, no money."[31] Korten argues for the same changes that Brown does: instead of creating money by banks lending at interest, we can create a system of non-debt government-issued money. With this system, he says, "The money would serve public purposes and reduce the need to collect taxes. Since the money is spent rather than lent, it need not be repaid, and it circulates permanently, thus eliminating the boom-and-bust cycles almost inevitable in the current system."[33]

The state of North Dakota has a wholly state-owned bank, established by the state legislature in 1919. It creates credit on its books just as private banks do. However, "this credit is used to serve the needs of the community, and the interest on loans is returned to the government. Not coincidentally, North Dakota has a $1.2 billion budget surplus at a time when 46 of 50 states are insolvent.[34] This prototype could be copied not only in every U.S. state but at the federal level."[35] In an article called "How Cash-Starved States Can Create Their Own Credit," Brown says that by law in North Dakota, the state and all its agencies, including schools and prisons, must deposit all their funds in the bank, and the state guarantees its deposits. Three elected officials oversee the bank: the governor, the attorney general, and the commissioner of agriculture. The bank's mission is to deliver sound financial services that promote agriculture, commerce, and industry in North Dakota.

The bank operates as a bankers' bank, partnering with private banks to loan money to farmers, real estate developers, schools, and small businesses. It loans money to students (more than 184,000 outstanding loans), and it purchases municipal bonds from public institutions. Again, like banks, the state is not limited to spending only the money it has. Like other bankers, they are allowed to create "credit" with accounting entries on their books.[36] The most common objection to this proposal is that it would be dangerously inflationary. But Brown says that will not happen as long as the money is used to create new goods and services. Price inflation results only when "demand" (money) exceeds "supply" (goods and services). When the two remain even, prices remain stable.

Other states are now thinking about creating state banks. State legislatures in Hawaii, Illinois, Massachusetts, Michigan, Missouri, New Mexico, Vermont, Virginia, and Washington have proposals for the formation of a state-run bank in one way or another. In addition, in 2010, candidates for political office in eight states—California, Florida, Idaho, Illinois, Missouri, Oregon, Vermont, and Washington State—advocated a state-run bank as part of their platform.[37]

We Can Use the Money System to Create Abundance

It is not just our view of human nature as selfish and materialistic that creates our sense of scarcity. Our money system reinforces scarcity, but it could be redesigned to create a feeling of abundance. Lietaer says when we use money to store value, rather than just facilitating exchanges, it pays to cut down trees, sell them, and put the money in the bank to earn interest. If we did not have interest, we would need to store value in something real, something that either increases in value over time or reduces future expenses. We could invest in insulation for our house because it would save heating costs; or when a child is born, to save for her education, the parents could plant a forest of trees, and harvest it when she starts college. However, because money itself earns interest, it makes sense to put it in the bank rather than pay now to plant trees for harvesting later.

This system forces competition between people to repay the interest so that they are not one of those who default. This leads Lietaer to say that it is no wonder that people so readily accepted Darwin's idea of the survival of the fittest as true. By definition, not every debtor can pay back his or her loans because there is never enough real money. However, *this scarcity is artificial.* Short-term thinking is not the result of human nature but of the design of our money system. Interest always favors today's money over tomorrow's resource.

We tend to think only in terms of the formal economy and using the national currency. But doing so limits the kind of society we can create. Lietaer says that we can choose a money system that creates a society without the extremes of

Ladybugs on Forget-me-nots

We can solve the debt crisis while cutting taxes.

To restore our power to create money from the private banking industry:

1. Nationalize the Federal Reserve. Make it truly federal, as its name implies and most people think it is now. The Constitution says Congress shall have the power to coin money and regulate the value thereof. A truly federal central bank could issue money by the people for the people, either directly or as loans. Interest on the loans would return to the government, reducing or eliminating taxes.
2. Nationalize bankrupt banks that are "too big to fail." Too big to fail does not mean too big to be put through a bankruptcy reorganization. In fact, that is the law: instead of being bailed out with taxpayer money, bankrupt banks are supposed to be put into FDIC receivership, purged of their debts, and, if they are too big to liquidate or sell off, taken over as public entities.
3. Set up state-owned banks. One state in the Union now owns its own bank — North Dakota. It is also one of only three states that are currently solvent. In fact, North Dakota is running the biggest surplus it ever had. A state-owned bank can generate credit on its books just as private banks routinely do now. In many states, a state-owned bank could be established through a voter initiative.

Ellen Brown "We Have Become the Servants of an Elite Group of International Bankers"
http://www.webofdebt.com/flyer01.pdf

poverty and privilege and a society that does not count Earth's destruction as healthy profit and economic growth. He says that our money system is the most powerful tool we have available on a collective level and that we should design it through conscious choices. We can redesign our money system for different collective purposes. Currently, we take for granted the functions money now serves as if nature had decreed the system. The transition to a sustainable world may be rocky; however, the future can be one of abundance. Lietaer believes it is possible to create a "Golden Age of Sustainable Abundance" within our lifetimes and asks,

> What if the scarcity is not mostly "out there" in nature, as we all have believed for centuries? What if the money system we have been using, by which we have been collectively hypnotized, was continuously creating that very scarcity that we most fear? Is there a limit to the amount of learning we can do, to the amount of passion, creativity or beauty we can generate and enjoy? What if every street in our cities could become a work of beauty? Could we not design a money system that works *for us* rather than having us work *for it*?[38]

Lietaer concedes that it is not quite *that* simple but that it *is* possible if we begin to think differently. To help people think differently, he uses the "Sufficiency of Money Game," played with a group of people:

> Just pretend that there is no scarcity of money. It has happened by magic. You have a foundation that must spend money in your community [however you want to define that]. . . . Ask a group of people these three questions:

1. What talent would you like to develop and offer to your community?

2. What is the vision of your newly formed community? What would you like to accomplish as a group?

3. Who are the other people or organizations you need to realize this vision?

Get everyone to explain his or her answers. In a second round, see how your different dreams can help each other, how some of your initiatives can mesh with those of others. You will often find that they strengthen and synergize each other in unexpected ways. . . .

There are people capable of and willing to do the work—people who have the skills and the knowledge to achieve these things. People want more meaningful things to do. Our problems are not caused by a scarcity of people or ideas. There are even organizations that have the skills to hire the people and put them to work. This could all be done. What is missing?

What is missing is money. *Everyone is waiting for money.* If one stops to think about it, it is a fascinating phenomenon. Imagine a Martian landing in a poor neighborhood and seeing rundown communities, people sleeping in the streets, children without mentors or going hungry, trees and rivers dying from lack of care, ecological breakdowns and all of the other problems we face. He would also see that many people willing to work are either unemployed, or use only a part of their skills. He would see that many have jobs but are not doing the work they are passionate about. And that they are all waiting for money. Imagine the Martian asking us to explain what is that strange 'money' thing we seem to be waiting for. Could you tell him with a straight face that we are waiting for an "agreement within a community to use something—really almost anything—as a medium of exchange"? And keep waiting?[39]

We don't need to wait. We can create our own "money" to act as a medium of exchange. Locally, we can do this with local currencies, which I will discuss later. However, nationally it can be done by making banking a public (government) function.

Lietaer says that we need to view economies as flow systems in which money's primary function is as a medium of exchange (not to store value). In a systems view, money is to the real economy as biomass is to an ecosystem, or blood to your body: it is an essential vehicle for catalyzing processes, allocating resources, and generally allowing the exchange system to work as a synergetic whole. He says the connection to structure is apparent: "The health of the whole in both economies and ecosystems depends heavily on the structure by which the catalyzing medium, in this case, money, circulates among businesses and individuals. Money must continue to circulate in sufficiency to all corners of the whole

Brown Bear Family

because poor circulation will strangle either the supply side or the demand side of the economy, or both."[41]

> The local currency and exchange movement may be the most important development in human liberation since the Magna Carta.
>
> Tom Greco
> The Co-Intelligence Institute
> *Improving Currencies*

As British commentator Ron Morrison observed in a 2004 article titled "Keynes without Debt," "[Today] bank credit supplies virtually all our everyday means of exchange, and this brings into sharp focus the simple fact that modern money is no longer constrained by outmoded intrinsic values. It is pure fiat [enforced by law] and simply a glorified accounting system. . . . *Modern monetary reform is about displacing the current paradigm of 'what can be afforded' with 'what we have the capacity to undertake.'*"[40] For example, assume the U.S. has agreed to create a wartime-speed shift to sustainability. One part of that would be a program to create as much mass transit as it is possible to do safely. Suppose U.S. companies have the capacity to create $100 billion of mass transit in year. If we had a national bank, Congress would create $100 billion to spend on mass transit. There would need to be regulations, quality control, and transparency, but supply and demand would be balanced. In this case, the money is spent into existence and there is no need to repay it.

Alternatives Include New Complementary Currencies

Lietaer defines money as "an agreement, within a community, to use something as a means of payment."[42] Therefore, currency is always about relationships, which means that money can connect people rather than control people. He names five functions of money but says only two are essential for an efficient currency.[43] The two essential functions are as follows:

- Money serves as a *standard of measure* in comparing the value of apples with oranges. Money is information with which to measure the exchange of goods. As such, it cannot be scarce, just as in building a house, you do not need to worry about running out of inches (a standard of measure).
- Money also is a *medium of exchange* and is more efficient than bartering or other methods. Money is not a thing; it is just information.[44]

Local currencies are not just an economic tool, they are also a cultural tool, and they can create work. Lietaer estimates that there are now more than 2,500 areas in the world using local currencies.[45] It is important to understand local currencies because when local economies are under stress, creating a local currency is one of the best ways to regain stability. Local currencies were common during the Depression in both the U.S. and Europe. We already use a corporate currency called frequent flyer miles. People now earn these by many things other than flying, and they can spend "miles" on many things: hotels and even long-distance phone calls. Coupons are another form of currency.

A community can have many currencies. A local currency is one that people can only earn and spend locally. When people pay some part—say 20%—of the price of goods sold at local stores in local currency, it helps to keep money circulating locally and increases the stability of the local economy.

For a currency to work, all we need is an agreement between parties to issue and accept the currency. Richard Douthwaite, in *Short Circuit,* gives the example of the owner of a delicatessen who had lost his lease. He needed to renovate a new store and pay for the move, but he could not get a loan from the bank. So, he issued "deli dollars." He sold $10 in deli dollars for $9 to his customers. (In this case, he chose to pay interest; generally, local currencies do not earn interest.) In one month, he raised the $5,000 he needed. (The "dollars" were dated with dates staggered over the next year so that customers would not all come in to redeem them for sandwiches at once.) The "deli dollars" were small, short-term loans from customers that became a form of community currency. This was a college town, and parents bought them for their kids. The dollars even ended up in the collection plate on Sunday because the parishioners knew that the pastor liked to eat breakfast there. It was also a good way to ensure that old customers would come to the new store.[46]

In Curitiba, Brazil, the shantytowns were full of garbage, and the city did not have the money to clean it up. The city offered one bus token for every bag of sorted garbage brought to the containers at the bus stop. Children soon picked the entire area clean and gave the bus tokens to their parents, who could use them to get to work. Garbage became a currency—a medium of exchange. Because the bus tokens and the garbage were not scarce, the system created reciprocity and cooperation.[47] The city created as much money (bus tokens) as they needed to pay local people for the garbage collection.

There are primarily two successful types of local currencies:

- A mutual credit system; or
- An actual physical currency.

Mutual Credit Systems

One form of a mutual credit system is LETS, a membership association in which each member has an account. The LETS system uses a ledger of accounts on a computer to keep track of exchanges. All accounts begin with a balance of zero. In this system, participants "create" the money when they make a transaction. When a member sells something to another member, her account receives a credit (plus), and when a member buys something, her account receives a debit (minus). A credit causes an account balance to increase, while a debit causes an account balance to decrease. In such a system of mutual credit, the total of credit balances must always equal the total debit balances. This type of local currency does not require any centralized money supply management.

With mutual credit systems, people can always create money in amounts that are sufficient; therefore, there is no scarcity. In a

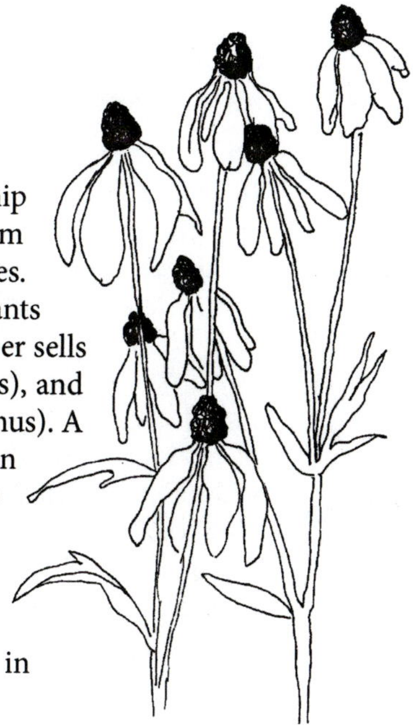

Purple Coneflower

mutual credit system, we can think of the "money supply" as being the total amount of debits or credits. Since these are two sides of the same coin, and since debits and credits are always equal, we need only count one of them. One of the elegant aspects of a mutual credit system is that the "money supply," within reasonable bounds, adjusts itself automatically in relation to the amount of trading that members wish or need to do. No central authority needs to decide whether to increase or decrease the supply of credits in order to keep the market value of the credits constant. The only thing to be decided is the maximum amount of debit balance allowed on any account to minimize the risk of default. In usual practice, account balances will be well below that amount.

Time dollars are another type of mutual credit, but in this case, it is a form of service barter. In a healthy economy, money is not the dominant value, nor even the dominant medium of exchange. One vital indicator of economic health is the presence of an active economy of affection and reciprocity in which people do useful things for one another with no expectation of financial gain. Voluntary sharing creates and sustains trust and mutual caring. The time dollar system matches resources with needs and is a way of providing recognition and exchange of the things people do for each other while allowing the exchanges to occur among a larger number of people. People earn time dollars by volunteering some service, and a computer records the time amount. This allows them to "purchase" another service from someone else. Time dollars creates a world that creator Edgar Cahn says "rewards decency and caring as automatically as the market economy now rewards selfishness and even ruthlessness."[48] For example, a teenager could earn time dollars by mowing someone's lawn and give his time dollars to his grandmother, who lives on the other side of town, so that she can use the dollars to have someone take her shopping. It becomes a way of facilitating bartering across a whole community.

> It is well enough that the people of the nation do not understand our banking and monetary system, for if they did, I believe there would be a revolution before tomorrow morning.
>
> Henry Ford

Time dollars are backed by moral obligation, a trait of community reciprocity. The IRS has ruled that they will not tax a transaction backed only by a moral norm as a "commercial exchange."[49] By connecting human needs and human resources, time dollars generate new wealth and build community. They highlight the value of people's contributions and encourage people to rely on one another.

A Physical Currency

example re K.L."

Local currency systems can also use circulating paper notes to give people a way of tracking their exchanges. For example, Ithaca HOURS (and other HOUR currency systems modeled on it) are paper currency notes that circulate in a local economy. Unlike LETS, Ithaca HOURS in Ithaca, New York, is not a membership organization. A free newspaper called Ithaca Money puts the HOUR notes into circulation. Businesses that advertise in this newspaper are given notes when they buy an ad. These advertisers usually accept HOURS for at least partial payment for products or services but have no obligation to do so. They publish this newspaper six times a year; it lists members' services and phone numbers, along with related articles, ads, and announcements. Every eight months, member stores may apply for two additional HOURS for their continuing participation. This is how they carefully and gradually increase the supply of currency in the community. In the case of Ithaca, the value of an HOUR is set at $10, the average hourly wage in Ithaca.

The U.S. government requires that communities design physical local currencies to look different from federal dollars, and local dollars must be counted as taxable income when accepted for trades or services that the government would normally tax.

An Example of Using Local Currency

With local currencies, every local expenditure cascades into a larger number of transactions that keeps the "money" local and enriches the community. These currencies do not benefit individuals directly, but rather indirectly, by increasing the number of exchanges that remain local. This strengthens and stabilizes the local economy and makes the community more resilient and less vulnerable to national or global recessions. By definition, people must spend a local currency locally (a town or city). Local money generates a cascading effect of other transactions, called the multiplier effect.

For example, a homeowner hires a painter for $500. She pays the painter $200 in national currency (needed to pay for supplies) and $300 in local currency. If the local grocer has agreed to take local money, the painter can spend the $300 at the grocery store. The grocer pays part of his clerk's salary in local money (say $300). The clerk spends $300 in local money to buy a share in a community-supported agriculture program from a local farmer. The farmer spends $300 in local money at the local hardware store.

Painter	$ 300
Grocer	$ 300
Clerk	$ 300
Farmer	$ 300
Hardware store	+ $ 300
	$1,500

The original $300 put $1,500 into the local economy because it was repeatedly exchanged for goods and services. Spending $300 on something that came from outside the community not only costs the community that $300 but the other $1,200, as well. To get the $300 back to the community, someone must sell something in the "out of town" economy.

These local currencies never have interest. In fact, they usually have *negative* interest, called a demurrage charge. If the owner of the money does not spend it within a certain period—usually a month—the

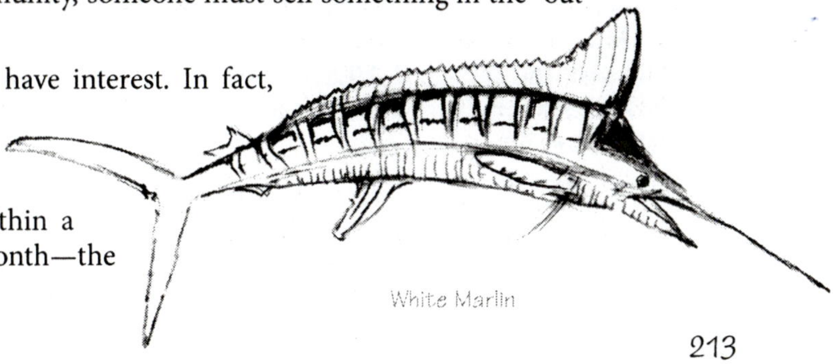

White Marlin

owner must pay a fee to revalidate the money. This charge helps to keep the money circulating. By designing the currency so that people must spend it quickly, it increases the impact on the local economy. In our example above, if the money had made the rounds of these five people in one week instead of two weeks, it would have put twice as much money into the local economy (assuming it continues to move at the same rate). By spending the money quickly (increasing the "velocity"), there is more money in the system.

> The risk that government will abuse its money-creation powers can be minimized through a combination of separation of powers and public transparency. An independent federal agency would decide how much new money to create. The Congress would decide how it would be spent.
>
> David Korten
> *Agenda for a New Economy*

Bernard Lietaer, in an interview in *Yes! Magazine,* says he has found only three periods in history in which there was negative interest (a charge for holding currency), and each time it had a profound effect. These three periods were classical Egypt, a period of about three centuries in the European Middle Ages, and a few years during the 1930s during the Depression in the U.S.

In ancient Egypt, people were given a token in return for storing grain. They could use this token for exchange, creating a "grain standard currency;" that is, they used grain as currency. However, when they returned to exchange their tokens for their grain, they would receive somewhat less grain than they had tokens because of the costs of storing the grain. This amounted to negative interest.

Egypt was the breadbasket for the ancient world because instead of keeping value in money (the tokens, which would decrease in value), everybody invested in assets that would last over time, like land improvements and irrigation systems. The Romans replaced the Egyptian currency with their own money system, which had positive interest rates. Lietaer said, "After that, Egypt ceased being the grain-basket, and became a 'developing country' as it is called today."[50]

In another example, Lietaer describes Europe during the Middle Ages between the tenth and thirteenth centuries, when local lords issued local currencies, periodically recalled them, charged a tax, and reissued them. This created a form of demurrage, making money undesirable as a store of value. This resulted in a blossoming of culture and widespread well-being, as well as the building of the great cathedrals. The cathedrals required an extraordinary investment for a small town. In addition to its religious value, the cathedrals attracted pilgrims, who had the same economic function as today's tourists. These cathedrals created a long-term cash flow for the community, and it is still working today: in Chartres, most of the city's businesses still depend on the tourists who visit the cathedral, 800 years after it was finished. Other productive assets were also investments, such as taking good care of the mills, with people replacing parts even before they wore out.[51]

In the early fourteenth century, with the invention of gunpowder, kings centralized their power and monopolized the money system—and few cathedrals were built after this. The economic incentive for collective long-term investments was gone.[52]

There were many local currencies during the Great Depression in the U.S.; these made communities much more stable and self-reliant, as well as insulating them to some degree from the national depressed economy. Some local governments allowed people to pay their local taxes in local currency, increasing the circulation of money locally.

> [Government banking] wouldn't be perfect, but it would be far less subject to abuse than the present system, under which the Federal Reserve functions as an arm of Wall Street to decide how much money to create and who gets it, creating and allocating trillions of dollars to favored private Wall Street institutions without public disclosure or oversight.
>
> David Korten
> *Agenda for a New Economy*

Creating Systems Solutions

Lietaer and others, in "White Paper on All the Options for Managing a Systemic Bank Crisis," argue that we also need larger-scale complementary currencies.[53] Having the government issue credit would stop banking crises, but there could still be monetary crises.

By using many different currencies, we can greatly increase the resilience of the system. Lietaer says that the technical justification for enforcing a monopoly of national currencies within each country is to optimize the efficiency of price formation and exchanges in national markets. Nobody questions the efficiency of these markets, but they obviously lack resilience. Even more ironically, whenever a banking crisis unfolds, governments invariably help the larger banks to absorb the smaller ones, under the logic that the efficiency of the system is thereby further increased. However, Lietaer argues that when a failing bank has proven to be "too big to fail," it needs to be broken up into smaller units that can compete with each other, similar to the breakup of the Bell telephone monopoly in the 1980s.[54]

A monopoly of a single type of currency, like our bank-debt money created with interest, is by definition lacking in diversity and therefore resilience. Lietaer says, "Imagine a planetary ecosystem where only one single type of plant or animal is tolerated and artificially maintained, and where any manifestation of diversity is eradicated as an inappropriate 'competitor' because it is believed it would reduce the efficiency of the whole."[55] He says that an overly efficient system is condemned to collapse, however well managed. Today, the Eurozone is facing a financial crisis over just this realization—that forcing all the EU countries to use the Euro creates instablity.

The same issue of diversity can also mean that a town that has only one very large employer will find it harder to adapt if that company goes under than a town with several medium-sized employers and many more small ones.

The fields of engineering, business, and economics are all focused almost exclusively on efficiency, which results in an overly efficient and dangerously brittle system. Placing too much emphasis on efficiency automatically tends to increase size and consolidation at the expense of diversity, connectivity, and resilience, until the entire system becomes

Fowler's Toad

unstable and collapses. These findings arise from the very structure of a complex network system; therefore, they remain valid for any complex network with a similar structure, regardless of what is being processed in the system: "It can be biomass in an ecosystem, information in a biological system, electrons in an electrical power distribution network, or money in an economic system. This is precisely one of the strong points of using a web-like network approach instead of machine-like metaphor."[56]

> [State banks can use] the "fractional reserve" lending principles followed by all banks. California, for example, owns about $200 billion in real estate, has $62 billion in various investments, and has $128 billion in projected 2009 revenues. Leveraged by a factor of 10, that capital base could support nearly $4 trillion in loans.
>
> Ellen Brown
> *Web of Debt*

Complementary currencies increase the resilience of the whole even though they reduce efficiency. They "facilitate transactions that otherwise wouldn't occur, linking otherwise unused resources to unmet needs, and encourage diversity and interconnections that otherwise wouldn't exist."[57] In particular, they encourage exchanges that are too small or inefficient to compete for national currencies in a global marketplace. These currencies could operate in parallel with conventional national money. The need now is to ensure that the collapse of the financial system does not create a breakdown in the "real" economy, where businesses produce and sell non-financial goods and services.

Alternative currencies can also be developed to help businesses as a group. Lietaer says that it is possible for companies to lead themselves out of this crisis, and he cites the example of the WIR currency system that began in Switzerland during the Great Depression and is still in use today. In 1934, 16 businessmen got together and decided to create a mutual credit system, inviting their clients and suppliers to join. They created their own currency, whose value was identical to the national money but didn't bear interest.

The country's banks tried to stop it but didn't succeed, and the new system saved the businesses involved. "A cooperative was set up among the users to keep the accounts dealing with that currency. Soon participants could also borrow from that cooperative in that currency at the remarkably low interest rate of 1% to 1.5%. All such loans need to be backed by inventory or other collateral. Over time, the system grew to include up to one quarter of all the businesses of the entire country."[58] Whenever there was a recession, the volume of activity in this unofficial currency would expand significantly, thereby reducing the recession's impact on sales and unemployment. When the national economy was stronger, the unofficial currency was used less.

Lietaer proposes that "businesses take the initiative of creating such Business-to-Business (B2B) systems at whatever scale makes most sense to them." This is easier now because of information technology tools. There are various software already avaialble that are designed to manage complementary currencies. For instance, the WIR cooperative has a large-scale system that is operational in Switzerland in four languages; it deals simultaneously with national money and WIR. There are also several other fully operational software packages available for specific complementary currency applications. Using open-source software would provide the flexibility to add new functions. For instance, the Strohalm Foundation in the Netherlands has open-source software for mutual credit systems used for social purpose applications that is already in operation in various countries.[59]

To gain support for the alternate currency, businesses could lobby the government to temporarily accept their B2B currency as partial payment of business taxes, perhaps 10% or 20%. This could apply only while the banking system cannot finance the real economy to the extent that is necessary. Partial payment of taxes would be the most effective incentive that governments could provide to accelerate the widespread acceptance of this currency. In addition, state and local governments, which are often hardest hit in a recession, could choose to accept the complementary currency in payment of the city or state taxes.[60]

In short, a whole new set of tools is available that has already been tested somewhere in the world. By using a diversity of currency systems, we can make the economy much more resilient.

The Financial Crisis, Global Warming, and the Future

With the financial meltdown, we are now forced to use tax dollars to bail out the wealthy. One-half of all U.S. corporations do not pay any tax, and the wealthy received large tax reductions under President Bush's administration. (Newly revised estimates from Citizens for Tax Justice show that the Bush tax cuts cost the government almost $2.5 trillion over the decade after they were first enacted, 2001–2010.[61]) The system has been designed for the wealthy—simply because they have so much political power and have been able to convince people that if everything is designed to maximize their wealth, wealth will "trickle down." This mindset originates in our shared understanding of how the world "works." It's difficult to realize how differently our social systems could be structured.

Now, with both global warming and the 2008 financial crisis, everything could change. These changes work best when people live in areas that have a strong sense of community because healthy economies are about relationships. We are reaching a critical crossroad: either we reinvent our social and economic relationships toward community and partnership or we continue on our path to mass suicide.

Warthog

Summary of the New Economy

- Focus on *true accounting* and *fair distribution.*
    A. Prices must include the full cost of the product.
    B. Create new indicators that measure the increase of well-being and the decrease in resources used or waste created.
    C. Create indicators that measure equity and social progress.
    D. Count as wealth real goods and services, not the "phantom wealth" created by Wall Street for gambling.
    E. Use tax and income policies to favor the equitable distribution of wealth and income.
- Redistribute *economic power and control* to local and regional economies, thereby restoring a country's ability to control its own economy.
    A. Rebuild communities with a goal of achieving local self-reliance in meeting basic needs. Implement policies that create a strong bias in favor of human-scale businesses owned by local stakeholders.
    B. Bring back the understanding that a corporation's charter means that it will operate in the public interest.
    C. Allow and fund the ability of stakeholders to buy out corporations to democratize ownership.
    D. Revise intellectual property rules to facilitate the free sharing of information and technology.
    E. Encourage companies to put technology patents into the public domain, allowing free sharing with developing countries and greatly expanding innovation in this country.
- Restructure *financial services* to serve Main Street.
    A. Transfer the responsibility for issuing money to the federal government.

Ants

Chart Summarizing the Contrasting Concepts of Economics

The left side of the chart gives the characteristics of the current worldview, and the right side has the characteristics of the ecological worldview.

Two Worldviews in Economics

CURRENT ECONOMICS	ECOLOGICAL ECONOMICS
Empty Earth economics	Full Earth economics
Continual economic growth	Steady-State economics
Cowboy economics	Spaceship economics
No limits	Recognizes limits
Capitalism as serving owners	Capitalism as markets
Globalization	Markets are primarily local and regional
Competition/greed	Community/mutual well-being
Materialism, centralization, rationality, inequity, hierarchy, industrialization, urbanization, technology, bureaucracy, and secularization	Sustainability patterned after nature, local economies, local ownership, decentralization, "bio-regionalism"
The best people are wealthy	The biases of the system create the wealthy
Scarcity	Abundance
Money values	Life values
Production for profit	Distribution of consumption for well-being
Quantity	Quality
Technology dominates	Interpersonal skills and knowledge dominate
Short view: the next quarter	Long view: seven generations ahead
Information and knowledge hoarded and made scarce	Information and knowledge freely shared
Financial capital	Social and human capital
Physical (manufactured) capital	Natural capital
Efficiency	Efficiency and resilience

Wolf Spider

Chapter 10

Reconnecting to the Sacred, to Nature, and to the Feminine

We will never transform the world unless we can create a renewed sense of the sacred and an experience of the Spirit and the soul. Many discussions of ecology do not include a spiritual dimension, which I feel is essential and is part of seeing life whole. Transforming the world begins with transforming ourselves through a personal spirituality because the beginning transformation is an inner one. Spirituality is about the whole—one interconnected divine reality in which mythic vision is present in the unconscious. Individuals and groups from many traditions are coming together to imagine, create, and share ways to connect to the "whole," to our collective and individual souls.

Our culture focuses on the outer world and ignores the inner world. Many of our current problems are the result of ignoring or denying any interior reality: imagination, art, spirit, and emotion. Therefore, when we feel empty, we look to the outer world and try to fill the emptiness with material things (consumer goods or drugs), but we can never get enough (since it is not really what we are seeking in the first place).

Reconnecting to the Sacred

We have restricted reality to the physical world and excluded from consciousness the awareness and power of spiritual realms. Spiritual disciplines teach that we are inseparable from the web of life. People are rediscovering ancient methods for achieving altered states of consciousness to connect to this wholeness, such as meditation, yoga, or shamanic journeys. Because we claim that the only reality is the physical world, we have been able to ignore any connection to and experience of our inner world.

> It is only when we see the invisible that we can attempt the impossible.
>
> Philip Scharper
> Catholic publisher and essayist

When the physical became the only reality, science became the only legitimate way to know reality, and non-sensory perception (mysticism) was denied any validity. In addition, Newtonian, reductionist thinking pushed aside deeper ways of thinking and knowing. For many, this logic led to the assumption that because God is neither matter nor observable through the (material) senses, God does not exist. Scientific reason replaced revelation and the experience of the Holy. This created a related shift from a focus on the inner world to one on the outer world.

Awareness of the sacred and the soul is an alternative mode of consciousness. Having a spiritual sensibility is like having perfect pitch, but modern science has rendered most people "tone-deaf" to the Spirit. Our current crisis is not just in our political and economic systems, it is also in the state of our souls. We need to consciously find a deep connection to our soul and the soul of others. Theodor Apt, in *Progress without Loss of Soul*, asked, "Is attending to the reality of the soul a personal or collective task?" I think it is both, but it starts with the individual.

As the poets and painters of centuries have tried to tell us, art is not about the expression of talent or the making of pretty things. It is about the preservation and containment of soul.

James Hillman

The transformation we need is fundamentally a spiritual transformation—one that Joanna Macy, a Buddhist, calls "the Great Turning." For her, it includes a shift in our understanding of where God is, from being transcendent (God is other, out there, above) to being immanent (God is within all things, within us, and Earth). She says:

> We are taking the sacred back into our lives. The swing is from transcendence to immanence. The most vital movement of our era involves making the sacred immanent again. I see it happening in every spiritual tradition—in the Jewish renewal movement, in Creation Spirituality [Christianity], in women's spirituality, and in the resurgence of Wicca, and the teachings of ancient indigenous peoples. We are reawakening to the sacredness of life itself, in the soil and air and water, in our brothers and sisters of other species, and in our own bodies.[1]

Seeing life as sacred means we are no longer isolated and lonely. Instead, Macy says, people are starting "prayer groups and healing groups, sacred circles and home churches. They're ordaining themselves."[2] By doing this, she says, we can undo centuries of thinking which has made us "so lonely and selfish and nuts and powerless. It all goes together, greed and powerlessness go together."[3]

When we seek to experience the sacred, we ask different religious questions. We ask *where* is God, (transcendent or immanent) rather than *who* or *what* is God; this has created a movement from belief (in who or what) to the experience (of where), that is, it creates a shift in focus from obedience to relationship. This shift assumes that what is important is our *relationship* with the immanent God. When the focus is on the acceptance or rejection of specific beliefs, people are faced with an either/or choice; this in itself defeats the soul, for it cuts off all imaginative experience of the divine. When we pay attention to the reality of the soul and resist the secularization of experience, we can create a new world.

As we raise our own consciousness toward this new spirituality, we raise the consciousness of those around us and help to heal the planet. Western thought developed a belief system in which only reason was important. But the soul is messy.

Great Blue Heron

The world is moving away from reason as the only way to know the world and has begun to embrace other ways of knowing: intuition, imagination, bodily knowing. We need to be able to integrate spirituality with reason. We need to trust the imagination, which is the language of the soul. If imagination is minimized, ignored, or devalued, the soul has no voice in our lives. Thomas Berry, theologian and cultural historian, explains it this way:

> We need to move: from a spirituality of alienation from the natural world to a spirituality of intimacy with the natural world, from a spirituality of the divine as revealed in words to a spirituality of the divine as revealed in the visible world about us. From a spirituality concerned with justice merely to humans to a spirituality of justice to the devastated Earth community. From the spirituality of the prophet to the spirituality of the shaman.[4]

The new belief system we need will honor all religions and spiritual practices because what is significant is what we share; we can learn from our differences, not demonize them.

> After several millennia of assigning the sacred to a transcendent dimension removed from ordinary life, the world around us begins to go dead and losses its luminosity and meaning. The Earth is reduced to a supply store of material resources and a sewer for our wastes.
>
> Joanna Macy
> *Yes! Magazine*

To restore our awareness of the soul, we need to make spirituality a more serious part of everyday life. In *The Care of the Soul: A Guide for Cultivating Depth and Sacredness in Everyday Life,* Jungian analyst Thomas Moore says our greatest problem is a loss of this experiential sense of the Holy or what he calls "loss of soul," which is implicated in all of our troubles and affects us individually and socially. He says when the soul is neglected, it doesn't just go away: "it appears symptomatically in obsessions, addictions, violence, and loss of meaning."[5] His book reframes issues of love, depression, jealousy, illness, and work as efforts of the soul to redirect lives. He calls his book "a program for bringing soul back into life."[6]

Moore describes much of this in terms of imagination; he says that the "increase of imagination is always an increase in soul."[7] It is imagination that allows us to resist the secularization of experience. For example, Moore says, imagination can find the message of Spirit even in illness: "In a very real sense, we do not cure diseases; they cure us, by restoring our religious participation in life. If the gods appear in our diseases, it follows that our lives may be too secular and in need of such a visitation."[8] He also re-images daily life in terms of the soul: "To the soul, memory is more important than planning, art more compelling than reason, and love more fulfilling than understanding."[9]

For the fundamentalist, the Bible is something to believe in literally; for the soul, it is a powerful stimulus for the religious imagination, for searching the heart for its deepest and most exalted possibilities. Moore says, "We go to church as much to subdue that spirit as to acknowledge it."[10] Reconnecting to the soul softens the judging and moralism that have been so common.[11] That is, the focus on sure and certain beliefs protects us against the power of the soul, which is not limited to easily controlled belief systems. Or, as psychologist Carl Jung said, religion can be a defense against a religious experience.

Because the spirit lives primarily in the unconscious, spiritual sensibility and artistic sensibility are closely connected. To help people reconnect, seminaries and churches need to teach courses and create experiences that can bring out the mystic in people. Matthew Fox, in *The Coming of the Cosmic Christ,* says, "Just as science has its laboratory hours for teaching the methods of scientific inquiry, so too theology requires laboratories in painting, clay, ritual, massage, and music to teach the art of mystical development."[12] Fox has been developing these courses and says, "Courses in the mystics, and above all courses that *bring out the mystic* in each minister, rabbi, or priest-to-be, must be taught in our seminaries. Such courses will necessarily include art as meditation courses; every seminary needs more of these to awaken the mystic within and to empower future spiritual leaders with skills with which they can train others in church and synagogue settings."[13]

Because scientific projections for Earth's future seem so relentlessly negative, and the task—transforming the world and creating a sustainable Earth—is so urgent and overwhelming, I'm sure most people feel as helpless and hopeless as I have at times. However, when I began to study painting about 10 years ago, my ideas and writings began to interconnect, coming together as a whole system. I have hung my abstract paintings all over my home. Participating in creative action helped bring ideas together and reassured me. For me, creating the paintings has been a good reminder that the changes we need are not linear but more weblike, too complex to measure or follow. Transformation can only happen with the engagement of the unconscious, and the unconscious (the soul) lives in the non-verbal, creative level with painting, music, or dance. Give the soul knowledge to inform the intuition, and the impossible becomes possible.

There are many artists involved in this effort to re-soul the world. In *The Reenchantment of Art*, art critic Suzi Gablik describes how artists are consciously working to "restore to our culture its sense of aliveness, possibility and magic."[14] She sees many artists' work today as a reaction to our modern world: "One of the peculiar developments in our Western world is that we are losing our sense of the divine side of life, of the power of imagination, myth, dream and vision. The particular structure of modern consciousness, centered in a rationalizing, abstracting and controlling ego, determines the world we live in and how we perceive and understand it; without the magical sense of perception, we do not live in a magical world."[15] Gablik says, "The

African Cape Buffalo

principal function of the shaman is magical healing and soul-retrieval; soul-loss, once regarded as the gravest of all illnesses, is never mentioned in Western medical books."[16]

We see science and spirituality as mutually exclusive because we misunderstand both. Creating an ecological sensibility requires that we understand and respect the role of both science and spirit—and do not confuse the two.

Science and the Sacred

> The felt presence of the sacred will be like fuel for the Great Turning. . . . In the breakdown of the industrial growth society, things will get a lot harder and scarier for a while. And when we get scared we get mean. We turn on each other. I think our greatest danger is fear and blaming and scapegoating that fear arouses.
>
> Joanna Macy

The discrediting of science by fundamentalist religion has been a major problem in ecology, for it has made it easier for people to deny the science of global warming. Now people are beginning to combine opposition to evolution with opposition to global warming and they argue that schools should teach "alternatives" to both.[17] I especially do not want to frame the counter-argument to that as a rejection of religion in the name of science. We need an alternative view of religion that doesn't require rejecting science, or choosing between science and religion.

We can name the sacred and connect with the Spirit without denying science or reducing the Spirit to the literal. Spiritual sensitivity neither denies nor confirms science because they each address two different kinds of reality. Confusing the two has led to a debate over evolution and creationism. Ecology and evolution are closely connected. Many universities have Departments of Ecology and Evolution. In science, ecology is the study of the interrelationships between organisms and their environment. Evolution is the study of the interrelationships between organisms and their environment over time.

People who deny evolution say it is "only a theory," by which they mean an opinion, and so insist that their opinions are equally valid. Therefore, such people want creationism taught in school as science. It is alarming that so many people are so scientifically illiterate. Science is not about conclusions, such as evolution. Science is a method of inquiry called the scientific method and it is the use of this method that makes something "science." Using this method, scientists develop theories by collecting data and developing a hypothesis that might explain the behavior of the data. A hypothesis is a tentative explanation, what colloquially we call a "theory." Then, scientists create an experiment to test the hypothesis. If they can show that the hypothesis is wrong, then they create a new one, perhaps collect more data, or modify the experiment used to test it, to gain more insight. If you can't construct a replicable experiment to test a hypothesis, it is not science.

"Peer-reviewed" research means that anyone with the same resources or equipment is able to perform the same test and get the same results. In addition, scientists never "prove" a theory; they work to *disprove* theories until there is one left that no one is able to disprove, and so it stands—for now. Therefore, "theories" are, by definition, state-of-the-art science, and yet always provisional. Scientists use new evidence to continually revise and amend all theories, including evolution. In contrast, arguments for creationism have no physical data, no evidence, no method, and no experiment to test their hypothesis. Fundamentalists refer only to the Bible and the reports in Genesis. This is a method of inquiry, but it is not the method used in science.

When they reject the purely scientific understanding of issues like evolution, they call them "materialistic." However, fundamentalists are also "materialistic" when they assume that for the Bible to be true, it must be literally, concretely (materially) true. Alfred North Whitehead called this kind of reasoning the "Fallacy of Misplaced Concreteness." However, although the Bible is not literally accurate, it is religiously meaningful.

> The most beautiful and most profound emotion we can experience is the sensation of the mystical. It is the power of all science.
>
> Albert Einstein

Science only addresses things that can be observed, quantified, and measured and that are subject to repeatable experiments. I believe science is accurate when we want to understand the physical world, and I accept the evidence of evolution. What I do not accept is the idea that science is the *only* way to know the world and that anything outside the realm of science is either illusion or superstition. As a religious person, I believe God created the world. However, I do not consider that to be a statement about physical data but a statement about meaning.

Science is powerfully instructive when used to examine the physical world, but science says nothing about meaning. To believe God created the world implies a value system that names the natural world, as God's Creation, as sacred. Our value system today names and defines the natural world in only economic or monitary terms. In this value system, trees have no value until they are lumber; wilderness has no value until it is "developed."

The religious voice is essential because it can introduce non-economic values into political decision-making. The religious belief that the natural world is created by God means that God holds us accountable for how we use it. The natural material world is not ours to use up and destroy.

We can value science and still name the world sacred. Science contradicts only fundamentalist religious beliefs that define the behavior of matter. Teaching creationism as science is not only bad science, but also bad theology. In a world understood only scientifically, our current theologies are inadequate if they require a rigid literalism to retain a sense of a sacred universe.

Malayan Tapir

In fact, twentieth-century science has begun to present a very different view. The newer sciences like quantum physics have demonstrated that the ultimate reality is not just matter but also energy and that there are legitimate nonempirical forms of knowing. Newtonian science has been for centuries the primary metaphor for the nature of reality, but twentieth-century quantum physics and complexity science offer new metaphors. Quantum physics now finds that on the atomic level, reality is not at all what we had supposed. Instead of a mechanistic, "clockwork" idea of separate parts, everything is now seen as fundamentally interconnected, a reality seen as Mystery rather than mechanism. Scientists also talk about living in an eleven-dimensional universe, where seven of these are unseen. In quantum physics, we cannot see ourselves as separate or as an "objective observer" of some other distinct reality. Physicists find that the very act of observing alters the actual physical properties of the reality perceived.

> But do we, as a Bible-based society, feel fully the reality of having been chased from Eden?
>
> James Hillman

Ecological theory includes a spiritual sensibility that defines reality as sacred interconnectedness where Earth and all of Earth's inhabitants are seen as manifestations of divinity. We cannot create adequate change until we can re-enchant the world and bring back a sense of the soul so that a concern for the sacred can again help determine social action. And we can do so without contradicting science.

Reconnecting to Nature

One specific form of denying the sacred has been denying nature any divinity. Before the Enlightenment, people saw all of life as organic, alive, mysterious, and powerful, and they saw themselves as participating in a sacred world. With this organic understanding, they experienced the world in terms of an all-encompassing Mother Earth. The emerging capitalist commercial system, along with Descartes, Newtonian science, and rising literacy, gradually destroyed this connection to nature. The new mechanistic philosophy of the mid-seventeenth century changed the metaphor of nature from a *living organism* to that of a *machine*. This machine metaphor calmed fears of chaos and disorder and offered the illusion of control. The resulting mechanistic world was often referred to as "the clockwork universe."

Historian Morris Berman also describes this medieval view of the world: "The view of nature which predominated in the West down to the eve of the Scientific Revolution was that of an enchanted world. Rocks, trees, rivers, and clouds were all seen as wondrous, alive, and human beings felt at home in this environment. The cosmos, in short, was a place of *belonging*. A member of this cosmos was not an alienated observer of it but a direct participant in its drama. His personal destiny was bound up with its destiny, and this relationship gave meaning to his life."[18] This medieval sense that the natural world was alive is called animism, from the Latin word for soul—anima. They believed that everything has a spiritual reality or soul, including trees, animals, plants, landforms, and bodies of water. Sociologist Max Weber called this shift from the world as living organism to the world as machine "the disenchantment of the world." Adequate social transformation will never be possible until we re-enchant and re-sacralize the world.

Environmental historian Carolyn Merchant, in *The Death of Nature: Women, Ecology, and the Scientific Revolution,* calls this shift from living organism to machine the "death of nature" and says it is the most far-reaching effect of the Scientific Revolution.[19] This shift served commercial interests. For example, people in the Middle Ages believed mining desecrated and destroyed the sacred Earth.[20] Merchant says that this image of nature created an ethics that said Earth must be treated and respected as a living organism. Secular economic concerns needed Earth's minerals, but mining could become legitimate or moral only after science declared nature "dead matter."[21] If nature is a machine, mining no longer harms a living Earth, and we've been raping Earth ever since.

> The human soul has not felt at home in the modern cosmos: the soul can hold dear its poetry and its music, its private metaphysics and religion, but these find no certain foundation in the empirical universe.
>
> Richard Tarnas
> *The Passion of the Western Mind*

Theologian David Ray Griffin also traces this historical shift. He describes the premodern view as a world created by a personal God involving a holistic, "magical" view of nature, in which divinity was immanent. He says that Descartes and Newton accepted these premodern views and assumed their new scientific view of nature would still support a belief in God. They thought that splitting Spirit and the rest of nature (matter) would increase the importance of the Spirit. God (Spirit) became the only creator, and nature became merely matter, having no creative or divine powers of its own. This view was meant to defend the existence of miracles, since anything unexplained by causal science was attributed to God. Therefore, events were either natural or the work of God, so that "when extraordinary events occurred (which no one doubted) they had to be considered genuinely miraculous."[22]

Griffin says this effort to make room for both science and God "backfired." Griffin calls this a "mistake:" the scientific method was turned into metaphysics. The change happened in two steps. Originally, nature was believed to be alive with divinity (pantheism or animism). In the first step of this radical change, the Enlightenment emphasized dualism—the world was split into spirit and matter. The second step was the evolution into materialism, in which the spirit is lost and only matter is real. Griffin calls what resulted "scientism," the "doctrine that nothing exists except what can be studied in terms of this [scientific] method."[23] The modern world came to have a worldview with two primary ideas: ontologically, all that existed was matter, and epistemologically, knowledge and "all perceptual experience of the world must come through one's (material) sensory organs."[24]

Christian theology also denied nature was sacred and identified any kind of pantheism or animism as heresy. The church's desire for power played a part in this denial of Spirit in

Horse Chestnut Tree

nature. The church denounced pantheism and animism because they were being used politically to support "leveling" social movements that challenged hierarchical views in society and religion.[25] The Church protected its authority by accepting this spirit/matter split and discrediting pantheism, for in pantheism authority comes from personal experience, rather than from the teachings of the church.

Despite centuries of mechanistic conditioning, we want to name, once again, this world as holy.

Joanna Macy

The loss of belief in a sense of Spirit naturally available to intuition had repercussions. As material*ism* began to define reality, society became more and more material*istic*. Griffin insists that the basic religious drive remains: to live in harmony with the ultimate reality or power of the universe. Atheistic materialism simply meant the replacement of one religious ultimate with another: matter is the ultimate reality of the universe, the creator of all that exists, the measure of value.[26] Materialism therefore resulted in the tendency to equate the real and the material, so that "we are 'in harmony with' matter by possessing and controlling it."[27] The shift to materialism meant the creation of a different kind of person. Earlier spiritualities were focused on self-control; the new science focused on the desire to control outer things. As Griffin says, "Developing virtues and overcoming vices is no longer a way to be in step with those dimensions of reality that are ultimately important. The way to do this is to use power to possess and control the really real thing: matter. *Having* becomes more important than *being*."[28]

This shift from an enchanted world to a mechanistic world has had many consequences. Medieval historian Lynn White, in his 1967 classic article, "The Historical Roots of Our Ecological Crisis," was the first writer to suggest that ecological problems began with religion. He examined the connection between our concept of God and our treatment of nature and traced ecological problems to the loss of animism in religions and to the acceptance of dualism. White says, "In antiquity every tree, every spring, every stream, every hill had its own *genius loci*, its guardian spirit. . . . By destroying pagan animism, Christianity made it possible to exploit nature in a mood of indifference to the feelings of natural objects."[29] White concluded that more science or technology couldn't solve our present ecological crisis. The solution could only come when "we find a new religion, or rethink our old one."[30] In other words, he felt that the crucial question is how "to create for humans a sense of sacredness within the natural world."[31]

Recreating this psychic connection to nature is the theme of David Abram's book, *The Spell of the Sensuous: Perception and Language in a More-Than-Human Word.* His book is about animistic awareness, which he says is a style of thinking that associates truth not with static fact, but with a quality of a relationship. Abram is an ecologist, a philosopher, and a sleight-of-hand magician. He says, "We have lived 98% of our evolutionary existence as hunter-gatherers carrying on an animistic conversation with every flapping form. This is what we are made of!" Yet now we think of the other animals and the land as mute. But in cultures with strong oral traditions, people still speak in ways that keep the world alive.

Short-billed Marsh Wren

Place awareness

Principally, tribal identity is held in stories that reside in the land. The story characters are animals, plants, and other beings that dwell in the land. But today, because the sacred is confined to sacred texts, it frees us to destroy Earth.

Abram wants us to become conscious of living on Earth by re-immersing ourselves in the land, in the soil, in breath, and in the interchange of air with all creatures on Earth. In many indigenous communities, the very air was considered sacred, connecting all living creatures with the world around them and with the Creator. Today, we no longer see the air as the breath of God but as a place to dump pollution. We believe that the power of language is solely the privilege of humans, but Abram says, "Everything speaks, just not in our language." He feels we lost our sense of reciprocity with the natural world with the invention of writing. Before writing, human speech, whether in songs, stories, or spontaneous sounds, was inseparable from exhaled breath: "For so many oral cultures, breath and the air have these deep associations with meaning." Abram continues, "The air is the repository of all those stories that are not being spoken, that are hanging out waiting to find someone who will begin to tell them." In traditional oral cultures people speak in ways that keep the world alive. Scientist Gerald Barney refers to this need, saying, "Every person must learn to think like the Earth, to act like the Earth, to be Earth."[32] We need to recover a sense of awe and wonder at the natural world. Many indigenous people live in deep relation to nature, and their sense of divinity is Earth centered. The world and our place in it changes when nature is reanimated and ensouled. What if trees are not just lumber, what if (as many Native Americans believe) they talk to each other?

There are many programs that are helping people reconnect to a sense of being embedded in nature, rather than being apart from it. In *Reconnecting with Nature: A Restoration of the Missing Link in Western Thinking,* Michael J. Cohen says we have 53 senses, not five. His project, Nature Connect, has published methods and materials for reconnecting with nature. We have learned to think in indoor, nature-disconnected terms. Even outside, most of us move from driveway to cement to parking lot and never even touch the Earth, let alone connect with it. When we reconnect, we can incorporate nature's wisdom in our thinking, regaining peace and healing. We can relate to nature on a spiritual level. His website teaches people how to reconnect with nature.[33]

In a related effort, Richard Louv's book, *Last Child in the Woods: Saving Our Children from Nature-Deficit Disorder,* created a movement with "Leave No Child Inside" initiatives throughout the country in cities and school systems.[34]

Verreaux's Sifaka

Reconnecting to the Divine Feminine

The denial of the sacred in nature is closely related to the denial of any feminine form of divinity. Cultures have traditionally linked women with nature, and the subordination and exploitation of both share the same history. The worldview that created our ecological crises has two roots: first, the mechanistic science of the seventeenth century created by Descartes, Newton, and others, which turned nature into an inert machine that we could destroy at will, and second, the much older patriarchal value system. In the patriarchal system, only the masculine is important, God is only male, and women and feminine values are depreciated and marginalized. This value system emphasizes control, dominance, power, and hierarchical authority. David Ray Griffin describes this shift: "Domination over nature replaced harmonious cooperation as the goal; dispassionate objectivity replaced the union of head and heart as the proper stance of the scientist, who therefore needed no personal transformation; and nature, understood as feminine, was emptied of all spirit and feeling, and especially all divinity."[35]

> The Reformers banned … the cult of the Virgin Mary and the saints, and so took these images away. Christianity became more of a male world than ever. . . . [This] made it a cerebral, emotionally impoverished and narrowly masculine affair, on a more basic level the only important people and "gods" now were all men.
>
> Karen Armstrong
> *The Gospel According to Women*

This discussion is not about men and women. Many women share a very patriarchal view of power and the world, while many men share the values of empathy and community. Feminism is not about men—it is about a concept of power called patriarchy, which is the hierarchical "rule of the father" or the patriarch, who rules the sons as well as women. In denying the legitimacy of domination, we also need to embrace "feminine" values, such as care, community, nurturing, and cooperation. To do that, we need to become conscious of how our culture uses ideas and to ask, What do we lose by living in a culture in which imaging God as female is still heresy? The idea that God is male is not about genitalia; it is therefore a social construction, and social constructions are never neutral. Consider what is being affirmed and what is being denied. This concept assigns the characteristics of half of humanity to God, and the characteristics of the other half of humanity are those that are emphatically *not* God. As theologian Mary Daly says, "When God is male, the male is God." But it is not just males, it's the entire authoritarian worldview that is reinforced and legitimized by this concept of God.

The Denial of the Divine Feminine

Nowhere has the denial of the feminine been more intense than in religion. The loss of the divine feminine has a long history and occurred in both Judaism and Christianity. As Judaism and Christianity developed, they both named the feminine irrelevant or dangerous and denied the validity of Goddess religions, beginning with a change in creation myths. Theologian Rosemary Radford Ruether says the creation myths in Goddess religions were replaced with Biblical stories that she calls "patriarchal reversal myths," based on the "war against the mother" and the eventual suppression of the mother symbol altogether.[36]

Originally, female gods or nature spirits represented the generative power of life. But in Genesis, the natural process was reversed, and two male figures (the male God and Adam) create new life, and this life is created not through a womb, but with a rib. Through this, Ruether says, the transcendent male ego claims "to be itself the creator, not the child, of the mother."[37] The creation story then goes on to depict the female as disobedient and therefore responsible for sin. Goddess religions, although based on female power, included both Gods and Goddesses. The Goddess and God did not compete for power but were worshiped for the role they played in the cycle of nature. Ruether comments, "The peculiarity of imaging God solely through one gender . . . is ignored because male monotheism reinforces the superiority of patriarchal rule."[38]

God writes the Gospel, not in the Bible alone, but also on trees and in the flowers and clouds and stars.

Martin Luther

We do not need to return to Goddess religions, but we do need to be aware of the vast difference between the logic of male-based religion and the logic of female-based religion. The two create very different worlds, with very different priorities. In this discussion, keep in mind that the word "Goddess" can refer to ancient pagan nature religions, but also, in English, "Goddess" is simply the feminine form of the word "God," and the term needs to be used in a discussion of the relationship between "the feminine" and God. Also keep in mind that our religious language does not change the reality of God, but new religious language does change our politics and our worldview. "The feminine" is such a red flag because feminine values inherently deny hierarchical and authoritarian power. Honoring the feminine is what will allow us to humanize the world and create societies of partnership, mutuality, and reciprocity. The loss of our connection to the feminine is intimately connected to the loss of the sacred in nature. When the female Mother Earth became dead, mechanically interacting matter, Earth was no longer a *She* but became an *It*. Our current worldview was created by discrediting everything female. The new worldview cannot be created until the feminine and masculine are brought back into balance.

Mary Condren and Leonard Shlain explain, from different perspectives, what happened when the feminine was lost in the shift away from Goddess and nature religions. Condren examines the history of when Judaism first replaced the Goddess religions and again when Christianity replaced the Goddess religion in Ireland. Scientist Leonard Shlain describes the effect of reading on the brain and contrasts the worldview of the pre-literate Goddess religions with the worldview of religions based on the written word.

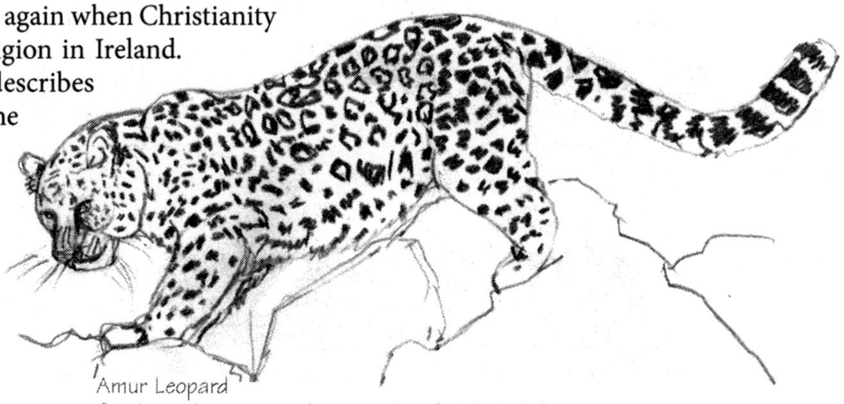

Amur Leopard
On the eight most endangered list (WWF & ICF)

The Loss of the Feminine with the Loss of the Goddess

One of the books we read in a seminary course called Power, Authority, and Abuse was *The Serpent and the Goddess: Women, Religion and Power in Celtic Ireland* by Mary Condren. Condren was a Carmelite nun for three years and is now Director of the Institute for Feminism and Religion in Ireland, which is an independent educational institute in Ireland.[39] She also teaches at Trinity College in the Centre for Women's Studies. Her book reviews the unexamined power relations and the unconscious logic behind replacing the Goddess religions with the Israelite monotheistic religion and again replacing the earlier goddess religions in Ireland with Christianity. Her discussion of this history tries to uncover the ways in which our religious images function politically. By placing these shifts in larger historical and social contexts, Condren is looking for the motives for shifting from a female-centered religion (and society) to a male-centered religion (and society) and trying to identify what was lost.

> We abuse the land because we see it as a commodity belonging to us. When we see the land as a community to which we belong, we may begin to use it with love and respect.
>
> Aldo Leopold

The story of Eve in Genesis, particularly the role played by the serpent, reflects this shift. In Goddess nature religions in the Middle East, the serpent had been a symbol of the Goddess and of transformation and rebirth in nature, both because the snake sheds its skin and because it hibernates in winter, to re-emerge in spring. The book of Genesis was probably written during the reign of King Solomon, when Serpent symbolism and polytheistic gods and goddesses were still worshiped.

In this era, Israel was changing from a tribal system of organization, without statehood, capital city, or central government, to a centralized system. Because they were frequently under attack, the tribes of Israel needed to consolidate authority in order to mobilize for war. Condren demonstrates that creating this new political organization required a correspondingly dramatic change in the religious and social organization of the people.[40] The serpent image threatened this new religion of Israel: "If Israel was to grow as a nation-state, with all the entailed political and military trappings, goddess religions would have to be overthrown. Alliance would have to be to one god, Yahweh, and the central symbolism of the new religion would be based on Promise and [linear and male-defined] History rather than on the Life and Cycle Regeneration of nature represented by the Serpent."[41]

In polytheism, the gods and goddesses were imaged as local deities, which led to competing loyalties and fostered a decentralized society. However, if the new God was to be able to command unquestioning obedience, his power had to be absolute and centralized so that political power in Israel could do the same. This shift to monotheism also forced a choice of sex (which comes in twos).

These same issues confronted every Western culture as it moved from decentralized forms of organization based on family ties or ties to particular territory to centralized forms of politics and state. Condren says, "The questions raised took many forms but essentially boiled down to a question of conflicting loyalties. To whom was one ultimately responsible: to biological family, tribe, country, political leader, or to ones God?"[42] The primary lesson in the Genesis story of the fall is a warning about the consequences of disobedience. This lesson is repeated in the willingness of Abraham to sacrifice his son Isaac. This "ultimate test of patriarchal ethics" demonstrated the willingness to obey the demands of God, no matter how arbitrary. Obedience, Condren says, becomes more important than life itself.[43] In this contest for loyalty, Abraham demonstrated his loyalty to Yahweh by being willing to break his kinship relations (which were paramount in Goddess religions), transcending his biological ties in an act of obedience to his God, Yahweh.[44]

Condren explains how this new social structure created by the logic of the Father God was entirely different from the logic of the Mother Goddess. Dismissing the Goddess ended the political, familial, and religious sources of power traditionally associated with women. Immortality was no longer found in the blood ties between parent and child.[45] In early nature religions, death was seen as a natural part of cyclical regeneration; but the story in Genesis made death the punishment for sin, which in turn was blamed on women, which was another way to discredit the Goddess.

We have no prescribed way to do the vision quest, no ceremonies for meeting the gods in the magic circle; the faculties with which we might have joined them have atrophied.

Suzi Gablik
The Reenchantment of Art

This same shift from a female-based religion to a male-based religion was repeated in the early Middle Ages, when Christianity replaced the original Goddess religions in Ireland. This, too, was a shift toward obedience and domination. In this case, domination again included unquestioning obedience to a higher authority but added the sexual control of women. Condren says this is related to the development of "male reproductive consciousness" because procreation is a profoundly alienating experience for men. In order to be sure a child is his, a man must control the woman: "giving birth to the hero [or an heir] can only take place under conditions where men are in control."[46]

In Ireland, local gods and goddesses again prevented political unification and therefore reduced the effectiveness of organizing military invasions. This shift in religions represents a fundamentally different way of organizing society, one in which conquest becomes the measure of greatness. Condren says that in this new religion, "patriarchal organization is based on territory rather than on relationship, on hierarchy rather than on consensus, on contract (or law) rather than on kinship. Military might rather than the rule of custom is respected."[47] She says that "as patriarchal forms began to develop both in the Church and the State, any form of religion in which women played any kind of symbolic or major role was gradually eradicated. Women were seen as the epitome of chaos, men were considered to be the repository of rationality, and so it was a gradual process, but over the period of about five centuries women's

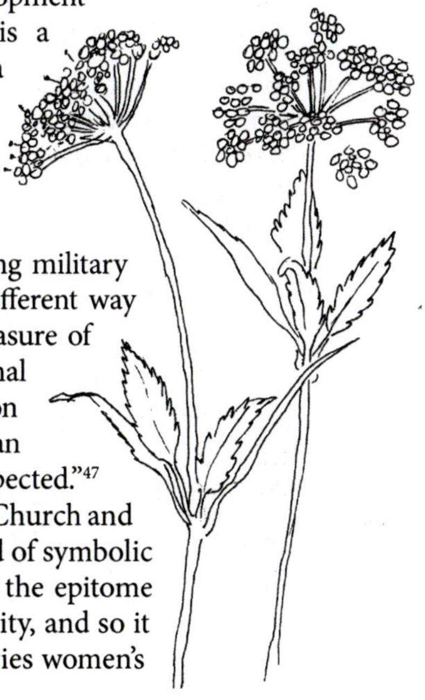

Golden Alexanders

religious authority was dismantled."[48]

To replace the indigenous religion in Ireland, Christians first created a unifying portable god (one not tied to the land), by naming one Goddess, Brigit, transcendent over territorial considerations.[49] Brigit was the female principle and was a synonym for the old Goddesses of Europe. Christianity first "merely claimed that its power was superior to that of the old religion; then it asserted that women only used these powers for evil ends; lastly it denied that women possessed any sacred powers in the first place."[50] Here, the cult of the warrior and hero replaced female creativity with male creativity, which in turn becomes connected with the defiance of death.[51] The only female figure in Christianity was Mary, who became important as a symbol of virginity and as a symbol of the control of female sexuality. Condren says you became a child of Mary if you were perfectly obedient, submissive, and passive.

> The collective psyche seems to be in the grip of a powerful archetypal dynamic in which the long-alienated modern mind is breaking through . . . to rediscover its intimate relationship with nature and the larger cosmos.
>
> Richard Tarnas
> *The Passion of the Western Mind*

Condren then relates the history of requiring celibacy for priests, which she calls "an extraordinary political achievement."[52] In the church, the control of sexuality coincided with the rise of patriarchal power. Priests vowed to become celibate so that they could prove their ability to be in control and that their "will" could overcome "nature" (i.e., the powers of women).[53] The power of "wise women" or the Goddess was closely related to fertility and sexuality, and to compete with the power of the Goddess, priests had to have power related to fertility.[54] The relationship between control of sexuality and of women and the rise of patriarchal power in the church began in the Middle Ages and has continued til today. These two concepts—obedience to one's superiors and the control of sexuality by men—established and maintained patriarchal dominance.[55]

By making sex sinful, the church created a powerful way to control the minds of the faithful because the church could not use physical force or economic power (since they were dependent on the contributions of the faithful).[56] Sexuality was denounced as evil, whereas in the Goddess religions it was sacred. Women had been revered for their ability to give birth. Redemption had been seen as the rebirth of natural cycles. Tragedy, ambiguity, and cyclical regeneration were assumed to be the natural way of life. But with the shift to the Father God, women, once revered as symbols of creativity, became signs of danger, temptation, and pollution. Once this shift was made, Condren says, "needless to say, women could not be bearers of the new 'life' in Christianity, for whereas in the old religion, the Goddess could be found in nature—the trees, bushes, and holy wells, now access to God the Father took place through the mediation of the male priests."[57]

Condren describes how the "birth-based" religion of the Goddess was replaced by the "death-based" religion of the Father God. This exchange meant the loss of female religious power and a shift to a very different logic. "The cult of the dead, relics and martyrs, replaced the veneration previously given to objects of life. The blood of the martyrs was known as the 'seed of the church', thus ensuring that symbols of death

death debt

rather than those of life would become common. For Christian theology, it was not birth but rebirth that mattered; the rebirth of the soul rather than the flesh."[58] This rebirth was imaged by baptism, which in Christian theology is made possible by the death of Christ. Baptism "represented a further denigration of natural birthing in favor of that which took place through death. Baptism represented an extremely potent form of fertility, far superior to natural birthing, but it was a rebirth that essentially occurred, or was made possible, through death."[59]

In these changes, there were three overlapping concepts that defined the shift from a female-based religion a male-based religion, shifts that still define much of our current society.

1) *Power.* To consolidate political power, the culture created one central god. Some of this drive was simply utilitarian, to enhance military power through political centralization. But in addition, men did not want to see themselves as dependent on the Mother (human or nature). In the goddess religions, people tried to live in harmony with the immanent divine. But the male-based religions created a unilateral, external power in order to legitimize a belief in power hierarchies (of which men were the earthly representatives) and of the use of this power to dominate.

   In today's theology, this glorification of absolute power has been questioned in liberation theology, feminist theology, and process theology. Liberation theology is about giving power to the marginalized. In process theology, we co-create the world with God because God's power is always limited or expanded by our response. This theology sees God as using only persuasion, never coercion, so that power is seen as power-shared rather than power-over.

2) *Birth.* The second primary motivation that Condren and Ruether mention is "male reproductive consciousness." The authority of men is closely related to men's control of women's bodies. Originally, the culture did not realize that men had any connection to birth. Fatherhood was a social construction, whereas motherhood is physiologically obvious. When people finally understood conception, men decided that the entire baby was present in the male seed, and women only contributed the incubator. Women were referred to as the "empty vessel." And because the whole baby is in the male seed, masturbation (by men) and homosexuality became seen as tantamount to murder and therefore a grievous sin.

3) *Death.* Fear of death led to a religion that both defies death and worships it. Death, unlike birth, was

Tropical Kingfishers
Fewer than 125 left

something men could "create." In her 1923 book, *His Religion and Hers,* the feminist economist and philosopher Charlotte Perkins Gilman describes the ways in which the crises of male and female experience result in very different orientations to life:

> To the [male] death-based religion, the main question is, "What is going to happen to me after I am dead?"—a posthumous egoism. To the birth-based religion, the main question is, "What is to be done for the child who is born?"—an immediate altruism. . . . The death-based religions have led to a limitless individualism, a demand for the eternal extension of personality. . . . The birth-based religion is necessarily and essentially altruistic, a forgetting of oneself for the good of the child, and tends to develop naturally into love and labor for the widening range of family, state, and world.[60]

Mel Gibson's movie *The Passion of the Christ* took the death-based religion to the extreme in his portrayal of a Christianity that worships death and suffering, rather than a Christianity of life and love. I know one minister who said he was going to devote the rest of his career to telling people that this movie is *not* what Christianity is about.

The Loss of the Feminine with the Written Word

Leonard Shlain, in *The Alphabet versus the Goddess: The Conflict between Word and Image,* describes this same shift from Goddess to God religions, again explaining how our current worldview developed. In his description of this loss of the feminine, Shlain also sees a shift from organic holistic societies to more mechanistic and male-oriented societies. However, he sees literacy and its effect on the brain as the primary cause of this shift. Shlain is Chair of Laparoscopic surgery at the California Pacific Medical Center in San Francisco and Associate Professor of Surgery at University of California–San Francisco. As a brain surgeon, Shlain became interested in the relationship between cultural changes and brain development.

> The historic mission of our time is to reinvent the human—at the species level with critical reflection, within the community of life systems, in a time-developmental context, by means of story and shared dream experience.
>
> Thomas Berry
> *Reinventing the Human*

Shlain's research revealed that acquiring literacy may "re-wire" thinking patterns and fundamentally change the way a culture understands reality. Using the terms *word* and *image*, he reviews the history of literacy and its effect on culture. He calls his findings circumstantial evidence but adds that the correlations are very strong. These findings demonstrate that learning to read brings a loss, along with its huge advantages and gains. Awareness of this loss can help us retrieve what was lost and create a balance.

Shlain contrasts characteristics of pre-literate (Goddess) cultures with those of post-literate cultures. Pre-literate cultures relied on images in which the brain grasps all parts of the whole simultaneously, integrating the parts; most images are perceived in an *all-at-once* manner. Reading relies on words, which the eye scans in a linear sequence; written words are understood in a *one-at-a-time* fashion.[61] Shlain says, "A feature of nonverbal communication is that no symbolization interferes with the direct appreciation of reality. The right brain perceives the world concretely. For example, a facial expression is 'read' without any attempt to translate it into words. The right hemisphere is also the portal leading to the world of the invisible. It is the realm of altered states of consciousness where faith and mystery rule over logic."[62]

However, in deciphering alphabetic writing, the brain uses reductionism, sequence, and analysis. Custom and language associate the holistic, simultaneous, and synthesizing worldview with feminine values, and associate linear, segmental, and abstract thinking, with masculine values.[63] However, he points out, all men and women have both.

Most people are on the world, not in it.

John Muir

The early Goddess religions, which relied on images rather than words, gave way to the three patriarchal Western religions, Judaism, Christianity, and Islam, all based the written word. As Shlain describes it, "Each monotheistic religion features an imageless Father deity whose authority shines through His revealed Word, sanctified in its written form. Conceiving of a deity who has no concrete image prepares the way for the kind of abstract thinking that inevitably leads to law codes, dualistic philosophy, and objective science, the signature of Western culture."[64] Pre-literate cultures were formed by holistic, right-brain modes that venerated the Goddess, images, and "feminine" values. Shlain contends that reading (using linear left-brain thinking) upset the balance between men and women, initiating the disappearance of goddesses, the abhorrence of images, the decline of women's social and political status, and the long reign of patriarchy and misogyny.

Hundreds of goddess-based sects thrived around 1500 B.C. in the Mediterranean world. Overwhelming archaeological and historical evidence shows that in prehistory and early history, both men and women worshiped goddesses, women functioned as chief priests, and property commonly passed through the mother's lineage. But by the fifth century A.D., Goddess worship had almost completely disappeared, and by then, women were prohibited from conducting a single major Western sacrament.[65]

Shlain contrasts the "feminine" right-brained oral teachings of Socrates, Buddha, and Jesus with the masculine creeds that evolved when their spoken words were committed to writing. For example, the first two of the Ten Commandments reject any goddess influence and ban any form of representative art, that is, images.

Shlain sees the church as using

Okapi Family

two weapons to fight for control: sexuality and literacy. Pope Gregory VII (who became Pope in 1073) decreed that corruption in the church was the result of priests breaking their vows of celibacy, so he ordered them to renounce their wives and children. Married men who chose to remain priests abandoned their wives to live in destitution in a society that now saw their children as illegitimate. Abbesses, who had been powerful and autonomous, were now tightly controlled by abbots.[66] Then in 1229, at the Council of Toulouse, church leaders decreed that "no lay person could possess any scriptural text in Latin, no scripture could be translated into the vernacular, and no university text written in Latin could circulate off campus. Since only males were admitted to the university, these edicts ensured that no woman could learn the language of religion, science, or philosophy. The punishment for disobedience was the stake."[67] This law also ensured that the men in control had virtually no input from women.

By ethical conduct toward all creatures, we enter into a spiritual relationship with the universe.

Albert Schweitzer

Studies of brain behavior reveal that the values that typify right brain activity include empathy with the plight of one's companions, generosity toward strangers, tolerance of dissent, love of nature, nurturance of children, laughter, playfulness, mysticism, nonviolence and the forgiveness of enemies. These aspects, in both men and women, express the feminine gatherer/nurturer side of human nature.[68] These gatherer/nurturer attributes are similar to the nurturing parent model from George Lakoff, the Earth Community model of David Korten, and the interconnected web/relationship model of ecological/systems science. In contrast, Shlain says, "People tend to exhibit left-brain attributes when absorbed in work, goals, focus, power, and money, attributes that would make a man or woman a successful hunter/killer."[69] The hunter/killer attributes echo the authoritarian model from Lakoff, the concepts of Empire in Korten's work, and mechanistic science.

Emphasis on masculine values increased after 1454, when Gutenberg invented the printing press and literacy rates exploded. Shlain contends that "the sudden inundation of society by alphabet letters caused a dramatic increase of left-brain hunter-killer values throughout Europe . . . . These trends became exaggerated in the wealthiest, all male, most literate segment of society—the papacy."[70] In medieval time, the church forbade people to read the Bible. In 1517, Martin Luther began the Reformation when he translated the Bible into German and published his sermons and writings. The primary religious conflict in the Reformation was over who should be allowed to read and interpret the sacred book. After Luther and after Guttenberg's printing press invention, violent religious persecution, called "the religious wars," began all over Europe and lasted for 150 years.[71] Germany's Thirty Years' War (1618–1648) began with Lutherans and Calvinists killing each other "over hair-splitting doctrinal disputes." Then, Protestants and Catholics fought each other. Shlain says, "Taken as a whole, the religious wars that wracked Europe for these 150 years after the printing press had transformed European culture can be viewed as a sort of mass madness."[72] This rejection of the feminine led to the witch-hunts in the fifteenth and sixteenth centuries. As Shlain says, "Witch hunting was *woman* hunting."[73] No group before or since has reached the "level of credulity and psychosis that seized the most educated male elite during the witch craze. As if in a deep hypnotic spell, men accepted as fact . . . that little girls in pigtails, pregnant women, and weak, elderly widows posed a mortal danger to society."[74]

Anne Llewellyn Barstow, in her book *Witchcraze*, estimated from court records that more than 100,000 people were killed. Some more recent investigators claim that the number was more than 1,000,000.[75] Shlain says this misogyny was so intense that all women and men who supported female values were completely eliminated:[76] "During the height of the witch craze, women were killed for being women, and women's rights and values plunged to their all-time low. Concurrently, Mary, the only surrogate Mother Goddess in all of Western culture, was under fierce attack by male members of Protestantism, a new alphabet-based religion that called for destroying every image of her they could find."[77]

Eventually, due in large part to the rise of science, the witch-hunts slowly abated and became what Shlain calls "a disturbingly repressed cultural memory." Yet even today, one way to discredit women, especially women's power, is to bring up witchcraft, evil, and the devil.

Shlain says that balance is now slowly being restored. Momentum is developing throughout modern culture to embrace feminine values and return balance to our lives and communities. In the nineteenth century, images returned with the invention of photography. Later, the discovery of electromagnetism led to images in film, television, computers, graphic advertising, and now the Internet. All of these have greatly increased the power of images. Shlain feels the imbalances are beginning to be corrected. He believes greater reliance on right-brain pattern recognition over left-brain linear sequence will move our culture toward equilibrium between the two hemispheres, between masculine and feminine, between word and image.

The Return of the Feminine

The paradigm shift from the current worldview to the new worldview is a shift to embrace feminine values and return balance to our lives and communities, undoing the repression of the feminine and the "death" of nature. In *The Passion of the Western Mind*, historian Richard Tarnas calls this shift "the return of the feminine." Tarnas says that centuries of masculine assumptions about reality, knowledge, nature, society,

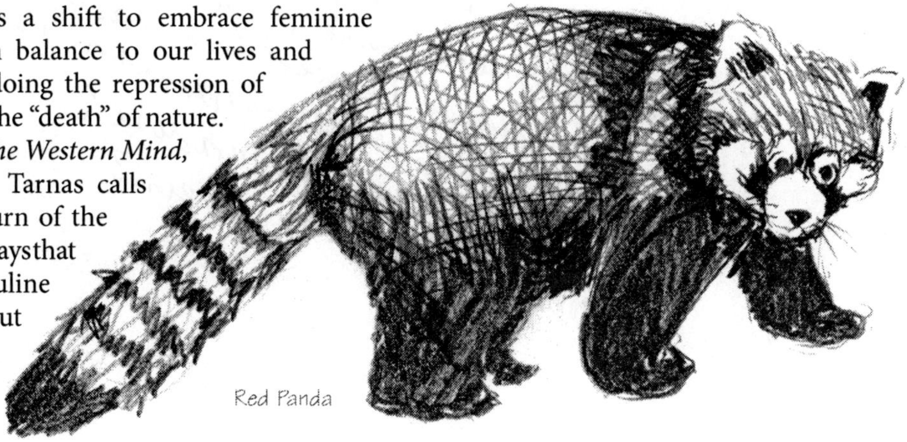

Red Panda

and the divine, along with parallel patterns and structures of domination, have resulted in the loss of living a full, "whole" life.[78] Therefore, he says, "*The crisis of modern man is an essentially masculine crisis.*"[79] This new attempt to create balance is going on in much of the culture, an attempt he believes is the culmination of history:

> And this is the great challenge of our time, the evolutionary imperative for the masculine to see through and overcome its hubris and one-sidedness, to own its own shadow, to choose to enter into a fundamentally new relationship of mutuality with the feminine in all its forms. The feminine then becomes not that which must be controlled, denied, and exploited, but rather fully acknowledged, respected, and responded to for itself. It is recognized: not the objectified "other," but rather source, goal and immanent presence.[80]

In recognizing the feminine as source, we also recognize nature as source. In fact, Tarnas believes that the uncertainty and confusion of our time is because this shift may be reaching a tipping point in this struggle "to bring forth something fundamentally new in human history."[81]

The sustainable world we need to create will include a politics of compassion and a world community of mutuality and reciprocity, an economy that works for everyone and preserves Earth, all based on a spiritual sensibility in which all people and all of nature are seen as expressions of a divine reality.

Ring-tailed Lemur

Velikovsky - Worlds in Collision
The neurological, genetic effects
unconsciously delivered them customs +
worldviews as well. Explanation of
ejection from Paradise?
other myths in this vein?

This chart summarizes the contrasting concepts of the two types of religion, discussed further in the next chapter. The left side gives the characteristics of the old worldview, and the right side has the characteristics of the new worldview.

The Two Worldviews in Religion

RELIGION OF EMPIRE	RELIGIONS OF EARTH COMMUNITY
Hierarchical relationships of authority and obedience	Interconnected relationships of compassion and empathy
Closed	Open
Linear man-made history	Cyclical, ongoing nature
God as transcendent, abstract, "out there"	God as immanent, often grounded in Earth
Pursuit of certainty	Ability to hold ambivalence and ambiguity
God must be male	God can be imaged as male, female, both, or neither
The masculine dominates	Feminine and masculine are in balance
Statements of certainty and absolutes	Our relationship with God is primary
Judgment	Politicizing compassion
Reality is secular	Reality is sacred
Structure	Process
God as strict Father	God as nurturing parent
Literal	Metaphorical and historical
Belief centered	Transformation centered
Word-based	Image-based
Reductionist	Holistic
Christendom paradigm	Emerging paradigm
State and religion merge	State and religion separate
Salvation in next world	Creating Kingdom of God in this world
Only Christianity is valid	All religions of compassion are valid
Prophetic and pastoral	Transformative for individuals and society

Chapter 11

Sustainability and Reinventing the Church

Much of the spiritual transformation we need will occur outside any organized religion. However, our global ecological crises require collective solutions, and these solutions need an institution to give them form and focus. Providing this institutional form is the primary missing link in the struggle to adopt the new values from ecology. Could the Church be the institution that makes a public religious call to the ecological vision? I believe the Church (and other religious institutions) can provide this focus. No other institution has the resources, ability, or interest in making a spiritually based argument. In the U.S., there are 100 million people connected to the Christian church; they are the largest grassroots organization in this country and could create a massive groundswell for change.

> For Creation waits with eager longing for the revealing of the children of God.
>
> Romans 8:19

The life and teachings of Jesus reveal a powerful example of the fight for radical social change. I once heard a description of Jesus that said, "It was as if history was a truck barreling down a hill out of control. Jesus threw himself in front of the truck, hoping to deflect the course of history. And he died, never knowing if it would make any difference." History is again out of control, and Jesus could again make a difference through a transformed Church. What would happen if the Church sacrificed its concern for doctrinal authority and control and instead threw itself against history, trying to deflect the coming crisis? And what would such a Church look like?

The effort to renew the Earth can also renew religion and in so doing bring new relevance to the message of Jesus for modern life. In addition, if the Church is not relevant to the world, the Church itself will not survive. Earth is dying. But so is the Church. I believe a transformed Church can return life to both.

Redefining the Church's Identity and Mission

Adequate transformations will never happen unless the mission and identity of the Church become saving God's Creation and creating the world anew in terms of sustainability. Loren Mead, who studies congregations, says, "*If you want to move the world, move a congregation.*"[1] I have long believed the Church could have a major impact on our response to our ecological crises, both conceptually and organizationally. The transformations needed include seeing our life together—nationally and globally—as one of mutual support and care, rather than a quest for dominance. It requires grounding our lives in the spiritual experience of the world and each other and creating a renewed valuing of "the feminine." It will also require a new understanding of economics. Today, our God is economic growth, and it is a jealous god that will have no other gods before it. Could the Church help create the political will for transformative change? That is, can the Church move from what Marcus Borg called a belief-centered paradigm to the transformation-centered paradigm described in Chapter 4.

Mainstream churches are losing members because their message has lost relevance to people's lives and to the problems facing the world. Theologian John Cobb Jr. says our mainline churches are empty because they don't have a convincing message. He said the Church should be the place we come to discuss the most important issues, a place where people can experience an opportunity to become what they need to be.[2] How would the Church actually become an agent of transformation in the public? Peter Senge's suggestion for institutional growth is, "Don't push growth; remove the factors limiting growth."[3] The major factor limiting Church growth is the lack of a convincing message. We need a vision that includes both a rich inner sense of the Spirit (mysticism) and an outer concern for creating the changes we need. If churches created dialogue groups and began to build a compelling shared vision of the world we need, membership would not be a problem.

The ecological crisis changes what it means to be religious, but people are always finding new ways to follow their faith. More than anything else people want meaning; they want to be empowered and inspired; they want to be called to commitment. Human survival requires a Church centered in the transforming experience of the presence of God that can give us the courage to change the systems of oppression, dominance, and power that are destroying Earth. Such a theology would carry God's message to live in harmony with all living creatures and with those who are still to come. The Spirit is leading society to a new world so that the Church can help God save the world. We now desperately need the Church to fight for planetary salvation rather than personal salvation.

Could the Church be the institution to make a public argument that, although grounded in its own beliefs and traditions, gives form to (incarnates) the ecological vision in a language that all religions can share?

I know that most churches are overwhelmed and understaffed and have few additional resources for yet another issue. One of the reasons for this overload is the large number of people who have dropped out of the Church and a society in which people have little time for volunteering. But addressing our ecological crises and changing our culture's worldview cannot just be one additional project among many. We need churches that understand their very identity and mission as saving Creation, allowing people to connect their faith with the efforts in ecology.

Will the Church become irrelevant, or will it be central to global transformation? The civil rights movement would not have

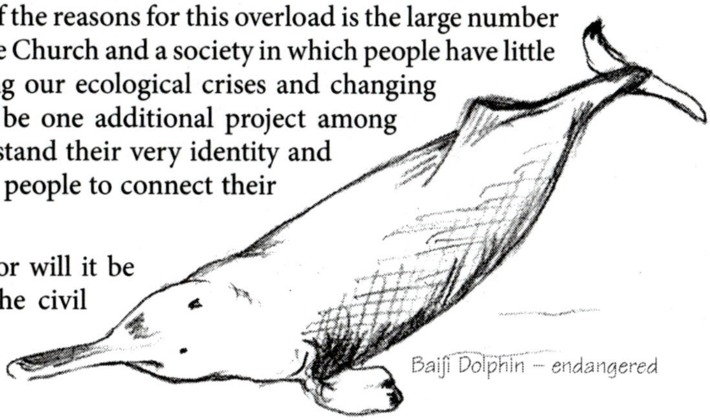

Baiji Dolphin – endangered

happened without the Church. Now we need the Church for an even larger task. We need the Church to witness that a new world is possible, a world that privileges people over profit and a world in which each of us accepts responsibility for the whole. But as we begin to see what the new life could be, it becomes clear that it requires the death of the old. It is the Church itself that needs to undergo a death and resurrection experience. The Church would have to "lose its own life," in the sense of any claim to an exclusive truth, in order to allow for the resurrection of a truly spiritual reality. What we need is a new Reformation.

In churches, there is a natural understanding of the need for the old to die before the new can be born. It has always seemed to me that the most interesting day in the Easter story is Saturday. How do we live when we've seen the crucifixion but do not yet know Easter? If we have the social upheaval anticipated by ecologists, we will be living through many Saturdays. Christianity has a powerful, constant image of hope, allowing people to live with a sense of radical trust. Christians can live creatively in the midst of uncertainty because they know how the story ends: transformation to a new life is ensured.

> If a revolution succeeds, it does so normally as a cathartic renewal of religion itself, [for no] true liberation is possible unless persons are religiously motivated toward it. To be religiously motivated is to be drawn from the depths of one's being.
>
> Aloysius Pieris
> *An Asian Theology of Liberation*

The language of the sacred and a deep sense of Spirit are fundamental to a meaningful understanding of ecology. What I am describing is the Church the world needs and the one I would like to join. In the Seminary, discussions often focused on the reasons mainstream churches were losing membership. However, these conversations took place between people who had spent their entire lives connected to the Church. They had little input from people who saw themselves as religious but were no longer in the Church.

I have two related goals here. I want the politics of liberals to include a strong religious dimension. I also want to define a religious vision that would draw back into the mainstream Church many of the millions of people who have left, so that together with those still there, the Church can take the lead in creating the world anew. To do this, I want to make clear how different the mainstream Church is from fundamentalist religion and therefore how much more room there is in it for many different views.

For the Church to matter again, the message of the Church must matter, and it must matter for this world, not the next. The Church is simply too important to give up on, and there is no substitute. Many churches have been very successful in working to serve the needs of their congregants and their local community, as well as organizing an outpouring of practical and emotional support for all levels of disasters, including the tsunami in 2004 and Katrina in 2005. The Church's ability to coordinate volunteers in massive efforts is awesome. However, these efforts are focused on specific problems. Would the Church be able to articulate the scale of the climate crisis and embody the new thinking required to solve it?

Three Paradigms in Church Life and Mission

To understand what the Church could be now, we first need to see what it has been in the past. The idea of mission has always shaped the life of the Church. Loren Mead, in *The Once and Future Church: Reinventing the Congregation for a New Mission Frontier,* says that over time, the Church has had three different understandings of its mission.

The Apostolic Paradigm

The first understanding of mission is what Mead calls the Apostolic Paradigm, the time of the apostolic period, or the first three centuries. During this time, the early Church existed in a hostile, antagonistic, and persecuting world. Mead says, "Each group of Christians was an illicit community, proscribed by the laws of the land. In many places, it was a capital offense to consort with or be a Christian."[4] Life for the early Christian communities was intense and personal. Jesus gave a powerful call for a people to be sent forth. Mead says that, at this time, each member was aware of their call to witness to God's love in that hostile world, in a mission that started at their front door. They understood that they had been sent to serve and convert the world, to care for the sick, the prisoner, and the widow, the fatherless and the poor, although what this meant varied in different locations.[5] They understood God's good news as a call to every member, while the congregation tried to give members the courage, strength, and skill to engage in that mission.

The Christendom Paradigm

The second understanding of mission, or the Christendom Paradigm, began in the fourth century, in A.D. 313, when the Roman Emperor Constantine converted to Christianity and made it the official religion of the Roman Empire. In this new paradigm, there suddenly was no distinction between the world and the Church, and the Church no longer existed in a hostile environment; rather, the law now identified the Church with the Empire.[6] Mission was no longer the responsibility of the ordinary person; it now happened at the edges of the Empire. Professionals—the soldier and the missionary—carried out the Empire's and the Church's mission. In this paradigm, mission and imperialism became identical.[7]

The congregation became a parish, a geographical region in which all the people living within that area were part of a congregation by birth. Being a good Christian was the same as being a

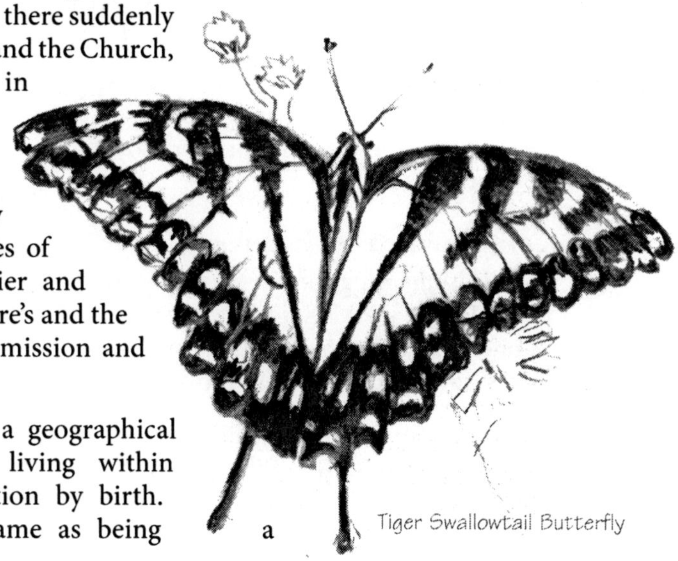

Tiger Swallowtail Butterfly

good citizen. Mead says, "One's sacred duty was to preserve the way things were and to avoid personal immorality, disloyalty, or disobedience to constituted authority."[8] Loyalty and obedience were the primary virtues. Mead argues that many people in contemporary society have grown up in the Christendom paradigm, and most of the structures of congregational, denominational, and ecumenical life remain oriented to this paradigm. This is still true because as the connections between Empire and Church came apart, the Christendom paradigm did not die.[9] Within this paradigm, it makes sense to fight for a Flag Amendment. The amendment says it is illegal to "desecrate the flag." Desecrate is a religious term meaning to violate the sacred character of something. Normally, it would be idolatry to suggest that the national flag is an image of deity or an object of worship. However, when religion merges loyalty to God with loyalty to the state, it is natural to see the flag in in religious terms.

> At the moment of commitment, the universe conspires to assist you.
>
> Goethe

Today, conservative religion emphasizes patriotism and support for any U.S. war because they still live in this paradigm. Even though many conservatives connect support of American wars with religion, it is extremely non-Biblical. Jesus preached against violence; as a bumper sticker asks, "Who would Jesus bomb?" During the Bush administration, conservative religion helped to create the imperial presidency (as in Empire). Conservatives see the president as all important, which becomes a modern version of the divine right of kings, and a fundamental part of a worldview of authority and obedience. This is a very undemocratic concept, but it is in keeping with the merger of religion and the state; in this view, there is no problem in merging church and state. The real separation of church and state is not about keeping displays of the Ten Commandments out of public buildings. It's about maintaining democratic institutions and culture without an authoritarian worldview that merges the power of God and the power of Empire. We can require the separation of church and state in the law, but to be effective, this separation needs to be done in theology.

When Christianity was co-opted and turned into the imperialist religion of the Roman Empire, we lost the religion of compassion, grace, and mutual care that Jesus preached. Ever since, the Church's primary loyalty has been to itself and perpetuating its own power and control.

The Emerging Paradigm

A third, emerging paradigm is replacing the second because the Christendom paradigm is no longer working. In fact, Mead says, "The era of Christendom is over."[12] This third understanding of mission is now developing but is not yet clear. In this paradigm, the mission frontier has again moved back to the local.

Once again, the Church and the individual person of faith are beginning to discover a sense of a new mission frontier. Mead writes, "We are surrounded by the relics of the Christendom Paradigm, a paradigm that has largely ceased to work. But the relics hold us hostage to the past and make it difficult to create a new paradigm that can be as compelling for the next age as the Christendom Paradigm has been for the past age."[13] Therefore, our task, Mead insists, is nothing less "than the reinvention of the church:"[14]

> We are at the front edges of the greatest transformation of the church that has occurred for 1,600 years. It is by far the greatest change that the church has ever experienced in America: it may eventually make the transformation of the Reformation look like a ripple in a pond. That transformation is occurring because of the persistent call of God that our whole world be made new, and that the church's mission in that world be itself transformed in new patterns of reconciling the world to God.[15]

Mead admits the new church may not be what we expect or want.[16] I believe that this new Church should be defined by the threat to God's Creation, which I, like many others, experience as a new revelation and a new call to covenant.

Mead says people dropped out of the Church because the Church stopped being important to them.[17] Membership in the mainline denominations has been changing profoundly. Church membership grew enormously after World War II and peaked around 1965. Since then, membership has declined in all mainstream denominations, with a loss of 20% to 40% of their members. However, during this same time, the population of the U.S. grew by more than one-third, so the losses in terms of a percentage of the total population are much worse.

The problem, Mead says, is that "the things we do best have little to do with who stays or who goes."[18] Yet this loss of membership "is not something we can generate a program to fix. A project to increase church membership will not work. Nor will a new improved stewardship program. This news is bad news precisely because what we have done in the past is no longer sufficient."[19] I have visited many churches from time to time, including ones with very good sermons, great choirs, and wonderful liturgies. I still leave feeling, So what?

However, like me, Mead says, "I deeply believe that the storm we are in presents the greatest opportunity the churches and religious leaders have ever had."[20] In fact, Mead says, "Never before have those in religious congregations had more—potentially—to give the other structures of society."[21]

Transforming Churches

The climate crisis is fundamentally a crisis of values and meaning. In *The Soul's Code,* James Hillman says, "Each person enters the world called." Søren Kierkegaard said, "We come to Earth with sealed orders." The ecological crisis is our call and our

Cross River Gorilla
On the eight most endangered list (WWF)

orders. We need to ask, What does God want of us at this time in history? I believe that God is calling the Church to transform society in order to restore Earth and ourselves to health. Psychologist Carl Jung said the big problems are never solved, they are only outgrown. The Church could be a major force in creating a society that has "outgrown" beliefs in materialism, radical individualism, and dominance.

In what may be the last years of the human race, the role of the imagination, I am driven to conclude, lies not in supplanting religion but in imagining how existing organized religious traditions might adapt their old resources to meet this new challenge.

Jack Miles
Global Requiem

Ministers speak of an individual church as being either prophetic or pastoral. A prophetic church emphasizes social justice and care for the marginalized. It focuses on doing good works in the local community and on supporting missions. A pastoral church emphasizes taking care of each other, of the elderly and the sick. Here, people come to church to feel better, which is understandable. In addition, most people feel enough stress; they don't want more at church. In the pastoral church, spirituality and community care become essential. Although many churches say they do both, it is very difficult to do so, because a prophetic church challenges and a pastoral church comforts. Yet even prophetic churches do not challenge the system; they generally only provide help for those who need it. The Earth crisis is outside the concepts of either prophetic or pastoral churches. People can serve the poor and those in need, but nothing in the church system challenges people to rethink their world and themselves.

In creating this new church, a fundamental characteristic is that it needs to be democratic. Mead says that in the Christendom paradigm, the role of the clergy was "strong, central, and unquestioned." Clergy had the power to veto every important issue. In that era, the clergy owned the Church.[22] Just like other areas of society, the Church needs to free itself from nonproductive power issues. However, despite the fact that the laity (the congregation) must now reclaim ownership of the Church, the clergy, Mead says, is still a key resource. "They are badly needed to ground the new structures in which lay people will gather to be formed and sent. They are critical training officers for the church of the future. The educational systems we have for training clergy, however, were invented to produce pastoral leadership for a Christendom Church."[23]

Mead says the clergy's main role now is one of assisting and empowering the laity to take responsibility for this new mission. Mead sees the training and re-training of clergy to do this as a major challenge of the churches. Having laity reclaim ownership of the Church is so essential, he says, because there will only be change if the laity takes the lead. However, a major problem is that the staffs of regional and national denominational offices have been cut back repeatedly due to budget problems. Therefore, there is little regional leadership left to help create these changes.

Peter Senge, in *The Fifth Discipline,* a book on systems thinking and institutional learning, says that the biggest problem in creating change is that "individuals learn all the time and yet there is no organizational learning." He says that the critical question in planning is, Can we accelerate institutional learning?[24] His idea for accomplishing this is to build a shared vision that can answer the question, What do we want to create? He says when people are drawn into a vision they can make their own, it is powerful because "a shared vision is not an idea. It is not even an important idea such as freedom. It is, rather, a force in people's hearts, a force of impressive power. It may be inspired by an idea, but once it goes further—if it is compelling enough to acquire the support of more than one person—then it is no longer an abstraction. It is palpable. People begin to see it as if it exists. Few, if any, forces in human affairs are as powerful as shared vision."[25] Senge says, "A group of people truly committed to a common vision is an awesome force."[26]

⇒ Shared vision— Cheryl

Churches can start the process of change even without knowing what to do. Everyone starts from a different place, and because everything is connected, wherever you start is the right place. Begin with people sharing their stories of their love of Earth and their stories of the changes they have noticed. I envision the Church as the organization that facilitates sharing information throughout their networks of churches. The Church can argue for an ecologically sustainable world, support each local congregation as they choose their place in this movement, and connect with others working on this issue. We need a shared vision that includes a rich inner sense of soul and spirit as well as an outer concern for creating meaning in the social choices we make.

No matter how uncertain the outcome, the Church can help people continue to work, all the while trusting in the loving presence of God. The Church can create rituals to celebrate God's Creation and to celebrate each step toward creating an Earth Community.

In this effort, there are three different needs. First, people will need emotional support and help because the coming transformation will be frightening. In order to avoid scapegoating and blaming, we need to address people's fears. As always, the Church can give people great emotional support and a sense of safety even as they work for change. When we are frightened, we can use the language of grace and trust in the loving presence of God.

Second, the Church can develop a process in which people can change their behaviors in their homes, at work, and in their own congregations. These efforts help raise awareness and build commitment. Working on energy, recycling, and other issues can introduce people to the issues and begin to create the political will needed for transformative change. Making these changes in our own lives can become a spiritual discipline. This will include finding non-material sources of fulfillment. Every time a person starts to be more intuitive, relational, and cooperative, the world changes.

Third, the real work is to create a new, shared "global project." To do so, the Church needs to become a learning community. Keep in mind that it is too late for transitional changes; what we

Jellyfish

need is whole-system transformation. However, the primary focus of change is on our understanding of the economy. The Church can be a major force in creating a moral economy, one that puts people and the planet over profit. Our spiritual reality is at issue, not sectarian beliefs. We need religious meanings that provide new ways of understanding the process of world transformation, religious meanings that can mediate the sacred dimensions of reality. Religious education should be about increasing our spiritual imagination and introducing people to the mystics. It would shift the educational tasks of the Church from supporting and sustaining congregational life to shaping the life and work of the congregation so that it can serve as an agent of transformation in the public. Saving Earth is a religious calling.

Mead says the new mission will take generations to take form.[27] I disagree—we don't have generations. Religion that matters has always had a sense of urgency, even danger, about it. We live in dangerous times and we need religious people to take up the challenge of creating the world anew. Grounded in this broader context, each congregation will find its own response.

For this message to be available in the Christian Church, it will be necessary to find a language that is true to the Christian view yet is sufficiently pluralistic to create an argument that can speak to and be affirmed by Jews, Muslims, Buddhists, and other religious people. To save Earth, we need the efforts of all religions, and it is essential that Christians work with other faith systems recognized as valid and complete. Although I focus on the Christian Church, all three major Western religions (Judaism, Christianity, and Islam) are working on our ecological crises. All agree 1) that God exists, 2) that one can be in relationship with God through both personal and collective means, 3) that this relationship places demands on us, and 4) that Earth is God's Creation. The Church, Synagogue, and Mosque can become allies in creating a common voice for a sustainable Earth. Interfaith efforts can create an inspiring vision of how to live together on a sacred Earth by making Earth their common cause.

This joint message must reject the social mechanisms of hierarchical dominance and control, reject materialism as the measure of either life or society, recognize our interconnectedness and need for community, and realize that public choices must be made with reference to our shared ultimate reality: that all people and Earth embody divinity in a sacred universe.

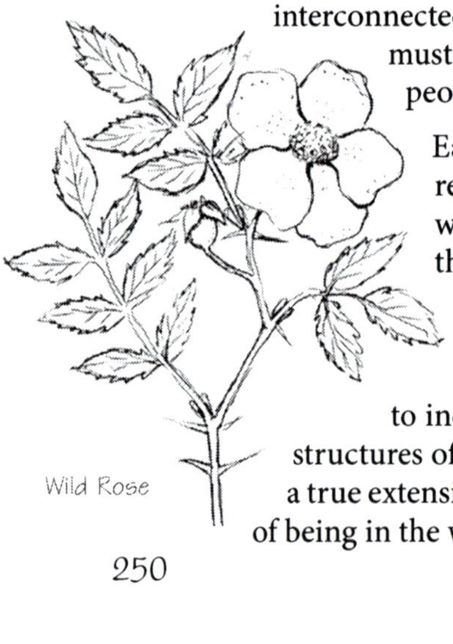
Wild Rose

Earth's powers of renewal of life are in jeopardy. An adequate response would create a new religious and political movement. If we can do this, the political power fight over the language used or the names given to the sacred will give way to a genuine shared sacred reality. If the Church can do so, the Resurrection will become the community of those who have said Yes to what Jesus preached: that is, a personal spirituality that is willing to incarnate divinity and a radical social message about overturning structures of power and dominance. In doing so, the Church would become a true extension of His life into history, and a witness, as He was, to a new way of being in the world.

Part IV

Specific Solutions

Trumpeter Swan Family – Endangered

Chapter 12

Sustainability in Energy

Global warming threatens the environment, the economy, and national security. America responds well to challenges and is capable of switching quickly to renewable energy—we just need to commit to doing so. The next three chapters give specifics about the changes needed for sustainability. The solutions are already available. The total response required by global warming is a political problem, not a scientific one or even a technical one.

> If we can create an honest energy market, the transition to renewables will accelerate dramatically.
>
> Lester R. Brown
> *World on the Edge*

Since the financial meltdown, economists have been saying that no one could have predicted this collapse, but in fact many people did warn us that our financial economy was a house of cards. Ecologists have warned that global warming could cause the collapse of civilization and perhaps warm Earth to the extent that human beings could not survive. But when ecologists say this, people call them alarmists and insist that that kind of collapse could never happen. It not only *can* happen, but it can happen almost overnight. With the financial collapse, we have now experienced how fast radical change can occur in complex systems.

The global warming website RealClimate.org says that climate change is a problem with a very high "procrastination penalty."[1] George Monbiot, a British journalist specializing in global warming and author of *Heat: How to Stop the Planet Burning,* says, "We are the last generation that can make [an adequate response] happen, and this is the last possible moment at which we can make it happen."

We Need to Reduce Emissions by 80% by 2020

Both Lester Brown of the Earth Policy Institute and Al Gore have defined the level of response needed. *Time for Plan B: Cutting Carbon Emissions 80 Percent by 2020* summarizes in eight pages the recommendations in Lester Brown's book, *Plan B 3.0: Mobilizing to Save Civilization.* Cutting CO_2 emissions 80% by 2020 will take a worldwide mobilization at wartime speed. Brown explains the steps needed to reach 80%:

- Energy efficiency could save enough energy to equal the increase in global energy demand.
- Replacing fossil fuels with renewable energy sources for electricity and heat production could cut carbon emissions by one-third.
- Restructuring our transportation systems and reducing coal and oil use in industry could save another 14%.

- Ending net deforestation worldwide could cut CO_2 emissions by another 16%.
- Planting trees and managing soils to sequester carbon could absorb 17% of our current emissions.[2]

I would add that reductions can also be made in our food system, moving to local food to cut transportation costs and greatly reducing beef consumption (see the next chapter). Brown also says these changes do not require new technologies and that what has been missing is leadership.

In June 2008, Al Gore challenged the U.S. to get 100% of our electricity from renewable energy (wind, solar, and geothermal) within 10 years. He said, "We're borrowing money from China to buy oil from the Persian Gulf to burn it in ways that destroy the planet. Every bit of that's got to change."[3] He might have also said that then that money goes to fund terrorism.

Gore's website, repoweramerica.org, lists four steps in shifting to renewable energy for electricity.

- energy efficiency
- using renewable energy
- creating a new unified national electrical grid
- transitioning to efficient, plug-in cars with rechargable batteries charged using renewable energy.

Everyone has been absolutely blindsided by global warming, and the challenge is now much harder because China and India have vastly increased their use of fossil fuels. After the 1995 IPCC report, Japan and Europe accepted global warming and began the Kyoto process. If the U.S. had joined them and we had leveled off our emissions at the 2000 level, we would be creating half the warming now anticipated. President Bush cost us the past 8 years, and this is why the cuts must now be so deep and fast.

We need a rapid worldwide shift to non-carbon energy—wind, solar, geothermal, tidal and wave power, and, ultimately, hydrogen fuels. NASA's James Hansen says that if we do not act decisively within the next 10 years, we will be living on a "totally different planet." He also said,

> We have to stabilize emissions of carbon dioxide *within a decade,* [or] temperatures will warm by more than one [additional] degree. That will be warmer than it has been for half a million years, and many things could become unstoppable. If we are to stop that, we cannot wait for new technologies like capturing emissions from burning coal. We have to act with what we

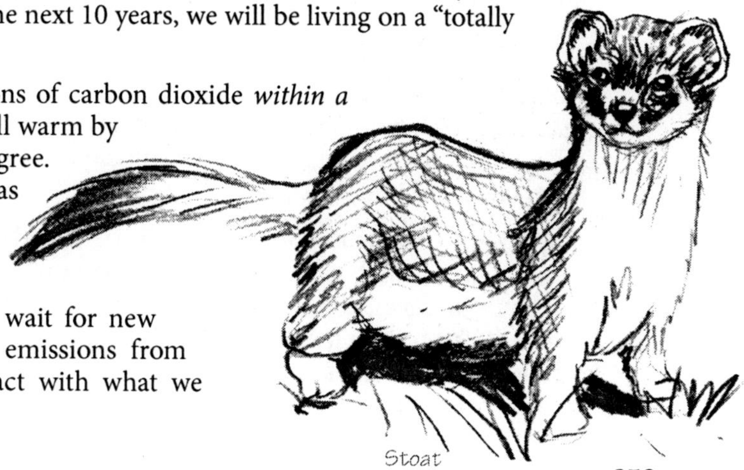

Stoat

> have. This decade, that means focusing on energy efficiency and renewable sources of energy that do not burn carbon. *We don't have much time left.*[4]

The issue of adequately confronting global warming is complicated, but I am making these arguments.

First, the consequences of global warming will cause deaths in the tens of millions and refugees in the hundreds of millions; therefore, it is now a moral imperative that we accept that our economy will have to change.

> Plan B also excludes the oft-discussed option of capturing and sequestering carbon dioxide (CO2) from coal-fired power plants. Given the costs and lack of investor interest within the coal community itself, this technology is not likely to be economically viable by 2020, if ever.
>
> Lester Brown
> *World on the Edge*

Second, arguing for drastic reductions in emissions is usually dismissed because it will "harm the economy." However, not responding will be catastrophic and could collapse the world economy. In addition, you need to ask, *whose economy*? The economy is not monolithic; it is an interconnected web of many "economies." It is true that the changes we need will harm the economics of some industries. It is true that, in time, it will destroy the economies of the fossil fuel industry and the industrial farming industry. But the fact is that both of these industries are on borrowed time already—one due to peak oil and the other because its practices are unsustainable. There are always winners and losers in economic shifts. The problem is that these two industries have enormous power to control the debate and limit changes.

Third, if we choose to respond at the rate the crisis demands, it could restart the economy by creating an explosion of new industries and millions of new jobs. Sustainable practices are generally more labor intensive, and these jobs are often high paying and highly skilled. In addition, most sustainable jobs are also place-specific, and so they cannot be shipped overseas. These jobs are embedded in the local economy, so towns will not be lowering business taxes to get large corporations to move to their area. Instead, being local, jobs will be decentralized, more stable, and under local control. If we respond as we need to, the new companies and jobs in the renewable energy sector will rival the number of jobs jobs created by the computer. In addition, much of what needs to be done will revitalize small towns and rural America.

Fourth, reducing energy use saves money for companies and homeowners. The organization Climate Group profiled 84 companies and found that, among others, DuPont reduced their greenhouse gas emissions by 60% and saved $3 billion with their energy efficiency improvements.[5]

Fifth, if we no longer import foreign oil, we would cut our trade deficit in *half.* And instead of giving all that money to foreign economies, it could be available for our own economy. Cutting our greenhouse gas emissions will also improve our national security. Because we would no longer have to protect our access to oil in other countries, we could have a much smaller military.

*Sixth, t*he complex, expansive, centralized command and control mechanisms of our energy system are inherently vulnerable to terrorist targets, and there is no way for us to realistically make them secure from attack. This includes both plants and distribution lines. Richard A. Clarke, who headed the cyber-terrorism efforts of the Bush administration, warned of an "Electronic Pearl Harbor." A combination of cyber-attacks and physical attacks could lay waste to the nation's oil and gas pipelines, power stations, and transmission lines and could devastate the economy and throw the country into chaos. Renewable energy and local production do not have these vulnerabilities.

In any case, the real issue is not cost, it is *time.* We are up against *nature's* deadline. We need to respond to the climate crisis at wartime speed and throw the full force of the economy into creating the shift to renewable energy. We are nearing thresholds that, if crossed, could trigger the melting of Greenland, the melting of West Antarctica, or runaway global warming. None of the global climate models include the effects of the greenhouse gases released by the melting of the permafrost in the Arctic, and this melting is accelerating.

Nobody really knows what it will cost to stop global warming, but the cost of *not* limiting it—in both lives and money—will be astronomical. We can shift to renewable energy at wartime speed. But before I discuss renewable energy, I first want to cover coal, nuclear power, and peak oil.

Moratorium on Coal

Both Al Gore and James Hansen say that we need to implement a moratorium on construction of new coal-fired power plants, as Denmark and New Zealand have already done. NASA's James Hansen, in his testimony before Congress, said that a moratorium on new coal-fired plants is "the most critical action for saving the planet at this time."[6] He says that it makes no sense to build coal-fired power plants when we will have to bulldoze them in a few years.[7]

There is growing grassroots opposition to coal-fired power plants. The Sierra Club "Beyond Coal Campaign" announced in December 2009 that it has stopped 100 planned new coal plants.[28] Beginning in 2001, there were 150 new proposed coal plants. Sierra Club's campaign worked in almost every state to fight local coal plant proposals, turning out to public hearings, holding rallies, and meeting with officials to push for cleaner and cheaper energy options for their area. Grassroots pressure, combined with rising costs and upcoming federal carbon regulations, contributed to ending plans for these 100 new plants. Most of the remaining ones are on hold.[8] In October 2007, regulators in Kansas were the first to turn down a permit for a new large coal-fired power plant solely because

Chameleon

of the global warming gases it would emit.[9]

There are about 600 U.S. coal plants. A typical (500-megawatt) coal plant burns 1.4 million tons of coal each year.[10]

The Union of Concerned Scientists compares coal with wind and lists the many major problems in our addiction to coal:

- Air pollution: Burning coal causes smog, soot, acid rain, global warming, and toxic air emissions.
- Wastes generated: Ash, sludge, toxic chemicals, and waste heat create more environmental problems.
- Fuel supply: Mining, transporting, and storing coal levels mountains and pollutes the land, water, and air.
- Water use: Coal plants need billions of gallons of cooling water and harm wildlife.[11] A typical 500-megawatt coal-fired power plant draws about 2.2 billion gallons of water each year from nearby water bodies, such as lakes, rivers, or oceans, to create steam for turning its turbines. This is enough water to support a city of approximately 250,000 people. [12]
- In addition to creating smog, acid rain, and air toxics, in an average year, *each* typical coal plant generates 3,700,000 tons of CO_2, as much CO_2 as cutting down 161 million trees.[13]

In August 2007, U.S. Senate Majority Leader Harry Reid of Nevada, who had been opposing three coal-fired power plants in his own state, announced that he was now against building coal-fired power plants anywhere in the world. Then in early November 2007, Representative Henry Waxman of California announced his intention to "introduce legislation that establishes a moratorium on the approval of new coal-fired power plants under the Clean Air Act until EPA finalizes regulations to address the greenhouse gas emissions from these sources."[14]

> One of the quickest ways to cut carbon emissions is to change light bulbs. Switching to more efficient lighting around the globe could save enough energy to close more than 700 of the world's 2,800 coal-fired power plants.
>
> Lester R. Brown
> *World on the Edge*

Investment houses are downgrading coal company stocks and recommend that their clients switch to other energy stocks.[15] In February 2008, Citigroup Inc., J.P. Morgan Chase & Co., and Morgan Stanley agreed that emissions-capping is inevitable. They will require utilities seeking financing for plants to prove the plants will be economically viable even under potentially stringent federal caps on CO_2.[16] Mark Brownstein, an Environmental Defense official, says if utilities have to pay for emission allowances, "the days of conventional coal really are over."[17]

The coal industry is advertising "clean coal" technologies such as carbon capture and sequestration (CCS) and Integrated Gasification Combined Cycle (IGCC). But the Sierra Club Beyond Coal Campaign says that CCS remains an unproven technology, experts disagree about how long it will take for this technology to be available for commercial and wide-scale use, and IGCC emits just as much global warming pollution as other coal plants. The Sierra Club says the coal industry is also pushing liquid coal as a clean alternative, yet liquid coal creates almost double the CO_2 emissions per gallon as regular gasoline; furthermore, replacing just 10% of our nation's fuel with it would require a more than 40% increase in coal mining.[18]

Sierra Club says, "Despite the industry's hype, there's no such thing as 'clean coal.'"[19] Environmental groups are split on backing research for clean coal. Some support CCS. CCS technology is designed to trap and store (either in the Earth's crust or the deep oceans) the massive quantities of CO_2 emitted from coal power plants.[20] Groups like the Natural Resources Defense Council and Environmental Defense Fund are already lobbying on behalf of CCS. Others, such as the Sierra Club and the World Wildlife Fund, insist that affordable and proven technologies, such as energy efficiency and wind or solar energy, should be more fully implemented before CCS is considered.[21] Greenpeace specifically opposes the technology. Kate Smolski, Greenpeace USA global warming campaigner, said, "The No. 1 reason is it's a way the dirty polluting coal industry can prop itself up. It's an unproven technology. And it takes resources away from solutions that we can use right now."[22]

A Massachusetts Institute of Technology report released in 2007, *The Future of Coal*, concluded that the U.S. CCS program is not on track to achieve large-scale commercial operation for at least a decade.[23] Worldwatch Institute President Christopher Flavin is also skeptical of CCS. "It will be many years before we know for sure whether large-scale carbon sequestration is practical and affordable," Flavin says. "The only thing that's certain today is that we shouldn't assume CCS will be a major solution to climate change—unlike solar, wind, and energy efficiency, all of which are being deployed on a significant scale today."[24]

Andrew C. Revkin reported that "Vaclav Smil, an energy expert at the University of Manitoba, has estimated that capturing and burying just 10 percent of the carbon dioxide emitted over a year from coal-fire plants at current rates would require moving volumes of compressed carbon dioxide greater than the total annual flow of oil worldwide—a massive undertaking requiring decades and trillions of dollars."[25] Because of massive cost overruns, on January 30, 2008, the Bush administration announced that it was cancelling the major pilot program with 13 utilities and coal companies to build a demonstration coal-fired power plant in Illinois with underground CCS.[26] The budget of this project nearly doubled, to $1.8 billion, and the government was responsible for more than 70% of the cost.[27]

Alligator

Nuclear Power

Using global warming as a reason, the nuclear industry has been pushing for a "nuclear renaissance" for a decade, but it would make only a limited impact in terms of global warming. Nuclear power only creates electricity, but a large part of the global warming problem comes from energy for heat and hot water and from the fossil fuels used in transportation. Currently, nuclear power is 7% of total U.S. power production (including heat) and 20% of electricity production. In the U.S., there are 104 operating commercial nuclear reactors in 39 states, and there are 442 reactors worldwide. There have been no new plants ordered in the U.S. since 1978. The U.S. leads the world with 101,000 megawatts of nuclear-generating capacity; second-ranked France has 63,000 megawatts.[29]

> At the end of 2009, there were 23,000 megawatts of PV installations worldwide, which when operating at peak power could match the output of 23 nuclear power plants.
>
> Lester R. Brown
> *World on the Edge*

Nuclear power is presented as the key replacement for coal-fired power plants because they emit almost no carbon. But so does energy from wind or solar, from recovered-heat industrial cogeneration, or energy saved by end-use efficiency, and these do it more cheaply and can be built faster. In addition, all of these carbon-free resources cost at least one-third less than nuclear power per kilowatt-hour, so they save more carbon per dollar invested.[30]

In any case, nuclear power actually uses considerable amounts of fossil fuels in the mining and milling of uranium, the construction of the nuclear reactor and cooling towers, the decommissioning of the radioactive reactor, and the transportation of massive quantities of radioactive waste.[31]

Increasing nuclear power is *not* the answer. Yet, in spite of the drawbacks, because of the threat of global warming destroying the planet, some environmentalists have begun to back nuclear energy as a "necessary evil." The argument is often presented in terms of a choice between coal and nuclear power, with the implication that wind, solar, and geothermal can't do the job. But they can. And they do not have the serious limitations that nuclear power does, including safety, waste disposal, water use, spread of nuclear weapons, and most of all, cost.

Safety. On February 16, 2002, the Davis-Besse nuclear power plant near Toledo, Ohio, shut down when a pineapple-sized section on the 6.63-inch-thick carbon steel lid that holds in the pressurized, fission-heated water in the site's sole reactor had been entirely eaten away by boric acid formed from a leak. The Nuclear Regulatory Commission shut down Davis-Besse for two years (from 2002–2004) in an attempt to fix the problem, address design flaws that had been in place since the plant started operation in 1977, and address what the agency described as a poor safety culture. All told, the repairs totaled $600 million, one independent contractor received a jail sentence, and FirstEnergy paid $33.5 million in civil and criminal penalties.[32]

Since 1979, after Three Mile Island partially melted down, U.S. nuclear reactors have had to shut down 46 times for a year or more for repairs or other safety improvements, according to the Union of Concerned Scientists. For example, in February 2000, a steam generator tube abruptly ruptured at the Indian Point nuclear power plant in Buchanan, New York. This was a potential problem first identified in 1997 that had not been fixed. Four incidents in the history of the nuclear power industry have been worse than the cavity at Davis-Besse, and two have been roughly equivalent.[33] According to the Nuclear Regulatory Commission's (NRC's) analysis of so-called "accident sequence precursors," or events that could lead to a nuclear accident, U.S. nuclear reactors averaged seven such events per year between 1993 and 2004—three per year at boiling-water reactors and 11 at pressurized water reactors, most of which could be attributed to the loss of auxiliary, nonreactor-generated power required to run the safety equipment.[34]

However, safety depends on people as much as technology. The greatest safety concern may not be problems with the nuclear reactors themselves. Steve Winn, executive vice president of the utility NRG Energy, said in 2007, "All of us are only as good as the weakest member of the group."[35] John Ricci, manager of specialized technical training at the National Research Council, says, "You cannot regulate against stupidity."[36]

A major concern now is that with deregulation, energy production has become more and more competitive, and many companies are cutting corners. In addition, power plants are aging, creating new safety issues.

The Chernobyl nuclear reactor was destroyed by an explosion and fire on April 26, 1986. *Chernobyl: Consequences of the Catastrophe for People and the Environment,* published in late 2009, examined more than 5,000 published articles and studies, most written in Slavic languages and never before available in English. The authors were Alexey Yablokov, of the Center for Russian Environmental Policy in Moscow, and Vassily Nesterenko and Alexey Nesterenko, of the Institute of Radiation Safety, in Minsk. They say, "For the past 23 years, it has been clear that there is a danger greater than nuclear weapons concealed within nuclear power. Emissions from this one reactor exceeded a hundred-fold the radioactive contamination of the bombs dropped on Hiroshima and Nagasaki."[37] Janette Sherman, M.D., the physician and toxicologist who edited the book, said, "We now realize that the consequences were far worse than many researchers had believed." Drawing on extensive data, the estimated number of deaths worldwide due to Chernobyl fallout from 1986 through 2004 was 985,000, a number that has since increased.[38]

The authors said, "One nuclear reactor can pollute half the globe." The Chernobyl fallout covered the entire Northern Hemisphere. Belarus Yablokov found that radioactive emissions from the stricken reactor, once believed to be 50 million curies, may have been as great as 10 billion curies, or 200 times greater than the initial estimate. Nations outside the former Soviet Union received high doses of radioactive fallout, in particular, Norway, Sweden, Finland, Yugoslavia, Bulgaria,

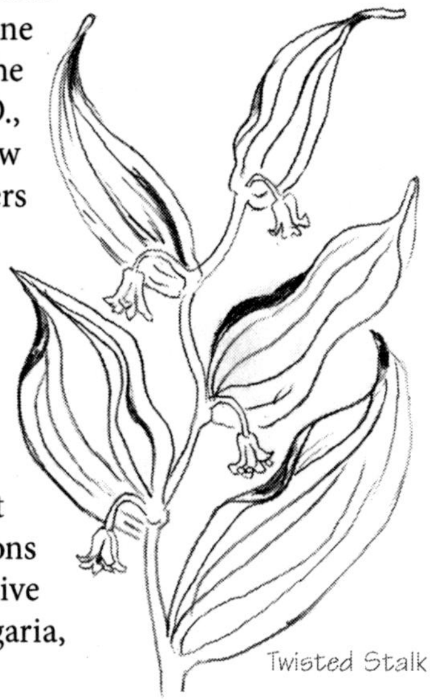

Twisted Stalk

Austria, Romania, Greece, and parts of the United Kingdom and Germany.[39]

This book explores effects of the Chernobyl fallout that arrived above the United States nine days after they entered the U.S. environment and food chain through rainfall. Levels of iodine-131 in milk, for example, were seven to 28 times above normal in May and June 1986. The authors found that the highest U.S. radiation levels were recorded in the Pacific Northwest, although a few years after the meltdown, thyroid cancer in Connecticut children had nearly doubled.[41]

Nuclear Waste Disposal. A major drawback of nuclear power is that in spite of decades of research, no safe solution for nuclear waste disposal exists anywhere in the world. The U.S. produces as much as 160,000 cubic feet (4,530 cubic meters) of radioactive material from its nuclear power plants annually—a number that spikes higher dramatically when old nuclear plants are decommissioned, such as Maine Yankee in Wiscasset, Maine, in 1997.[42] To store the nuclear waste, in 1987, Congress chose Yucca Mountain, 90 miles northwest of Las Vegas.

> Just let all ways to save or produce energy compete fairly—no matter which they are, where they are, what technology they use, how big they are, or who owns them.
>
> Rocky Mountain Institute
> Profitable Solutions to the Business, Climate, and Nuclear Proliferation Problems

The final approval of Yucca Mountain has been an issue for 32 years. President Obama has stopped the funding for the nuclear repository at Nevada's Yucca Mountain, putting the project on hold, but has not offered an alternative. This leaves nuclear waste in storage in 121 temporary facilities in 39 states—sites that are vulnerable both to leakage and to terrorist attacks.[43] Taxpayers have already spent nearly $11 billion on the project, including early planning stages. This includes fines paid to utilities for delays when the site did not begin accepting shipped waste in 1998 as planned—even if the plans went ahead, it would not have opened until 2017 (20 years later than originally planned). If finished, the originally estimated cost of $58 billion in 2001 is now estimated to be $96 billion.[44]

One reason that the Yucca Mountain site will probably be abandoned is that it wasn't going to work. In the original plan, the key reason the site was approved by Congress was that it allowed the high-level nuclear waste to be stored in a geological repository, where the rock would shield the radiation and for thousands of years not allow radiation to leak.[45] However, it was found that the rock will not act as they originally thought it would; therefore, scientists have now come to accept that the geology of the mountain itself can't keep moisture away from stored nuclear waste as planned. Department of Energy scientist Larry Rickertsen said, "The answer is clearer than ever. Engineering has to do the job."[46] But then there were arguments that engineering cannot do the job, either. In an article in *Science*, geology Professor Peter Burns, of the University of Notre Dame, said the engineers tend to think that no matter what happens underground, the buried fuel can be put "in a bigger and stronger box, and that'll solve the problem." However, Burns says, "This is just my opinion, and I'm not an engineer, but I personally take the attitude that 10,000 years is beyond engineering scale."[47]

Whatever alternative is found, if it is a repository, the fossil fuel emissions from transporting it there will be enormous. In shipping the nuclear waste to Yucca Mountain, the Department of Energy has said there would be 108,500 shipments over a period of 38 years. Although the Department of Energy prefers rail shipment of wastes, 24 nuclear reactors do not have access to rail lines and would need to use barges or heavy haul truck shipments. There is also no railroad to Yucca Mountain once the waste gets to Nevada. The Department of Energy's Final Environmental Impact Statement (FEIS) for Yucca Mountain includes a proposed 319-mile rail route to Yucca Mountain from the nearest rail line. Construction of these rail lines would be the largest federal transportation undertaking since World War I and would cost billions of dollars.[48]

There would inevitably be accidents in moving the spent fuel to Yucca Mountain. An Energy Watch Group report says, "Everyone agrees there will be accidents if nuclear waste is transported by train and truck though 45 states for 38 years to the repository at Yucca Mountain in Nevada. The Department of Energy (DOE) predicts that there will be about 100 accidents over the life of the project. The State of Nevada predicts about 400 accidents during the same time period."[49] Fifteen states would experience *daily* shipments of deadly high-level radioactive waste for 38 years. In addition, while being transported, the nuclear waste will be vulnerable to terrorist attacks. If Yucca did open, it would be filled to its statutory limit of 70,000 metric tons by the wastes generated through 2010.[50]

In the absence of a long-term solution, nuclear waste is being stored at the power plants. At present, the nation's nuclear facilities store spent fuel on-site in pools or dry casks, at a cost of roughly $1 million per cask.[51] But nearly all of the nuclear power plants in the U.S. have already run out of storage space because these pools were not designed to be long-term containers and enough room needs to be preserved in case of a crisis such as a meltdown.[52]

The other solution that has been presented is reprocessing the nuclear fuel; this is currently being done in France but has proved prohibitively expensive; there have also been concerns raised about the proliferation of plutonium for nuclear weapons.[53] The National Research Council, the research arm of the National Academy of Sciences, in 1996 estimated that reprocessing of existing used nuclear fuel could cost more than $100 billion.[54] Physicist Edwin Lyman, a senior staff scientist at the Union of Concerned Scientists, said, "We do not think any reprocessing scheme existing or proposed can mitigate the serious concern of proliferation and nuclear terrorism."[55] Some 250 metric tons of plutonium—enough for 30,000 nuclear weapons—has already been reprocessed according to the group.[56] Craig Nesbit, Vice President of Communications for Exelon, says, "There has to be some resolution to the high-level waste issue." He continues, "We don't think it's appropriate to build another plant in another community knowing that we are saddling that community for an indeterminate amount of time with high-level waste."[57]

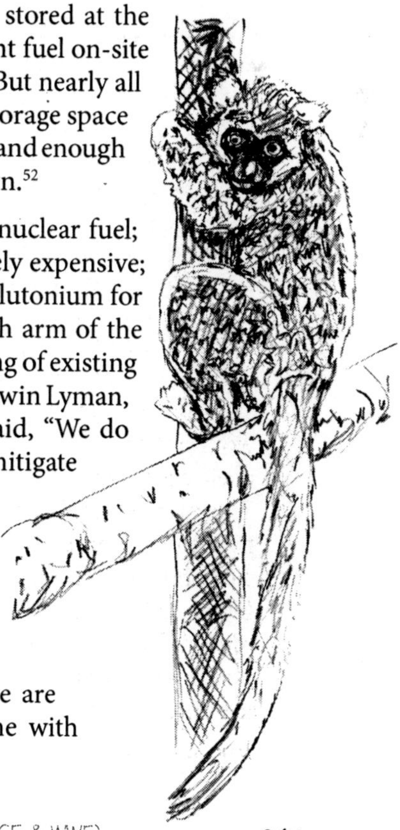

Greater Bamboo Lemur
On ten most endangered list (ICF & WWF)

The only real solution to the problem of nuclear waste is not to produce the waste in the first place. No form of renewable energy has this problem.

Water. In any energy system, the use of water is a significant concern. The world will soon face a severe water shortage, and nuclear power is dependent on water and is therefore vulnerable to drought. Nuclear power plants use massive quantities of water for steam production and cooling. Heat waves, which are expected to become more frequent and intense as a result of global warming, can force the shutdown or reduce the power output reduction of reactors.[58]

> In looking at the broad shifts from the reference year of 2008 to the Plan B energy economy of 2020, fossil-fuel-generated electricity drops by 90 percent worldwide as the fivefold growth in renewably generated electricity replaces all the coal and oil and 70 percent of the natural gas now used to generate electricity.
>
> Lester R. Brown
> *World on the Edge*

This problem of water became clear during the extremely hot summer and drought in Europe in 2006. The drought reduced the volume of water in the rivers and could have forced some power plants to shut down. In addition to there being less water, authorities in France, Germany, and Spain had to override their own environmental rules on limits to the maximum temperature of water released into rivers from cooling systems in nuclear power plants. Germany has shut down reactors at Kruemmel, Brunsbuettel, and Brokdorf, situated along the river Elbe, which flows through eastern and northern German. In Spain, the nuclear power plant at Santa Maria de Garona, one of eight Spanish reactors, was shut down for a time due to the high temperatures recorded in the river Ebro from draining the water used for cooling into the river. The power plant, Spain's oldest, provides 20% of the electricity generated in the country.[59] The European heat wave in the summer of 2003 caused cooling problems at French reactors that forced engineers to tell the government that they could no longer guarantee the safety of the country's 58 nuclear power reactors.[60] During the 2006 heat wave in the U.S., reactors in Michigan, Pennsylvania, Illinois, and Minnesota were affected.

The Spread of Nuclear Weapons. Another major problem with nuclear power is that around the world, the spread of nuclear weapons material is caused by the spread of nuclear energy. The same process used to manufacture low-enriched uranium for nuclear fuel also can be employed for the production of highly enriched uranium for nuclear weapons.[61]

Countries have used energy production by nuclear power as a cover for producing nuclear weapons. Nuclear weapons treaties safeguard only 1% of the world's highly enriched uranium and 35% of the plutonium.[62] Many nations have already used civilian nuclear power programs to cover attempts to make weapons, including Pakistan, India, Iran, and North Korea.[63] These risks have increased since the breakup of the Soviet Union. Nuclear power programs have provided the material, equipment, skills, and the civilian "cover" for bomb programs.

Amory Lovins insists that we cannot limit the threat of nuclear war as long as we have "an energy policy that insists on donating bomb kits to developing countries."[64] He explains,

> If there was no longer an international commerce in nuclear technology, the ingredients needed to make bombs by any of the twenty or so known methods would no longer be casual items of trade. People or countries might still be able to get the ingredients on the black market, but necessary components would be much harder to get, more conspicuous to try to get, and politically much more costly to be caught trying to get. For the first time, the reason for wanting the items would be *unambiguously* military. Such 'unmasking' wouldn't make the spread of bombs impossible, but it would make it far more difficult.[65]

Lester Brown points out that "if we say that expanding nuclear power is an important part of our energy future, do we mean for all countries or only for some countries? If the latter, who makes the A-list and the B-list of countries? And who enforces the listings?"[66]

Cost. The major limit to nuclear power is that it is spectacularly expensive. Nuclear power is the most expensive approach possible to reduce greenhouse gas emissions. *The Economist* observed in 2001, "Nuclear power, once claimed to be too cheap to meter, is now too costly to matter." It is cheap to run but very expensive to build. Most efficiency improvements have been priced at 1¢ to 3¢ per kilowatt-hour, whereas new nuclear energy will cost 15¢ to 20¢ per kilowatt-hour. Avery Lovins says since 2001, it has become several times more costly to build, and as old fuel contracts expire, costs will increase further.[67] In *The Nuclear Illusion*, Amory B. Lovins and Imran Sheikh say the cost of electricity is 14¢ per kilowatt hour from a new nuclear power plant and is 7¢ per kilowatt hour from a wind farm. This comparison includes the costs of fuel, capital, operations and maintenance, and transmission and distribution, but not the additional costs for nuclear of disposing of waste, ensuring plants against an accident, and decommissioning the plants when they wear out.[68]

Banks no longer want to make loans for nuclear power plants. Investors don't want to invest in nuclear power. Moody's has concluded that the decision to build a nuclear reactor is a "bet the farm" decision. And James Hempstead, a nuclear analyst at Moody's, says that any company that is even thinking about building a new nuclear plant faces the possibility of a credit downgrade.[69] In fact, in 2006, distributed renewables got $56 billion of private risk capital. Nuclear got nothing. In 2007, $71 billion was invested in renewables globally by private capital, while nuclear energy again got zero.[70] The only way the plants can be financed is at the taxpayer's expense.

Dr. Mark Cooper, in his November 2009 report "All Risk, No Reward for Taxpayers and Ratepayers," says, "The

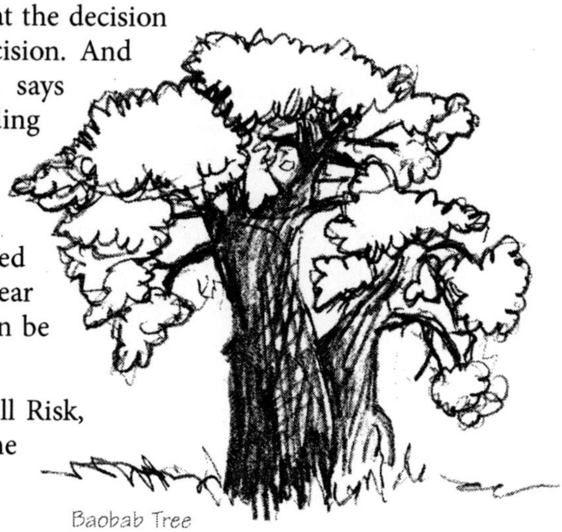

Baobab Tree

rejection of new reactors by financial markets is not a case of market failure, it is an example of market success, markets properly assessing risk and acting accordingly by refusing to underwrite unacceptable risks."[71] So Amory Lovins, an energy expert at the Rocky Mountain Institute, asks, "What part of this do people who take markets seriously not understand?"[72] This refusal of private markets to invest in nuclear power happened because in the 1970s and 1980s:

- one-half of the reactors ordered were cancelled or abandoned;
- those that were completed took, on average, twice as long to build as originally planned and cost twice as much as originally estimated;
- four-fifths of the utilities that undertook nuclear construction suffered large financial downgrades and all suffered substantial financial distress; and
- investments in new reactors resulted in spectacular bankruptcies of both investor-owned and publicly owned utilities.[73]

As the Cooper report notes, "The last time the nuclear industry circumvented the judgment of the marketplace, it resulted in what *Forbes* magazine called the 'largest managerial failure in American history.'"[74]

The taxpayer has always had to subsidize nuclear power. In 1957, Congress passed the Price-Anderson Act, creating taxpayer-backed liability insurance for the industry in the event of an accident.[75] Under the act, utilities are required to maintain private accident insurance of $300 million per reactor—the maximum the insurance industry will provide. In the event of a catastrophic accident, every nuclear utility would be required to contribute to a pool of up to $95.8 million. The collective cap on nuclear operator liability is $10.2 billion. However, Sandia National Laboratory estimated that a worst-case accident could cost $700 billion. So anything above $10.2 billion would be covered by taxpayers.[76] The U.S. government also subsidizes the cost of uranium enrichment. And from 1948 to 2003, the federal government spent $73.4 billion on nuclear energy research and development.

The Energy Policy Act of 2005 established a loan guarantee program (Title XVII) that enables nuclear reactors to get 100% taxpayer-backed loans for up to 80% of the cost of their reactor construction. According to the Congressional Budget Office, the default rate for new reactors is "very high—well above 50 percent."[77] The loan guarantees mean that taxpayers will bail out the nuclear utilities by paying back their loans when the projects fail.[78] So far, Congress has authorized $18.5 billion in loan guarantees specifically for nuclear energy, which could cover the expenses of two to three reactors. The industry is seeking numerous modifications that would enable them to get larger, riskier, unchecked loan guarantees.[79]

For the proposed new plants, the first detailed cost estimate was filed by Florida Power & Light for a large plant off the Keys of $12 billion to $18 billion. Progress Energy announced a $17 billion plan for a similar Florida plant, which was triple its estimate of a year earlier.[80] Many proposals use "Overnight Cost" estimates. For a project (such as solar or wind) whose construction period may be as short as several months, the difference between an "overnight" cost and the full cost to complete the project is not significant. However, for a nuclear project, which can take up to a decade to complete, costs can increase substantially during this long construction period. When added to the financing costs during construction, the total cost of a project can easily double, compared with its "overnight" cost.[81]

An added problem is the aging of the reactors currently operating. In an article in the *Bulletin of the Atomic Scientists*, nuclear consultant Mycle Schneider projects a decline in world nuclear generating capacity. Currently, 439 reactors are operating worldwide. To date, 119 reactors have been closed, at an average age of 22 years. If we assume a lifespan of even 40 years, then another 93 reactors will close between 2008 and 2015, another 192 will close between 2016 and 2025, and the remaining 154 will close after 2025.[82] Schneider points out that plant closings will soon exceed plant openings—and by a wider margin in the years ahead.[83] Only 36 nuclear reactors are currently under construction worldwide—31 of them in Eastern Europe and Asia.[84]

Nuclear energy is not the answer. The bottom line is that any money spent on nuclear power is money not available for efficiency modifications and developing renewable energy. Amory Lovins says, "If climate is a problem, we need the most solution per dollar and the most solution per year. We can get two to 10 times more coal displaced per dollar buying stuff other than nuclear. Every time I spend a dollar on an expensive solution I forgo a lot more that I could have bought of a cheaper solution."[86]

In addition to cost, nuclear energy requires the old emphasis on massive centralized power plants. Renewable energy represents not just a new form of energy, but a new distribution of social power, for renewable energy allows for the democratization of energy.

Peak Oil

Shifting quickly to renewable energy is also important because of peak oil. Peak oil occurs when the maximum rate at which oil is extracted reaches a peak because of technical and geological constraints, with global production going into decline from then on.[87] Over the past decade, oil production has been falling in 33 of the world's largest 48 oil-producing countries, including six of the 11 members of OPEC (Organization of the Petroleum Exporting Countries).[88] Oil extraction has exceeded discoveries by a factor of three during the past two decades.[89]

The respected International Energy Agency (IEA) in Paris assesses future energy supplies, and they did the first detailed assessment of more than 800 oil fields in the world, covering three-quarters of global reserves. The IEA estimates that the decline in oil production in existing fields is now running at 6.7% a year, not the 3.7%

Long-Beaked Dolphin

decline it had estimated in 2007. In constructing the first assessment of the world's major oil fields, the IEA concluded that the global energy system was at a crossroads and that consumption of oil was "patently unsustainable," with expected demand far outstripping supply.[90]

> In the old energy economy, pipelines and tankers carried oil long distances from oil fields to consumers, including a huge fleet of tankers that moved oil from the Persian Gulf to markets on every continent. In the new energy economy, pipelines will be replaced by transmission lines.
>
> Lester R. Brown
> *World on the Edge*

Making accurate estimates of remaining oil supplies is difficult because state-owned companies control three-fourths of the world's oil. Their reserve figures are never audited, are highly secretive, and are based as much on politics as on geology. The reserves of large private companies, such as ExxonMobil and Shell, are required to be regularly monitored by outside experts.[91]

A study by the German-based Energy Watch Group, released in October 2007, relies more on actual oil production data, which are more accurate than estimates of reserves still in the ground.[92] The study said that global oil production is currently about 81 million barrels a day and is expected to fall to 39 million by 2030.[93] It says that global oil production peaked in 2006—much earlier than most experts had expected, and they predicted that production will now fall by 7% a year.[94] The author, Joerg Schindler, said, "The world is at the beginning of a structural change of its economic system. This change will be triggered by declining fossil fuel supplies and will influence almost all aspects of our daily life."[95] The Energy Watch Group report quotes the British energy economist David Fleming as saying, "Anticipated supply shortages could lead easily to disturbing scenes of mass unrest . . . . For government, industry and the wider public, just muddling through is not an option any more as this situation could spin out of control and turn into a complete meltdown of society."[96]

The first half of the world's oil extracted and consumed was the easy half, that is, the oil that is found on shore or near to shore; oil close to the surface and concentrated in large reservoirs; and oil produced in friendly, safe, welcoming places. The second half, what is left of the world's petroleum supply, is the *tough* oil, that is, oil that's buried far offshore or deep underground; oil scattered in small, hard-to-find reservoirs; oil that must be obtained from unfriendly, politically dangerous, or hazardous places; or oil that is too expensive to extract.[97] There are large reserves of non-conventional oil, such as the tar sands of Canada. But this oil is dirty and will produce vast amounts of CO_2.[98] The price of oil will have to be much higher to cover the increased costs of obtaining this oil.

But continually focusing on oil is what caused this problem in the first place. Scientists have said that to keep warming at 2°C we can use only 50%–60% of the known remaining oil. We need to stop looking for yet more oil and concentrate on efficiency and renewable energy. With China's growing economy, the Chinese want to have U.S. consumption patterns. If they did, by 2030, China would need 98 million barrels of oil a day. Currently, the world produces 85 million barrels a day.[99] The western economic model—a fossil fuel–based, automobile-centered throwaway economy—is simply unsustainable.

Peak oil will cause energy prices to skyrocket. It is irresponsible for governments to ignore this coming crisis. Shifting to renewable energy is essential to address global warming, but it will also ease us out of fossil fuel use before peak oil hits.

Efficiency and Conservation

Shifting to renewable energy begins with efficiency and conservation. In building a sustainable energy system, it is critical that all efforts start with conservation and increased energy efficiency. All future increases in energy demand can be met by increasing efficiency. Amory B. Lovins and L. Hunter Lovins, at the Rocky Mountain Institute, say, "An estimated three-fourths of US electricity could now be saved through efficiency techniques that cost less than generating that power, even in existing plants."[100]

The Bush administration cut budgets for efficiency research in half, weakened cost-effective air-conditioner standards, and emphasized a supply-side strategy, particularly by trying to get congressional approval to exploit Alaskan National Wildlife Refuge oil.[101] Oil there would not be available for 10 years, it would be extremely expensive to produce, and there is relatively little of it. Even the conservative Cato Institute has called President Bush's claim that Arctic oil would reduce gas prices or American dependency on foreign oil "not just nonsense, but nonsense on stilts."[102] Every increase of *1 mile* per gallon in auto fuel efficiency would yield more oil than two Alaskan National Wildlife Refuges could yield.[103]

We need whole-system efficiencies that are not dependent on voluntary conservation. Amory Lovins says, "Unlike short-term behavioral changes, efficiency investments are irreversible: you don't scrap fuel-frugal furnaces or remove roof insulation when fuel prices drop, so efficiency ratchets up. And elegant frugality will outdo incrementalism: efficiency often yields expanding returns."[104] The Sierra Club, as well, says that efficiency has a proven track record: "The U.S. economy was 42% more efficient in 2000 than in 1970 as measured by energy per dollar of economic activity. Energy efficiency has added more total energy capability to the U.S. than all fossil, nuclear and renewable energy resources combined over the same period."[105] In fact, when the Bush administration released their energy plan in April 2001, it called for building 1,300 new power plants by 2020. The Alliance to Save Energy issued a plan that could eliminate the need for

Pinta Island tortoise
Lonesome George is the last of his species

all these new plants and save money doing it.[106]

In *Time for Plan B*, Lester Brown says there are enormous financial and technical benefits of investing in efficiency, but efficiency on a large-enough scale requires a massive mobilization. Efficiency is created through "demand-side management." It will address lights, air conditioning, heaters, appliances, computers, electronics, water heaters, motors, and fans. Each of these categories could consume 20% to 70% less energy (and much less, in some cases). Creating enough efficiency will require new building standards, updated appliance and equipment standards, regulatory reforms, and regulations giving utilities an incentive to sell *less* electricity, rather than more.[107] This means we should change the way utilities make money so that they can make money from selling efficiency rather than more energy. We also need commercial and industrial incentive programs for becoming more efficient and retrofit and replacement programs for replacing inefficient appliances like old refrigerators or industrial motors. Retrofitting buildings to improve energy efficiency will create tens of thousands of new jobs in a variety of sectors, including electricians, heating/air conditioning installers, carpenters, construction equipment operators, roofers, insulation workers, carpenter helpers, construction managers, and building inspectors.

Saving energy is cheaper than buying it, so firms are starting to buy energy efficiency whether or not they worry about climate.

Rocky Mountain Institute
Profitable Solutions to the Business, Climate, and Nuclear Proliferation Problems

The tax code should be changed to help homeowners buy energy-efficient household appliances, insulate walls, and install well-insulated windows. One way to increase efficiency is to use smart meters, devices installed in homes or businesses to enable information to flow back and forth between a utility and its electricity customers on electricity usage, giving rates that vary with the time of day. Enabling customers to help in balancing load reduces the total number of power sources needed. Homeowners can set a timer on a dishwasher to run it at 3 a.m. using cheaper electricity. This will reduce electricity bills and reduce peak demand.[108]

Energy used in buildings can be greatly reduced. In the year 2035, three-fourths of the built environment in the U.S. will be either new or renovated, presenting a historic opportunity for architecture. The American Institute of Architects (AIA) has developed an ambitious program to lower emissions called the "2030 Challenge," which was adopted by resolution by the U.S. Conference of Mayors (for municipal buildings). The resolution calls for an immediate 50% reduction in fossil fuel use in the building design and performance of new and renovated buildings. The goals then increase: AIA's energy reduction goals are to reduce the energy use in buildings by 60% in 2010; by 70% in 2015; by 80% in 2020; and by 90% in 2025. The goal is that by the year 2030, all buildings designed will be carbon neutral.[109]

The AIA says that these goals are readily achievable and that most buildings can be designed to use only a small amount of energy (at little or no additional cost) through proper siting, building form, glass properties, location, and material selection and by incorporating natural heating, cooling, ventilation, and day-lighting. Any additional energy a building would need to maintain comfort and operate equipment can be supplied by renewable energy sources.

However, meeting these goals will require mandatory energy-efficiency standards. The U.S. Conference of Mayors has endorsed that goal, but we cannot do it on a city-by-city basis; it requires national standards. It is possible to see buildings as "power plants." Jeremy Rifkin says, "These buildings will collect and generate energy locally from the sun, wind, garbage, agriculture and forestry waste, ocean waves and tides, hydropower and geothermal power sources—enough energy to provide for their own power needs as well as surplus energy that can be shared."[110]

The U.S. Department of Energy, in partnership with building professionals and organizations, is working on developing a Zero Energy Home (ZEH), which combines state-of-the-art, energy-efficient construction and appliances with commercially available renewable energy systems, such as solar water heating and solar electricity.[111] The combination results in a home that produces its own energy—as much or more than it needs. Even though the home might be connected to a utility grid, it has net zero energy consumption from the utility provider because part of the time it is supplying energy to the utility provider.

Increasing efficiency has created a new industry of energy-service companies (ESCOs). The National Association of Energy Service Companies estimates that ESCOs have provided $2 billion in energy-efficiency hardware, software, and services nationwide. ESCOs oversee energy-conservation projects for businesses, schools, and public agencies for a period of seven to 10 years. They contract to save a certain amount of energy and energy expenditures, and then the ESCO conceptualizes, implements, and finances the project. ESCOs base their payment on achieving or surpassing promised efficiency targets.[112]

Lighting accounts for 20% of electricity use. If each household replaced incandescent light bulbs (which waste most of the energy as heat rather than light) with compact fluorescent (CFL) bulbs, we could cut electricity for lighting in half.[113] CFL bulbs cost more initially, but they last 5 years rather than a few months and use 75% less energy. Research is intensifying to offer a wider range of light. If the world made the switch and turned to high-efficiency home, office, industrial, and street lighting, total world electricity use could fall by 12%, equivalent to the output of 705 coal-fired power plants.[114]

Individuals voluntarily choosing CFL bulbs cannot create sufficient widespread use of efficient lighting; it requires a global ban on the sale of incandescent light. The world has finally reached a social tipping point, and many countries are now banning incandescent light bulbs, with many countries annoucing target dates for phasing out inefficient bulbs: Cuba and Venezuela, 2005;

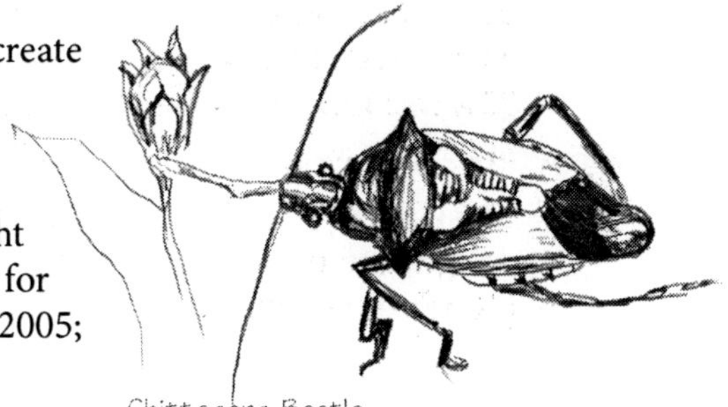

Chittagong Beetle

Ireland, 2009; Australia, Argentina, and the Philippines, 2010; the United Kingdom, 2011; Canada and Taiwan, 2012; and China, 2017.[115] The European Union began in September 2009, and the U.S. will start in 2012 and finish in 2014.

LEDs (light-emitting diodes) are the next generation in home lighting. LED bulbs last up to 10 times as long as CFLs and they do not contain mercury. They are available in 'cool' white light, which is ideal for task lighting, and 'warm' light, commonly used for accent or small area lighting.

> At the heart of Plan B is a crash program to develop 4,000 gigawatts. . . (4 million megawatts) of wind generating capacity by 2020, enough to cover over half of world electricity consumption . . . (requiring) a near doubling of capacity every two years, up from a doubling every three years over the last decade.
>
> Lester R. Brown
> *World on the Edge*

Many cities in the U.S. are replacing their incandescent traffic lights with LED arrays because electricity costs can be reduced by 80% or more. The high cost of producing LEDs has been a roadblock to widespread use. However, researchers at Purdue University recently developed a process for using inexpensive silicon wafers to replace the expensive sapphire-based technology. This promises to bring LEDs into competitive pricing with CFLs and incandescents. LEDs may soon become the standard for most lighting needs.[116]

Another way to conserve energy is through cogeneration, which is a method of using waste heat to generate power. Power plants, factories, and refineries vent steam and hot gases through smokestacks, which is wasted energy. Factories can put in a recovery device on the stack and use the steam to drive a turbine. A 2005 Lawrence Berkeley National Laboratory study of 16 major industries found enough waste heat to generate 96,000 megawatts of power, which is nearly one-fifth of nationwide electricity demand. Another method of cogeneration is to build a mini-plant, usually fueled by natural gas, to power large industrial or commercial properties; this cuts out transmission costs.[117] To help this process, we need an investment tax credit for cogeneration plants. [118]

The main point to remember is that we can meet all future increase in demand for energy with massive increases in efficiency. But then we also need conservation so that increased efficiency is not just an excuse to use more energy. There are many new ways to be more efficient, and all of them create new jobs and new industries.

Micropower: Creating Locally Based Decentralized Energy Systems

The fight over energy production is not just over renewable versus non-renewable energy, it is also over centralization versus decentralization, or what has been called the "democratization of energy." Twentieth-century energy systems used larger and larger facilities with greater distances between energy production and use. Originally, we created these huge, centralized systems to avoid wasteful duplication of equipment. At one time, planners assumed decentralized energy production would be chaotic; now, it is seen as far more resilient.

Micropower's advantages come from economies of production (producing more units to lower costs), rather from economies of scale (lowering costs by building bigger units). Shifting to micropower units is a shift as profound as the move from mainframes to personal computers. This is because when power is created and stored where it is used, as in micropower, it eliminates distribution costs and vulnerabilities. Cogeneration, or capturing the waste heat from microgenerators, can create total thermal efficiencies of 80% to 90%, compared with the 30% efficiency typical of today's centralized power system.[119]

In *Small Is Profitable: The Hidden Economic Benefits of Making Electrical Resources the Right Size,* the Rocky Mountain Institute describes 207 ways in which the size of "electrical resources" that make, save, or store electricity affects their economic value. They say that properly considering the economic benefits of "distributed" or decentralized electrical resources typically raises their value by a large factor, often approximately tenfold, by improving system planning, utility construction and operation, and service quality and by avoiding societal costs.[120] With new micropower units, homes and businesses can produce their own electricity with far less pollution. Buildings canmeet most of their own energy needs with rooftop solar systems, fuel cells, and flywheels—even becoming net energy generators that can sell excess power back to the grid.[121] New micropower technologies, which include fuel cells, microturbines, and solar roofing, can be one-millionth of the scale of today's coal or nuclear plants and produce little air pollution. Seth Dunn and Christopher Flavin, in "Sizing up Micropower," say, "Power systems of this scale are small enough to be factory-built in modular units and transported in one piece by truck or rail to the site where they are installed. Installation often takes a few hours or less."[122]

A report titled "Facing the Hard Truths about Energy," by the National Petroleum Council, an oil-industrial association, was submitted to the U.S. Department of Energy in July 2007. This report estimated that $20 *trillion* in new investment will be needed between now and 2030 to ensure sufficient energy for anticipated demand.[123] They say these funds will be needed for new investments for "building new, multi-billion-dollar oil platforms in water thousands of feet deep, laying pipelines in difficult terrain and across country borders, expanding refineries, constructing vessels and terminals to ship and store liquefied natural gas, building railroads to transport coal and biomass, and stringing new high-voltage transmission lines from remote wind farms."[124] Because future oil projects are likely to be more complex and remote, they will result in higher costs per unit of energy produced.[125] This approach is madness.

To begin with, we can meet new demand with efficiency. We don't need more oil; we need renewable energy, mass transit, and plug-in hybrids. But finding more oil will just increase global warming. Our

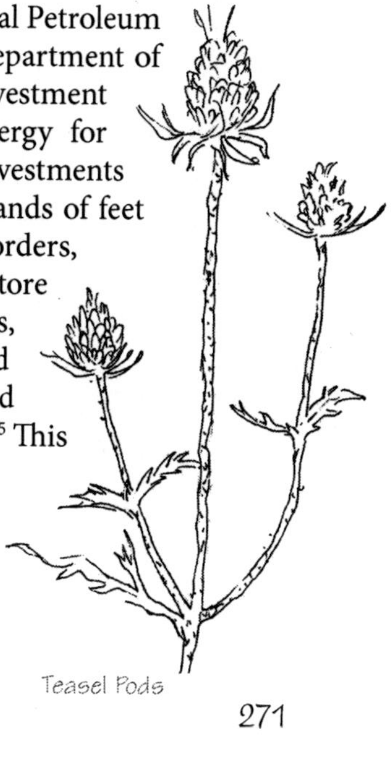

Teasel Pods

atmosphere is warming dangerously, and we cannot afford to use the oil that is left. It would cost far less to simply convert to renewable energy.

The shift to renewables has enormous implications for politics and democracy. Oil and coal energy are not only destroying the environment, they have also concentrated power in the hands of a few. The nature of fossil-fuel energy systems is, in large part, what has created today's concentration of power. Jeremy Rifkin says this old power structure is inherent in fossil-fuel energy: "Unevenly distributed, difficult to extract, costly to transport, complicated to refine and multifaceted in the forms in which they are used, fossil fuels, from the very beginning, required a highly centralized command-and-control structure to finance exploration and production, and coordinate the flow of energy to end users."[126] Only a limited number of nations, corporations, and individuals benefit from ownership of fossil fuels at the expense of everybody else.

> A strong, efficient national grid will reduce generating capacity needs, lower consumer costs, and cut carbon emissions.
>
> Lester R. Brown
> *World on the Edge*

A very different organization of social and political power is possible when using renewable and decentralized energy. These also allow for decentralizing authority and control, making the control more democratic and sustainable, rather than concentrating wealth and power. In Denmark, wind power promotion is based on a decentralized, community-based model in which the machines are built by local companies, financed by local bankers, and owned and installed by local farmers. The Danish approach has raised incomes and created jobs within communities.[127] One of the reasons why large companies like British Petroleum are interested in solar and wind is that they want to keep energy production centralized and in the hands of large corporations.

Unlike oil, renewable energy resources can be found almost anywhere in the world. The resources of sun and wind and even geothermal are more evenly distributed than oil, and no one owns the wind or the sun. Therefore, countries with technical knowledge, a skilled labor force, and openness to innovations can be the most competitive. This democratization of energy creates a less hierarchical and more networked society and would have profound effects on societies around the world.

National Unified Smart Grid

Renewable energy requires a new national unified smart grid. The U.S. electricity transmission and distribution system, called the "grid," is old, fragmented, and too limited in its reach. It is a series of independently operating regional grids that needs to be modernized and expanded into a national unified smart grid. This new infrastructure would allow us to efficiently carry large amounts of electricity over long distances in a network that is resistant to failure.[128]

A national unified electrical grid would include meters that spin both ways, so that it would not only provide power, it would also accept power, making it a "smart" grid. With a smart grid, homes and businesses with solar panels on their roofs can act as a

Renewable Energy Potentials

Wind~
A 2009 survey of world wind resources published by the U.S. National Academy of Sciences reports a wind-generating potential on land that is 40 times the current world consumption of electricity from all sources.

Solar~
The Algerians note that they have enough harnessable solar energy in their vast desert to power the entire world economy. This is not a mathematical error.

The American Solar Energy Society notes that solar thermal resources in the U.S. Southwest can satisfy current U.S. electricity needs nearly four times over.

Geothermal~
An interdisciplinary team of 13 scientists and engineers assembled by the Massachusetts Institute of Technology in 2006 assessed U.S. geothermal electrical generating potential. Drawing on the latest technologies, including those used by oil and gas companies in drilling and in enhanced oil recovery, the team estimated that enhanced geothermal systems could help the United States meet its energy needs 2,000 times over.

Lester R. Brown, *World on the Edge*

small utility company because when they have surplus power, they can sell it back to the larger system.

A national unified smart grid must also include energy storage. This means a combination of central and distributed storage technologies, such as thermal (e.g., working fluids in solar thermal plants), chemical (e.g., batteries, hydrogen), mechanical (e.g., compressed air, hydropower), plug-in cars, or other storage media. Some of these are proven technologies, while others, with the right policies, will be commercialized.[129]

Plug-in electrical cars can recharge their batteries from this clean energy. But these cars can also act as battery storage for the grid. And with plug-in cars, even stored energy in the cars' batteries could be sold to the grid at optimal times when the price is high and recharged when the price is low.[130] Al Gore says that if 3% of cars were plugged into the grid at optimal time, wind power could support up to 50% of the U.S. electricity mix.[131]

People often ask wheather renewable energy can provide enough power to run a national or global economy. Jeremy Rifkin says that with Internet grid technologies, businesses can connect thousands of desktop computers, creating distributed computing power that far exceeds that of the most powerful centralized supercomputers. In the same way, intelligent utility networks would allow millions of local producers of renewable energy to produce and share far more distributed power than the older centralized forms of energy.[132]

Northern Lapwing

Producing Electricity with Renewable Sources

In July 2008, Al Gore challenged the U.S. to produce 100% of its electricity from renewable sources, wind, solar, and geothermal within 10 years. The most common renewables are wind and solar, but there are other inexhaustible energy supplies: biomass; geothermal; ocean energy from the tides, currents, and waves; ocean thermal energy; and small-scale hydopower.[133] Brown says that in the new energy economy of 2020, fossil fuel–generated electricity drops by 90% but is more than offset by the fivefold growth in renewably generated electricity.[134] This shift is quite possible but will be fought by the oil and coal industry. However, Spain, Germany, and China are becoming world leaders in renewable energy, which, by 2020, will be one of the world's biggest industries, totaling as much as $2.3 trillion.[135]

> With no investment on their part, farmers and ranchers typically receive $3,000–10,000 a year in royalties for each wind turbine on their land. For thousands of ranchers in the U.S. Great Plains, wind royalties will dwarf their net earnings from cattle sales.
>
> Lester R. Brown
> *World on the Edge*

Wind Energy

Lester Brown's call for creating a new energy economy by reducing our emissions by 80% by 2020 includes the development of 5,000 gigawatts (5 million megawatts) of new renewable generating capacity by 2020, more than one-half of it from wind. This would be more than enough to replace all the coal and oil and 70% of the natural gas now used to generate electricity.[136] In *Time for Plan B,* Lester Brown says, "Wind is the centerpiece of the new energy economy because it is abundant, widely distributed, clean, climate-neutral, inexpensive, and inexhaustible. No other energy source has all these attributes."[137] In addition, wind power generates electricity with

- No air emissions
- No fuel to mine, transport, or store
- No cooling water
- No water pollution
- No waste

The main cost for wind power is the upfront capital expenditure for construction and installation. The only ongoing cost is for maintenance. The fuel is free; therefore, its price will never fluctuate.

Brown says we could develop 3 million megawatts of wind power capacity by 2020. He says energy consultant Harry Braun points out that "since wind turbines are similar to automobiles in the sense that each has an electrical generator, a gearbox, an electronic control system, and a brake, they can be mass-produced on assembly lines. Indeed, the slack in the U.S. automobile industry is sufficient to produce a million wind turbines per year."[138] Developing 3 million megawatts would require investing $4.5 trillion over the next dozen years, or $375 billion per year (at $3 million per installed turbine).

Although this seems high, world oil and gas capital expenditures are projected to reach \$1 trillion per year by 2016.[139]

For the U.S., a 1991 inventory by the U.S. Department of Energy estimated that North Dakota, Kansas, and Texas together had enough harnessable wind energy to satisfy national electricity needs. However, using today's wind turbines (which are twice as tall and more efficient than those used at the time of the survey), the wind resources from these three states would enable us to meet not only national electricity needs, but also *total national energy needs* (including heat).[140]

Texas, the oil state, is now the leader in producing electricity from wind. Brown says that "in 2006, Governor Rick Perry announced a public-private collaboration between the Texas Public Utility Commission and wind farm developers and transmission line builders to link wind-rich west Texas to the state's population centers. The initiative could lead to the development of 23,000 megawatts of wind-generating capacity, enough to meet the residential needs of more than half the state's 24 million residents."[141]

Danish firms now supply more than half the global wind power market—an achievement made possible by a two-decade-long strategic partnership between government and industry.[142] Lack of support in the U.S. of this alternative energy means our energy industry will fall behind as Europe moves to renewables. All renewable energy provides more jobs per unit of capacity or output and per dollar spent than conventional energies do. Many of the jobs are high-wage and high-tech and require a range of skills, often in areas that are rural or economically depressed.[143] An added benefit is the additional revenue and jobs that wind power will bring to rural areas, since most wind turbines are put in farmers' fields, allowing them to earn extra income. The money from this wind energy stays in the community, creating much stronger rural economies.[144]

Solar Energy

Enough solar energy falls on the surface of the earth every 40 minutes to meet 100% of the entire world's energy needs for a full year. Brown in *Time for Plan B* says we can harness the sun's energy for both heat and electricity generation. Solar cells are used either in stand-alone systems or in systems that can feed on to the grid.[145]

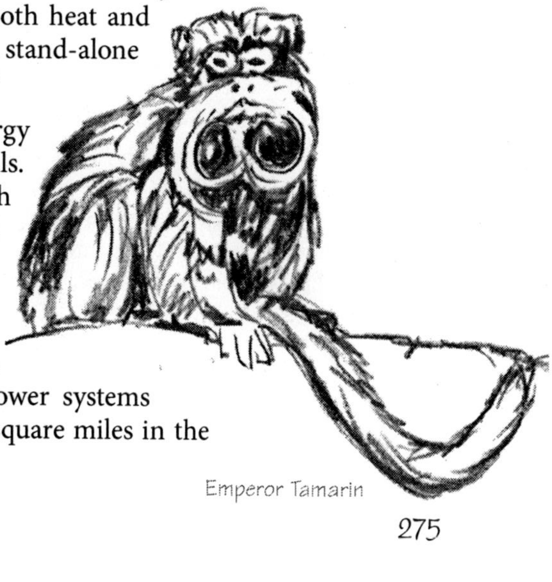

Emperor Tamarin

Solar energy can be used to create solar thermal energy (STE) with either power plants or photovoltaic cells. Concentrated solar thermal power projects—which capture heat from sunlight to generate steam, used to drive a turbine—show that producing electricity from the sun on a large scale can be profitable.[146] High-temperature collectors concentrate sunlight using mirrors or lenses and are generally used for electric power production.[147] Solar thermal power systems covering a parcel of land measuring less than 100 square miles in the

Southwest could theoretically supply 100% of America's electricity needs.[148]

This is different from solar photovoltaic (PV) cells, which convert solar energy directly into electricity. Solar PV can be mounted on rooftops, integrated into roof tiles, or placed in empty fields and can produce electricity even on cloudy days. There are currently thousands of companies developing, producing, installing, and maintaining PV systems in the U.S.[149]

Production of solar cells that directly convert sunlight into electricity is doubling every two years.[151] Brown says that one Plan B goal is to multiply the number of rooftop solar electric systems so that cumulative installed capacity in 2020 exceeds 1 million megawatts. Solar electric power plants and solar thermal power plants could add another 300,000 megawatts.[152] Engineers and architects are designing PVs to work as the "skins" of buildings—roofs, walls, skylights, and window glass—meaning that there would be no extra costs in new buildings for PV installation. A Swiss company, Atlantis Energy, is already selling PV roof shingles, which simultaneously generate electricity and capture emitted heat for space heating.[153] Solar cells can be built along highways, over parking lots, and on top of municipal buildings and transit stations.[154]

> Since no two wind farms have identical wind profiles, each one added to the grid makes wind a more stable source of electricity. With the prospect of thousands of wind farms spread from coast to coast and a national grid, wind becomes a stable source of energy, part of baseload power.
>
> Lester R. Brown
> *World on the Edge*

Variability in Combining Solar and Wind

One of the biggest drawbacks with solar and wind power is that they are "variable," or dependent on the sun shining and the wind blowing. Rocky Mountain Institute's Energy Resources Team has led a series of research projects to rethink the variability of wind and solar. Because the sun often shines when there is no wind and the wind blows differently in different locations, the team found that the key is diversification, both in combining wind and solar and in doing so over a wider geographic area. By combining multiple wind sites with solar generation, the total power generated for the grid becomes much more stable. The team found that they could obtain stable energy amounts with as few as eight sites.[155]

Amory Lovins concludes that using just wind power and solar energy can deliver reliable power without significant cost for backup or storage: "These variable renewable resources become collectively reliable when diversified in type and location and when integrated with three types of resources: steady renewables (geothermal, small hydro, biomass, etc.), existing fueled plants, and customer demand response."[156] He says both nuclear and centralized fossil-fueled power plants typically fail about 8% of the time, so utilities must install a roughly 15% "reserve margin" of extra capacity, some of which must be continuously fueled.[157]

The report by the Rocky Mountain Institute assumed our current electricity grid. One part of shifting to renewable is creating a unified national smart grid, which will make this even easier.

Geothermal Energy

Geothermal is the world's largest energy source. At Earth's core (4,000 miles deep), temperatures can reach over 9,000°F. When this heat bubbles up nearer the surface, the rock and water temperatures there sometimes reach 700°F. At Earth's surface, this heat occurs as hot springs or geysers. This natural collection of hot water is called a geothermal reservoir. The heat in the upper 6 miles of Earth's crust contains 50,000 times as much energy as is found in all the world's oil and gas reserves combined. Yet, only 9,300 megawatts of geothermal generating capacity have been harnessed worldwide.[158]

Lester Brown says that the U.S. is already the world leader in geothermal electricity generation. Geothermal development is projected to expand to provide 15 to 30 times as much power over the next few decades as a result of recent advances in enhanced geothermal systems (EGS), which can harness heat energy stored up to 10 km below the surface. According to an MIT study, 100,000 megawatts of electricity could be installed by 2050 with EGS technology, enough to power more than 70 million homes. In fact, the U.S. government recently issued permits for geothermal project development on 190 million acres of federal land in 12 states. This potential just scratches the surface of a massive amount of recoverable heat energy.[159]

Geothermal heat is used in three ways: in geothermal power plants, direct use, or geothermal heat pumps.

Geothermal Power Plants. All U.S. geothermal power plants are located in California, Nevada, Utah, and Hawaii, where reservoirs of steam or hot water are available. Engineers drill wells into the geothermal reservoirs to bring the hot water to the surface. There are three designs for geothermal power plants, all of which pull hot water and steam from the ground, use it, and then return it as warm water to prolong the life of the heat source. Power plants generate power 95% of the time because they are seldom offline for maintenance or repair. The power from these plants is referred to as baseline power, in the sense that the available power does not fluctuate with the time of day or the seasons—it is always there.

Direct-Use Piped Hot Water. The second use of geothermal energy is widespread through the western one-third of the U.S., where the underground water is 68°F–302°F. Hot water is piped to greenhouses, spas, fish farms, and homes to fill space heating and hot water needs.

Geothermal Heat Pump. The third use does not create electricity, but it greatly reduces the amount of electricity needed. Almost everywhere, the upper 10 feet of Earth's surface maintains a nearly constant temperature between 50°F and 60°F (10°C and 16°C). Geothermal energy is gaining wide acceptance for both residential and commercial buildings. The ground acts as a heat source in winter and a heat sink in summer. A series of pipes, called a "loop," is buried in the ground near the building, either vertically or horizontally. Fluid (either water or a mixture

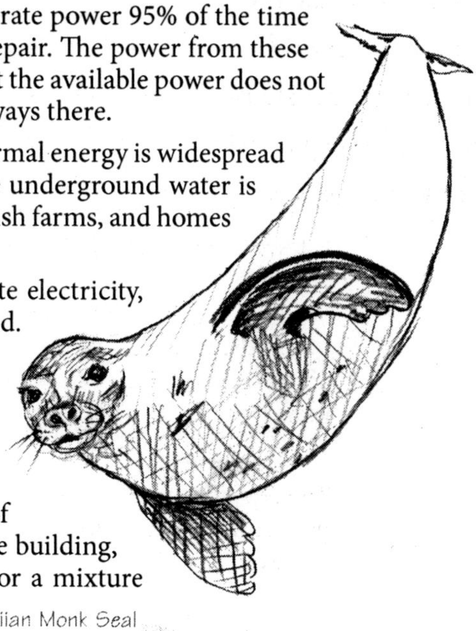

Hawaiian Monk Seal
On ten most endangered list (ICF)

of water and antifreeze) circulates and absorbs heat from (winter) or relinquishes heat to (summer) the surrounding soil. A geothermal heat pump system consists of pipes buried in the shallow ground near the building, a heat exchanger, and ductwork into the building. In winter, heat from the relatively warmer ground goes through the heat exchanger into the house. Since the water is already about 55°F, it takes much less energy to raise the temperature to the desired level. In summer, hot air from the house is pulled through the heat exchanger into the relatively cooler ground. Heat removed during the summer can be used as no-cost energy to heat water.[160]

A 2006 Massachusetts Institute of Technology study found that for the U.S., an investment of $1 billion in geothermal research and development—roughly the cost of one coal-fired power plant—could yield 100,000 megawatts of electricity-generating capacity from enhanced geothermal systems by 2050, the equivalent of 250 coal-fired power plants. The Plan B goals for the world involve increasing geothermal heat capture by a factor of five and geothermal electricity production twenty-twofold.[161]

Hydroelectric Power

Hydropower, including tidal and wave energy, will also be part of renewable energy. Directing, harnessing, or channeling moving water creates mechanical energy. The flow or fall of the water determines the amount of available energy, and a storage system accumulates water in reservoirs created by dams and then releases it when the demand for electricity is high. Only a small percentage of all dams in the U.S. produce electricity; most dams provide only irrigation and flood control.

> With financing from oil and tire companies, National City Lines, a bus company controlled by General Motors, bought up more than 100 electric streetcar lines in 45 cities between 1938 and 1949 and proceeded to dismantle them.
>
> Michael Renner
> Worldwatch Institute

The number of large dams worldwide grew from 5,000 in 1950 to more than 45,000 today—an average construction rate of two large dams a day for 50 years. The report of the World Commission on Dams, *Dams and Development: A New Framework for Decision-Making*, says there is a debate on dams, based in the wider, ongoing debate on equitable and sustainable development.[162] For a time, only the benefits of these engineering projects were considered, but there are also considerable social and ecological costs in terms of people displaced from their homes, fisheries destroyed, soils contaminated by salts, and aquatic species threatened.[163] Opponents say the benefits of large dams have often been deliberately exaggerated and that the services they provide could be obtained by other, more efficient, sustainable means.[164]

It is the gigantic dams that create so much damage. Smaller dams can be useful, and the hydropower from new wave and tidal power can provide a great amount of energy. For hydroelectric power, Lester Brown projects that the 850 gigawatts in operation worldwide in 2006 will expand to 1,350 gigawatts by 2020, including a large number of small hydroelectric facilities, tidal projects (some of them in the multi-gigawatt range), and numerous smaller wave power projects. If the interest in tidal and wave power continues, the additional capacity from hydroelectric, tidal, and wave power by 2020 could easily exceed 500 gigawatts.[165]

Ocean Power

Our oceans are full of usable energy sources, which we have barely begun to use. These sources include offshore wind, tides, waves and currents, aquatic plant biofuels, and kinetic energy.[166]

Energy for Transportation: Cars and Mass Transit

Transportation is the fastest-growing source of carbon emissions. There are many ways to reduce energy use in transportation. Cars as currently designed could be much more efficient. We can shift to hybrid, electric, and/or hydrogen cars. Cities can be designed to reduce the need for cars, and we can greatly increase mass transit.

Even without hybrid or other new cars, we could save a lot of energy just through raising fuel standards. President Jimmy Carter did so by creating CAFÉ (Corporate Average Fuel Economy) standards, when in the late 1970s OPEC policies created an oil shortage. Within six years, this raised fuel efficiency in American cars by 7.6 miles per gallon (mpg). During Carter's term and the five years following it, oil imports from the Persian Gulf region fell by *87%*. From 1977 to 1985, total U.S. oil imports fell by 42%, even while the GDP rose by 27%.[167] By 1986, the CAFÉ standards worked so well that they produced an oil *glut*. Reagan's administration rolled back the CAFÉ standards to the 1984 level in order to protect America's domestic oil industry from gasoline price collapse. There has not been an increase in mileage standards since.[168] In 1995, the Republican Congress made it illegal for the EPA to study higher CAFÉ standards.[169]

The 1986 price crash began a decade of cheap oil. After the 1986 oil-price collapse, the U.S. government stopped any promotion of efficiency and made deep budget cuts to technological innovation in energy productivity. Amory B. Lovins and L. Hunter Lovins of the Rocky Mountain Institute say, "If the United States had continued to conserve oil at the same rate that it did in 1976–85 or had simply bought new cars that got 5 mpg more than they did, it would no longer have needed Persian Gulf oil after 1985. Instead, policy in the 1980s discouraged energy efficiency, which was officially characterized as an intrusive, interventionist burden of curtailment and sacrifice."[170]

Various fuel efficiency technologies exist in industry and university labs, even for the conventional internal-combustion engine. In 2002, John DeCicco, a senior fellow at Environmental Defense, said that if all new cars and light trucks adopted available and emerging gas-saving technologies, the average fuel economy of U.S. cars would surge to 46 mpg, up from today's 27 mpg. He says, "The industry doesn't lack the technology, it lacks the priority."[171]

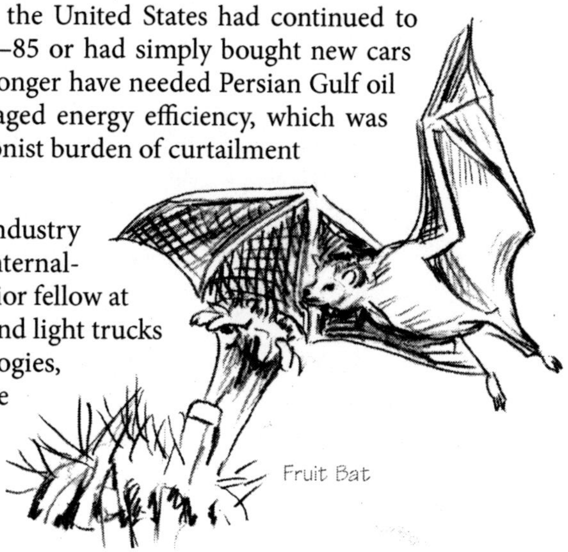

Fruit Bat

The next step is to have cars pay their true costs. Amory Lovins says transportation has the least true competition and the most untruthful prices. Automobile use is the most subsidized and centrally planned sector of most of the world's economies. In addition to problems of carbon emissions, transportation by car creates the social costs of congestion, traffic delays, accidents, roadway damage, and smog. Lovins says these "social costs approach a trillion dollars a year—about an eighth of America's gross domestic product. Because that figure is not reflected in drivers' direct costs, the expenses are in effect subsidized by everyone."[172] Therefore, he says, we need to create public policies with market mechanisms that would end the subsidies.[173]

In the United States alone, the total hidden social costs of driving, not paid by the motorist, total nearly $1 trillion, including such expenses as building and repairing roads, economic losses due to congestion, ill health caused by air pollution and medical costs for the victims of the 2 million accidents each year.

Paul Hawken, Amory Lovins, and L. Hunter Lovins
Natural Capitalism

Land use planning and transportation planning must be conducted together; in particular, we must stop subsidizing and mandating sprawl. How much and what kind of transportation a community has available is greatly impacted by how the community uses land. Creating better land use will help reduce the energy used by cars and will also improve the livability of communities. Richard Register, in *Ecocities: Building Cities in Balance with Nature*, says we need to fundamentally rethink the design of cities. He wants pedestrian cities—communities designed so that people can walk to most of the places they need to go or take public transportation.[174] One-third of the U.S. population is too young, too old, or too poor to drive, which leaves these people stranded. For instance, about 98% of Boston's welfare recipients live within walking distance of public transit, but only 32% of potential employers are close to a transit station.[175]

A major part of subsidizing the cost of driving cars has been the subsidies given to parking. In *Plan B 3.0*, Lester Brown says that Donald Shoup, in *The High Cost of Free Parking*, estimates that off-street parking subsidies in the U.S. are worth at least $127 billion a year, which obviously encourages people to drive.[176] Building regulations often require developers to provide as much parking for each shop, office, or apartment as people would demand if parking were free. In *Natural Capitalism*, the authors say this also diverts investment from buildings into parking spaces, making affordable housing scarcer.[177] They list many approaches that are already being used to reduce the use of cars:

- A San Jose, California, city council member proposed that developers of workplaces and multi-unit downtown housing be forbidden to provide a parking place but instead be required, at far lower cost, to provide a perpetual transit pass with each unit.
- In Frankfurt, Germany, an office *cannot* be built with associated parking: workers must buy their own.
- Britain is authorizing local taxes on firms that provide free or below-market employee parking.[178]

- ❖ Metropolitan Sydney taxes many nonresidential parking spaces to fund suburban railway-station parking and other transit improvements.
- ❖ In Tokyo, you can't buy a car without proving that you own or rent a place to park it. This allows the cost of parking to become part of the cost of a car.
- ❖ Stockholm proposed issuing a monthly permit to allow residents to drive downtown—but the same permit would also serve as a free pass to the regional transit system (which the permit funds).

Brown relates that in 1999, to address the lack of sufficient parking and to reduce traffic congestion on campus, Penn State University (in a small town of 40,000 in central Pennsylvania) offered $1 million annually to the bus-based local transit system in exchange for unlimited free rides for its students, faculty, and staff. Bus ridership jumped by 240% in one year and created a far more pleasant, attractive campus.[179]

The National Complete Streets Coalition (a coalition of citizen groups, including the million-member-strong Natural Resources Defense Council and numerous local and national cycling organizations) is challenging our current cars-only model of transportation. This "complete streets" movement developed because of concern over the obesity epidemic, rising gasoline prices, the urgent need to cut carbon emissions, air pollution, and the mobility constraints on aging baby boomers.[180] Scores of cities in Europe are declaring car-free areas, among them Stockholm, Vienna, Prague, and Rome. Paris has a total ban on cars along stretches of the Seine River on Sundays and holidays and wants to make much of the central city traffic free starting in 2012.[181] Many cities are also beginning to make subways attractive, even cultural centers with shops and restaurants. *Natural Capitalism* lists other ways to reduce use of cars. These include the following:

- ❖ In many American cities, allowing residents to rent out their daytime parking spaces could yield enough income to pay their home property tax.[182]
- ❖ Singapore is rarely congested because it taxes cars heavily, auctions the right to buy them, imposes a U.S. $3 to $6 daily user fee on anyone driving downtown, *and* channels the proceeds into an excellent transit system.
- ❖ Amsterdam banned cars gradually from its scarcer and costlier parking and uses an 18-mile-per-hour urban speed limit.
- ❖ In Denmark, where bikes outnumber cars two to one (four to one in current sales) and where walking and buses are widely used, there's no need for "huge roads and parking lots. This keeps towns and villages walkable, bikeable, and transit-reachable."[183]
- ❖ Pasadena, California, has even found it cost effective to give free bicycles to city workers who promise to commute with them and

Ponderosa Pine

plans to expand this to the general public, imitating the heavily used 2,300-free-bike ($3 deposit) program in Copenhagen.

- Palo Alto, California, requires office buildings to offer lockers and showers for bike commuters.
- In the 1970s, Portland, Oregon, estimated it could cut gasoline consumption by 5% merely by resuscitating the concept of the neighborhood grocery store.
- Car-sharing networks, popular in Europe since the late 1980s, provide easy access to a car without the costs of owning or the hassles of renting. Each member pays for a card that opens lockers holding keys to cars parked around the city. Members who call the network to reserve a car are directed to the closest one. European Car Sharing Network now has participants in more than 50 organizations in over 300 cities in Germany, Austria, Switzerland, and the Netherlands.[184] This same concept is now being used in Chicago.

We still need to design communities for people, not for cars, and to provide more access with less driving. Public transportation should be primary, with an emphasis on land use, so that city streets are pedestrian and bicycle friendly and parking lots are replaced with parks. Currently, the U.S. transportation system is focused on the construction and maintenance of highways and streets. Instead, urban transport systems need to be based on a combination of rail lines, bus lines, bicycle pathways, and pedestrian walkways that can provide mobility, low-cost transportation, and a healthy urban environment.[185]

Hybrid-Electric or Fuel Cell Cars

U.S. passenger vehicles (cars and trucks) contribute 20% of our global warming pollution. The gasoline for these cars is almost entirely from petroleum, nearly 60% of which is imported. It is now essential that we develop clean, affordable cars that 'fuel' by plugging in to a national unified smart grid carrying clean energy. Clean plug-in passenger vehicles will reduce dependence on foreign oil, provide transportation for as little as $1 per equivalent gallon, create price certainty with renewable energy sources that are abundant and free, and help solve the climate crisis.

Amory Lovins and the Rocky Mountain Institute have been working for years on the redesign of cars. Currently, cars and trucks are incredibly inefficient. In *Natural Capitalism*, Lovins says that "of the fuel it consumes, at least 80 percent is lost, mainly in the engine's heat and exhaust, so that at most only 20 percent is actually used to turn the wheels."[186] In addition, current cars are extremely heavy, and completely redesigning cars could save at least 70% to 80% of the fuel they now use and still make cars safer and more comfortable.[187]

Lovins argues that cars must be hybrid-electric.[188] We can do this with an on-board stack of fuel cells or a gas-electric hybrid engine. But with electric cars, the electricity that fuels the battery must be generated by renewable energy to make any real difference. Lovins is primarily interested in using hydrogen, which requires redesigning the automobile and creating a hydrogen-refueling infrastructure.[189] The redesign of the car could also apply to plug-in electric cars.

To make cars work with fuel cells, cars must be much lighter than they are now. Lovins says the design changes involve

- making the vehicle ultra-light, with a weight two to three times less than that of steel cars;
- making it ultra-low-drag, so it can slip through the air and roll along the road several times more easily; and
- After steps 1 and 2 have cut by one-half to two-thirds the power needed to move the vehicle, its propulsion system must be made hybrid-electric.[190] The electricity in this car needs to be produced by an on-board stack of fuel cells, which turn hydrogen and air into electricity and hot water.[191]

Amory Lovins and the Rocky Mountain Institute have been working for years on the redesing of cars. They want to make cars so efficient they need only one-third of the current amount of power, allowing fuel cells to be small enough to be affordable and the compressed hydrogen tanks small enough to fit.[192] The redesign the Rocky Mountain Institute created is called the Hypercar, which they say would use about a *tenfold* lesser flow of such consumable fluids as oil, antifreeze, and brake and transmission fluids (fourteen kinds in all are used in a standard automobile), and there would be a similarly decreased flow of the twenty-one most routinely replaced automotive parts.

> The rust-free, fatigue-free, nonchipping, nearly undentable composite body would last for decades until it was eventually recycled. Together with reduced materials flows in the processing industries upstream, each Hypercar could thus represent a total saving of materials dozens of times its own weight—a total of billions of tons per year. Best of all for the owner, the complex mechanical systems of the traditional automobile would be largely replaced by solid-state electronics and software.[193]

Today's advanced-composite materials are strong enough to absorb five times more energy per pound than steel. Although these composite materials are more expensive than steel, the primary cost of the steel in a car is not for the steel itself (which is only about 15%) but the pounding, welding, and finishing of it. The manufacturing of these new cars will be radically different because these new cars, made of composites and other molded synthetics, emerge from a mold already shaped and finished.[194] Therefore, the Hypercar is shippable overnight. The design of the Hypercar "exploits small design teams, low production runs, very low break-even volume per model, rapid experimentation and model diversification, and greater

Mountain Gorilla
On the ten most endangered list (ICF & WWF)

flexibility."[195] The Rocky Mountain Institute says that the U.S. potential of hydrogen-fueled cars "is to save 8 million barrels a day. It's as if we'd gone drilling in the Detroit Formation and found a Saudi Arabia down there."[196]

This new hydrogen and fuel cell technology could reorient the global energy system as much as the discovery of oil and the invention of the internal combustion engine did more than a century ago. In *Natural Capitalism*, Lovins describes the scope of these changes:

> These innovations are the forerunners of a technological, market, and cultural revolution that could launch an upheaval not only in what and how much we drive but in how the global economy works. Such Hypercars could ultimately spell the end of today's car, oil, steel, aluminum, electricity, and coal industries—and herald the birth of successor industries that are more benign. . . . [The cars] are likely to be leased as a service, even as part of a diversified "mobility service," rather than sold as a product.[197]

Lovins believes that since ultralight hybrids are not just another kind of car, they will probably be made and sold in completely new ways perhaps by computer companies. Car-industry jobs will shift, although their total number could well be sustained or increased.[198] Highly decentralized, local manufacturers could produce these cars. The inexpensive tooling of Hypercars could make the optimal production scale as small as that of a regional soft-drink bottling plant. Lovins says that cars could be ordered directly from a local factory, made to order, and delivered to a customer's door in a day or two. This just-in-time manufacturing eliminates inventory and its carrying and selling costs and eliminates the discounts and rebates required to move stock that is not matched to current demand.

Fuel cell cars can also be used to generate electricity for other uses since a car is parked an average of 96% of the time. Lovins says, "The vehicle becomes, in effect, a clean, silent power station on wheels, with a generating capacity of around 20 to 40 kilowatts."[199] He explains:

> When you are not using it, rather than plugging your car into the electric grid to recharge it—as battery cars require—you plug it in as a generating asset. While you sit at your desk, your power-plant-on wheels is sending 20-plus kilowatts of electricity back to the grid. You're automatically credited for this production at the real-time price, which is highest in the daytime. Thus your second-largest, but previously idle, household asset is now repaying a significant fraction of its own lease fee. It wouldn't require many people's taking advantage of this deal to put all coal and nuclear power plants out of business, because ultimately the U.S. Hypercar fleet *could have five to ten times the generating capacity of the national grid.*[200]

He emphasizes that this generating capacity would be completely decentralized, completely dependable, and terrorist proof.

The other way to go is to use gas-electric hybrid cars like the Toyota Prius, which gets 55 mpg. Without any change in miles driven or number of vehicles, if all our cars were gas-electric hybrids, it would reduce the use of gasoline in cars by 50%. A second battery could be added for increased storage and a plug-in capacity that would allow the batteries to be recharged from the grid. A second battery would reduce the use of gasoline by another 20%, or 70% in total. Lester Brown says this would require a strong, integrated national grid because the regional grids we now have are antiquated.[201] These changes would allow the batteries in these vehicles to become part of the storage system for wind energy. This would give us wind-generated electricity to power automobiles.[202]

Increasing Train Travel

Alternative transit options are increasingly cost-effective because more riders are served per dollar of investment. For decades, interstate highways and local roads have been "free," that is, paid for by taxes and not by those driving the cars. We now need to shift the money for building roads to increasing local public transportation, buses, subways, and light rail and make them "free." This would greatly increase ridership in public transportation and encourage people not to drive.

We need to take medium-distance trips by train, with air traffic used only for long distances and only occasionally. However, there is not one high-speed train (over 200 mph) project in the U.S. The Florida Department of Transportation had worked on a project for 14 years, but Governor Bush canceled it in 1999. We continue to give priority to highways and airports and neglect an entire mode of transportation—passenger trains. People often think that rail systems are too expensive to build; however, although it is a major undertaking, it is not costly, compared with other types of transportation. For example, Boston's Central Artery/Tunnel—a local highway project—cost $11 billion, while the Florida high-speed rail project, linking Miami, Orlando, and Tampa, was expected to cost about $6 billion.

A major block to the development of train transportation is that Congress has refused to allow states to use flexible surface transportation funds for intercity passenger rail, even though states can use these funds for most other types of transportation, including trails for hikers and bikers. In 1998, the federal government spent $30 billion on highways but less than $1 billion on railroads. Both highways and aviation have trust funds that ensure they have enough money. There is no trust fund for rail systems. Congress also needs to provide adequate funding for trains. We now need to shift transportation funds from highway construction to urban transit and intercity rail construction.

Biofuels

Biofuels, as currently created, are not the answer. Biofuels give the illusion that we can keep our gas-guzzling cars and still reduce car emissions. They also make a great deal of money for those producing it—which makes biofuels very

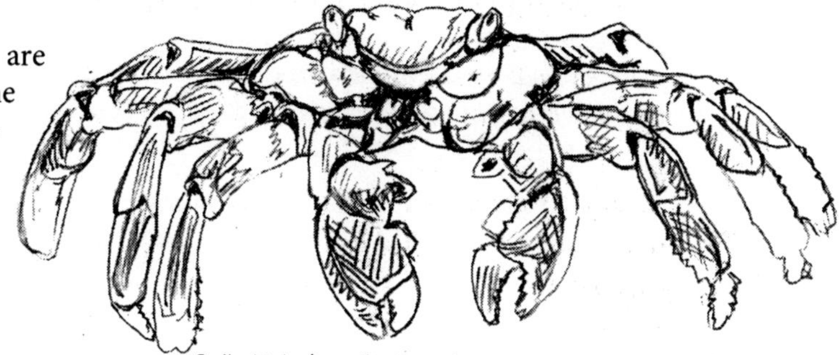

Sally Lightfoot Crab

attractive politically. *Biofuel* is a broad term used to describe all fuels derived from organic matter. *Bioethanol* is produced by processing starchy or sugar-rich crops, such as sugar cane, wheat, or corn. *Biodiesel* is the fuels produced by processing a variety of vegetable oils, including soybean, palm, and rapeseed (canola), and animal fats.[203]

Biofuels are now the main cause of deforestation in Indonesia and Malaysia, and they are becoming responsible for the extinction of the orangutan in the wild there as well. Biodiesel made from palm oil and soybeans is also causing rain forest deforestation in Brazil, Colombia, and Ecuador. As people burn forests to clear them to grow plants for biodiesel, both the trees and the peat they sit on are turned into CO_2. A report by the Dutch consultancy Delft Hydraulics shows that every ton of palm oil (for biodiesel) results in 33 tons of CO_2 emissions, or *10 times* as much as petroleum produces. The British government has said it will require transport fuel suppliers to report on the carbon saving and sustainability of the biofuels they supply. However, they cannot require them to do anything, because if they try to impose environmental standards on biofuels, it will conflict with world trade rules.[204]

In a world of blatantly increasing food insecurity — driven by population, dietary trends, rising oil prices, and growing climate instability — America's policy of burning one third of our corn crop in our engines (soon to be 37% or more) is becoming increasingly untenable, if not unconscionable.

Editorial, *Nature*

Governments around the world are subsidizing crops to produce biofuels, leading some to call biofuels a modern-day gold rush. The problem is that the use of biofuels has many unconsidered consequences. These problems led George Monbiot to say, "We need a five-year freeze on biofuels, before they wreck the planet. It used to be a matter of good intentions gone awry. Now it is plain fraud. The governments using biofuel to tackle global warming know that it causes more harm than good. But they plough on regardless."[205]

Brown says that growing food crops for biofuels sets up direct competition between affluent motorists and low-income food consumers for the same foodstuffs, including wheat, rice, corn, soybeans, and sugarcane. Using land for biofuels limits the land available for food, causing food prices to rise. He describes the boom in bioethanol as a competition between the 800 million people in the world who own automobiles and the 3 billion people who live on less than $2 a day, many of whom are already spending more than one-half their income on food.[206] The corn required to fill a sport-utility vehicle (SUV) tank with bioethanol just once could feed one person for a year. One-fourth of the 2008 U.S. grain harvest went to ethanol. Because of this, since the beginning of 2006, the price of maize has doubled. The price of wheat has reached a 10-year high, while global stockpiles of both grains have reached 25-year lows. Already there have been food riots in many countries (see Chapter 14).

In a paper published in the *Proceedings of the National Academy of Sciences*, Jason Hill, of the University of Minnesota, calculated that even if the U.S. diverted its entire current corn harvest to biofuels, it would meet only 11% of its current gasoline demands. In any case, if we have inefficient vehicles, we will never have enough land to create biofuels.[207] Our approach to transportation emissions needs to be holistic. It begins with mandating much higher gas mileage—a 30% reduction is easy to do with current technology. Even a 3% increase in fuel-economy standards for vehicles would save more gas than the entire 2006 production of corn ethanol.[208]

Help other Countries by Transferring Technologies

Ross Gelbspan says all developing countries would love to switch to renewable energy, but none can afford it. He suggests financing this energy modernization fund by a small tax on international currency transactions, which total more than $1.5 trillion every day. A tax of a quarter penny per dollar on those transactions would yield about $300 billion a year "for wind farms in India, solar assemblies in El Salvador, fuel cell factories in South Africa, and vast solar-powered hydrogen farms in the Middle East."[209] He says that since currency transactions are electronically tracked by the private banking system, we can avoid the need for a large, new bureaucracy by paying banks a fee to administer the fund. In addition, the involvement of the banks in administering the fund would help minimize corruption and ensure that the money goes directly to clean energy projects.[210]

We need a worldwide crash public works program to rewire the planet, which could change many destructive dynamics and create a new sense of purpose. Construction and deployment of many of these technologies is much more labor intensive than coal and oil extraction, both of which are highly automated. Gelbspan says we hear many complaints about the costliness of addressing the climate crisis, "but the real economic issue in rewiring the world with clean energy," he says, "is not cost, but whether the world has a large enough labor force to accomplish this task in time to meet nature's deadline."[211]

In dealing with the developing world, perhaps the most important issue is to open immediate negotiations with China. We must work out how much it would cost to decarbonise its growing economy, and help to pay that cost.[212] There are solutions—but they require unprecedented global coordination to address this problem.

Most important is making the initial cuts as deep and as fast as possible. The IPCC report makes it clear that this will be our last meaningful chance: anything less than an all-out assault on carbon in our economy will not be enough. However, by now our economy is only part of the problem. Even if we start to get our own house in order, we will need to figure out how, with desperate speed, to lead an equally sweeping international response.[213] As Gelbspan says, "Our only hope lies in a rapid and unprecedented mobilization of humanity around this issue."[214]

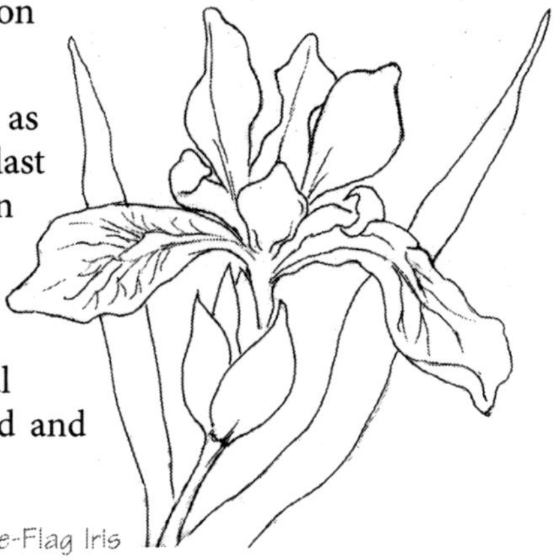

Blue-Flag Iris

Chapter 13

Sustainability in Our Food System

Globalization has created a massive restructuring of the food industry. It has created industrial agriculture, which sees farming as a factory with inputs and outputs and is focused on exporting food around the world. In the U.S., 2% of farms produce 50% of all agricultural products in the country.[1] Industrial agriculture depends on just a few dozen highly standardized seeds and relies on massive applications of pesticides, fertilizers, and fossil fuels. We need to reform the entire food system and create diversified regional food networks. The industrial agriculture system is fundamentally unsustainable.

Worldwide, small farms and farmers are desperate. Giant agricultural conglomerates such as ConAgra and Novartis have structured the market so that there is only one buyer for the crops and the same conglomerate is the only source for seed and supplies. This system either forces the farmer out of business or forces him to become a virtual "serf" on his own land.[2] This loss of small farms is occurring worldwide.

Our Food System Is a Major Contributor to Global Warming

Our food system contributes greatly to global warming. The 2006 report titled *Livestock's Long Shadow*, by the U.N. Food and Agriculture Organization, said that raising livestock creates 18% of greenhouse gas emissions.[3] The food system as a whole creates as much as 37% of emissions.[4] Therefore, reforming our entire food system would have a major impact on global warming.

Fifty million "environmental refugees" will flood into the global north by 2020, fleeing food shortages sparked by climate change.

Cristina Tirado, Professor at University of California, Los Angeles

Industrial agriculture is very energy intensive. Farms use energy for farm machinery, and rising prices from peaking oil production will have direct consequences for world food security.[5] Irrigation uses energy. In the U.S., close to 19% of farm energy use is for pumping water, and this energy use will increase as water tables continue to fall.[6]

Simply clearing the land and tilling the soil releases large amounts of carbon into the air. Fertilizer accounts for 20% of U.S. farm energy use.[7] Emissions come from chemical fertilizers (made from natural gas) and pesticides (made from petroleum). The heavy use of nitrogen fertilizers impacts global warming. When nitrogen fertilizers break down in the soil, nitrous oxide is released into the air. This nitrous oxide traps 200 times more heat per molecule than CO_2 does. Alan Thein Durning says, "American feed takes so much energy to grow—counting fuel for farm machinery and for making fertilizers and pesticides—that it might as well be a petroleum byproduct."[8]

Only one-fifth of the energy used in the U.S. food system is actually for farming.[9] Another 16% of food system energy use is devoted to canning, freezing, and drying food.[10] Packaging accounts for 7% of food system energy use.

Transportation of food from farmer to consumer uses 14% of the energy used in the food system.[11] Almost every state in the U.S. buys 85% to 90% of its food from somewhere else. Most food travels an average of 1,300 miles, much of it from across the world, whereas a typical food item grown locally travels 200 miles. Grains have always been shipped long distances, but now fresh fruits and vegetables are shipped over vast distances in refrigerated airplanes or trucks, which is extremely energy intensive.

Grazing cattle for beef is the main reason for tropical deforestation, which in turn makes global warming worse. The Center for International Forestry Research said that demand for Brazilian beef is causing the deforestation rate in Brazil to skyrocket. Their report showed that Brazil's cattle herd doubled over the past decade to 175 million head in 2002 and that the Amazon accounted for 80% of that rise, with 57 million cows.[12]

Farming Could Reduce Global Warming

Sustainable local agriculture without fertilizers and pesticides would greatly reduce the energy used in our food system. But in addition, a new type of agriculture could create a significant way to sequester carbon. It could also reverse desertification and restore soils and water tables. Sequestering terrestrial carbon in the soil is a natural process. Geological and oceanic sequestration have been suggested, but these will be much more expensive and have unknown ecological impacts.[13]

Soil could be a carbon sink, but currently it is a source of CO_2 emissions. The world's soils have released about 100 billion tons of carbon over centuries of farming. Most of that loss has occurred within the past 100 years.[14] There are two ways in which soil could sequester large amounts of CO_2. One is to restore degraded land, and the other is to shift to minimum tillage farming. *Natural Capitalism* says that the potential for change worldwide is enormous: "The earth's 5 billion acres of degraded soils are particularly low in carbon and in need of carbon-absorbing vegetative cover. Increasing degraded soil's carbon content at plausible rates could absorb about as much carbon as human activity emits. This would also improve soil, water and air quality, agricultural productivity, and human prosperity."[15]

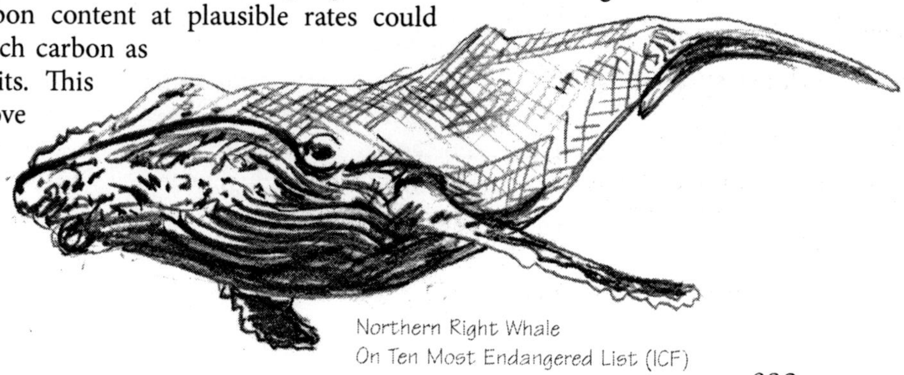

Northern Right Whale
On Ten Most Endangered List (ICF)

In the U.S., 48 million acres are prone to moderate or severe erosion by wind and 65 million acres by water erosion. An additional 0.7 million acres are degraded by salinization and 5.2 million acres of land by mining.[16] Surface mining of coal in the U.S. affected 5 million acres between 1978 and 2002, of which only half, 2.5 million acres, have been reclaimed. Similar potential exists in restoring salt-affected soils.[17] Protecting wetlands will also increase sequestration of carbon. Approximately 15% of the world's wetlands (100 million acres) are in the U.S., of which 5 million acres are in need of restoration.[18] If restored, all these lands could sequester enormous amounts of carbon.

> The current surge in world grain and soybean prices, and in food prices more broadly, is not a temporary phenomenon. We can no longer expect that things will soon return to normal, because in a world with a rapidly changing climate system there is no normal to return to.
>
> Lester Brown
> The Great Food Crisis of 2011

In addition to restoring degraded land, sustainable farming could greatly reduce the loss of carbon. The primary problem with farming is that plowing opens up the soil and mixes oxygen into the soil, which oxidizes the carbon in the soil, creating CO_2. Plowing also decreases the structural stability of the soil so that it becomes more susceptible to erosion by water and by wind.[19] Conventional agriculture also degrades the soil by leaving it bare and failing to return plant residues to the soil.

Dr. Rattan Lal, Director of the Carbon Management and Sequestration Center at Ohio State University, says that to create significant carbon increases in the soil, the soil must be shielded from the oxidation that occurs when it is plowed. Therefore, public policies should emphasize conservation tillage or no-plow farming techniques.[20] These methods are called 'reduced tillage' and are used in conjunction with preventing soil erosion by using an abundance of mulch and manure and planting crops to enhance the soil.[21] It also requires organic farming without pesticides and fertilizers, as well as planting more cover crops during the off-season, rotating crops, and using more perennials instead of annuals.

Many farmers are already using minimum tillage because it is good agricultural practice. This process could become part of trading carbon credits, although there will be issues of monitoring and compliance.[22] Soil carbon sinks are not a permanent solution, for soils can only hold so much carbon, but with these changes, the average soil will reach its maximum carbon content within 50 years.[23]

Demand for Food Is Increasing

The World Food Program has described the current global food crisis as a silent tsunami.[24] In addition to impacting climate change, our global food system has now resulted in severe food shortages and rising prices. From mid-2006 to mid-2008, world prices of wheat, rice, corn, and soybeans roughly tripled.[25] In six of the past nine years, world grain production has fallen short of consumption, forcing a steady drawdown of stocks. When the 2008 harvest began, world carryover stocks of grain (the amount in the bin when the new harvest begins) were at 62 days of consumption, a near record low.

Many earlier civilizations have declined and collapsed due to shrinking food supplies, causing Lester Brown to say, "Food is looking like the weak link in our civilization." These potential food shortages could be catastrophic; in fact, Brown says they could bring down civilization, for they are creating increasing political instability.[26]

These shortages and price increases are happening because demand is growing faster than supply. All of these factors—warming, scarcity of water, increasing drought (from increasing evaporation), loss of cropland, and increasing populations—are creating a devastating reduction in the world's food crops and, therefore, a threat of greatly increased malnutrition. And this malnutrition will be combined with new diseases and increased toxic pollution from global warming. The number of hungry people rose to 1.02 billion in 2009 or nearly one in seven people, despite a 12-year concerted U.N. effort to reduce hunger.[27]

First, demand is increasing as a result of population increases: 79 million people are added every year. This means that the world will add the equivalent of another China in 15 years, and almost all of it will be in developing countries, which are already struggling. Population is expected to level off at 9.1 billion in 40 years. Direct competition could develop between China and the U.S. for grain. If China begins to import grain, it will have to buy from the U.S. The U.S. will not be able to restrict exports because Chinese investors hold more than 1 trillion U.S. dollars.[28]

Second, demand is increasing because the diets of developing countries are changing. Per-capita meat consumption has more than doubled in the past half-century. When combined with increases in population, the overall demand for meat has increased fivefold.[29] Three billion people around the world want to eat grain-intensive livestock products.[30]

And third, demand is increasing as a result of the diversion of grain harvests to ethanol. In a *Scientific American* article titled "Could Food Shortages Bring Down Civilization?" Lester Brown says that one-fourth of this year's U.S. grain harvest will go to fuel cars (enough to feed 125 million Americans or one-half billion Indians at current consumption levels). But if the entire U.S. grain harvest were used to make ethanol, it could meet at most 18% of U.S. automotive fuel needs. The grain required to fill a 25-gallon SUV tank with ethanol could feed one person for a year.[31] Diverting grain to ethanol has also greatly increased food prices, especially for the poor around the world.

Using grain for ethanol merges the food and energy economies. If more money can be made by selling grain for fuel than can be made selling it for food, the market will move the grain into the energy economy. This creates a fierce competition between cars and people for the grain supply and leads to what Brown calls "a political and moral issue of unprecedented dimensions." He says this is generating global food insecurity on a scale never seen before.[32]

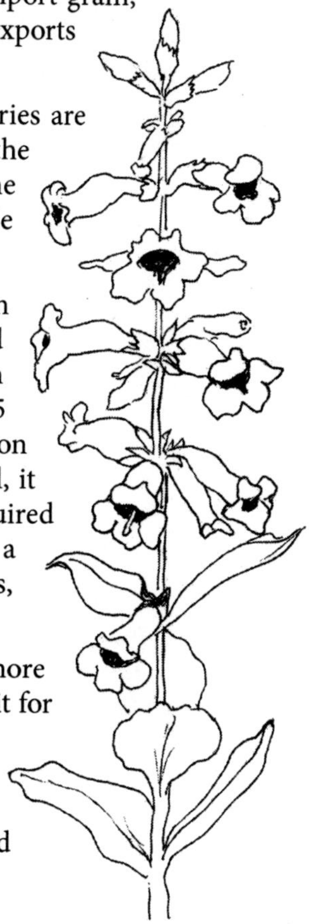

Large-Flowered Beardtongue

Food Supply Is Not Keeping Up With Demand

Producing enough food for the increasing global population will become a major problem. Between 1950 and 2000, world production of grain nearly tripled, allowing food production to keep ahead of population growth. However, after 1984, growth in production slowed and fell behind population growth.[33]

Water Scarcity

The biggest threat to food supply is water shortages, for 70% of freshwater is used for agriculture. There is no substitute for water. Global warming will make water scarcity even worse. Water availability limits food production even more than land availability, for it takes 1,000 tons of water to produce 1 ton of grain.[34] These water shortages will create food shortages. We need a focused effort on increasing water productivity, including more efficient irrigation systems and more water-efficient crops.

> The water crisis is the most pervasive, most severe, and most invisible dimension of the ecological devastation of the earth.
>
> Vandana Shiva
> *Water Wars: Privatization, Pollution, and Profit*

Water tables are falling in the three big grain producers—China, India, and the U.S.[35] In India, water tables have dropped in millions of irrigation wells in almost every state.[36] One-half of India's traditional hand-dug wells and millions of shallower tube wells have already dried up. Electricity blackouts are occurring in many states where one-half of the electricity is used to pump water from depths of up to 1 km (3,300 feet).[37] A World Bank study reports that 15% of India's food supply is produced by mining groundwater. This means that 175 million Indians consume grain produced with water from irrigation wells that will soon be exhausted.[38]

In addition, the melting of the Greenland and West Antarctic ice sheets is raising sea levels. If the Greenland ice sheet were to melt entirely, sea levels would rise by 23 feet. Sea levels are projected to rise by up to 6 feet during this century. The world rice harvest is particularly vulnerable to rising sea levels. In Bangladesh, even a 3-foot rise in sea levels would cover one-half of the riceland in a country of 160 million people. It would also inundate one-third or more of the Mekong Delta, which produces one-half of the rice in Vietnam, the world's No. 2 rice exporter. And it would submerge parts of the 20 or so other rice-growing river deltas in Asia.

The worldwide melting of mountain glaciers is of even greater concern. In China and India, most of the irrigation is from river water. However, the glaciers in the Himalayan and on the Tibetan Plateau are melting, and they feed to major rivers in Asia. This will limit the water availability in the dry season. This could become a humanitarian catastrophe. China is the world's leading producer of wheat, India is second, and the United States is third. In contrast to the United States, most wheat grown in China and India is irrigated. Combined with rice production, these two countries totally dominate the world harvest. Brown says that "the projected melting of these mountain glaciers in Asia represents the most massive threat to food security the world has ever seen."[39]

In addition to rivers running dry and mountain glaciers melting, aquifers are being depleted. Usually, aquifers can be recharged, but some of the most important ones cannot, for example, the "fossil" aquifers, so called because they store ancient water and are not recharged by precipitation. For these—including the vast Ogallala Aquifer, which underlies the U.S. Great Plains; the Saudi aquifer; and the deep aquifer under the North China Plain—depletion would spell the end of pumping. In arid regions, such a loss could also bring an end to agriculture altogether.[40]

Because it takes 1,000 tons of water to produce 1 ton of grain, exporters of grain are exporting water. Canada exports 18 million tons of grain, mostly wheat, which represents 18 billion tons of water. The U.S. annual grain exports of 90 million tons of grain, representing 90 billion tons of water, an amount that exceeds the 67-billion-ton annual flow of the Missouri River.[41]

Land Scarcity

In addition to water scarcity, the second major threat to food supply is available usable land. The loss of cropland is now being combined with water shortages to reduce food production worldwide. Land is limited because of soil degradation and conversion of agricultural lands to other uses. Over the past 50 years, the world has lost one-fourth of its topsoil from soil degradation. We will need a worldwide effort to conserve soil.

The Soil Conservation Service estimates that more than 3 billion tons of topsoil is lost from U.S. croplands each year, seven times faster than the natural build-up rate.[42] The Soil Conservation Service says that American farms are suffering from the worst soil erosion in history.[43] Although artificial fertilizers give higher yields for a number of years, eventually the yields decline because the fertilizers destroy the soil's structure and make it more vulnerable to erosion. This loss of soil reduces yield.[44] Haiti was largely self-sufficient in grain 40 years ago. Since then, nearly all its forests and much of its topsoil have been lost, forcing the country to import more than one-half of its grain.[45] This topsoil loss is occurring in one-third of the world's cropland.

Pygmy Marmoset

Cropland is also being lost to desertification resulting from overgrazing, overplowing, and deforestation. Dry-land ecosystems cover over one-third of the world's land area. Overgrazing and cultivation of marginal lands, along with water scarcity, are turning vast areas of land into desert, threatening the livelihoods of more than 1 billion people in 110 countries. In northern and western China, advancing deserts have forced the abandonment of 24,000 villages and their surrounding cropland.[46]

China's cattle, sheep, and goat population tripled between 1950 and 2002. The land available for grazing in China is about the same as in the U.S.; however, the U.S. has 8 million sheep and goats and China has 298 million. The sheep and goats, concentrated in the western and northern provinces, are destroying the land's protective vegetation.[47] When vegetation is lost, the strong winds of late winter and early spring can remove millions of tons of topsoil in a single day.[48] China is now planting one-tenth of its grainland with trees to try to stop the advancing deserts.[49]

The Secretariat of the United Nations Convention to Combat Desertification has projected that, without concerted efforts to arrest and reverse desertification, Asia could lose one-third of its arable land, South America one-fifth, and Africa two-thirds.

It takes 2.5 billion gallons of water per day to irrigate the world's golf courses. It would take 2.5 billion gallons of water per day to support 4.7 billion people at the UN daily minimum.

Worldwatch
Matters of Scale: Planet Golf

Urban sprawl has also greatly limited the amount of agricultural land. Most cities have developed near rivers and waterways, which is also where most of the country's prime farmland is located. The areas near rivers and waterways currently produce more than one-half of the total value of U.S. farm production, and per-acre yield is 2.7 times that of areas that are not near cities near waterways. The American Farmland Trust reported that from 1982 to 1992 we lost an average of 400,000 acres every year of "prime" farmland with the best soils and climate.[50] Urban sprawl has also destroyed more than 100,000 acres of wetlands each year. In *Shrinking Fields: Cropland Loss in a World of Eight Billion*, Gary Gardner says nearly every country has allowed its grain-producing land to be paved over or degraded by erosion.[51]

Since the mid-twentieth century, grain land area per person has fallen by one-half, primarily because of urban sprawl. For every five cars, an area the size of a football field is paved. This land war between cars and crops will affect food security everywhere.[52] There is not enough land in China, India, and other densely populated countries like Indonesia, Bangladesh, Pakistan, Iran, Egypt, and Mexico to support automobile-centered transportation systems in those countries and still feed their people. Brown says that the competition between cars and crops for land is becoming a competition between the rich and the poor, between those who can afford automobiles and those who struggle to buy food.[53]

Increased Temperatures Lower Yield

The third major threat to food supply, besides water and land scarcity, is rising temperatures from global warming. A report titled *Quantifying the Costs of Agricultural Adaptation to Climate Change*, by the International Food Policy Research Institute, warned that "of all human economic activities, agriculture is by far the most vulnerable to climate change."[54] Warming temperatures will decrease yield, will increase the number of pests, and will create new diseases in both plants and animals. Farmland can also be lost as a result of rising sea levels or water loss from melting glaciers because of global warming. Developing countries will be hit hardest by climate change and will face bigger declines in crop yields and production than industrialized countries.[55] A study by the U.S. National Academy of Sciences has found that for every degree of temperature over the optimum, there will be a 10% decline in wheat and rice yields.[56]

As the atmosphere gets warmer, weather patterns will change. Farmers need to know when to plant the seeds, but now, heat waves, more destructive storms, and higher temperatures are making planting more and more unpredictable. Agriculture has evolved over 11,000 years of climate stability. The crops we grow were bred to flourish in this climate. All of that is going to change.

Collapsing Fisheries

Fourth, food supply is being limited by collapsing fisheries. The U.N. Food and Agriculture Organization estimates that three-fourths of the world's oceanic fisheries are being fished at or beyond their sustainable yields. The capacity of the world's fishing fleet is now double the sustainable yield of fisheries. To ensure a future supply of fish, 75% of the world's fisheries require immediate steps to freeze or reduce fishing. In fact, the global fish catch may need to be cut in half to prevent additional collapses.[57]

The WWF says that the population of breeding tunas has been declining steeply for the past decade and will be wiped out completely in 3 years if fisheries managers and decision-makers keep ignoring the warnings from scientists that fishing must stop.[58] Bluefin tuna is now at risk of commercial extinction. WWF forecasts that breeding stocks of the fish that migrate from the Atlantic to spawn will be wiped out in the Mediterranean by 2012.[59] Scientists estimate there were 5 million snapper off the U.S. Atlantic Coast in the 1950s. However, overfishing has reduced the snapper population by 90%, to about 500,000, despite rules that limit the number and size of fish that anglers can keep.[60]

Professor Boris Worm, of Dalhousie University, published a detailed study in the peer-reviewed journal *Nature* that reports that at the current rate, all global fish populations will have collapsed by 2048. Worm writes, "This isn't some horror scenario, it's a real possibility. It's not rocket science if we're depleting species after species. It's a finite resource. We'll reach a point where we run out." Another study found that whenever the vast industrial trawlers are sent in, it takes just 15 years to reduce the fish population to a 10% shadow of its former self.[61]

Fish farming increases fish yields but damages coastal areas. Since 1992, global aquaculture production has more than doubled, from 15 million metric tons of fish to 36 million metric tons in 2000. About 40% of farmed fish are raised in coastal and marine areas. These fish-farming operations have destroyed mangrove swamps, polluted the water with antibiotics and other chemicals, and displaced traditional fisheries.[62] In addition, 36% (30 million tons) of the world's total fisheries' catch each year is ground into fishmeal and oil to feed farmed fish, chickens, and pigs.[63] The practice is not

White-Throated Sparrow

only impacting fish populations but also local food security. Farm animals impact our seafood supplies much more than seafood certification programs because globally, pigs and chickens consume six times the amount of seafood that U.S. consumers do and twice the amount of Japanese consumers.[64]

Food Shortages Are Creating Social Unrest and Failed States

Rapidly increasing demand and diminishing supplies caused global food prices to increase by almost 50% in 2008.[65] This is an explosive situation and threatens political stability. It is creating enormous pressure on governments and threatening to bring down governments in many counties, as has already happened in Somalia and Haiti. In the Western Hemisphere, Haiti—one of the first states to be recognized as failing—was largely self-sufficient in grain 40 years ago. In the years since, however, it has lost nearly all its forests and much of its topsoil, forcing the country to import more than half of its grain.[66]

> In this unstable situation, a new mechanism to stabilize world grain prices is needed—in effect, a World Food Bank (WFB). This body would establish a support price and a ceiling price for wheat, rice, and corn.
>
> Lester Brown
> *World on the Edge*

As the world's food security unravels, a dangerous politics of food scarcity is coming into play: individual countries acting in their narrowly defined self-interest are actually worsening the plight of the many. The trend began in 2007, when leading wheat-exporting countries such as Russia and Argentina limited or banned their exports, in hopes of increasing locally available food supplies and thereby bringing down food prices domestically. Vietnam, the world's second-biggest rice exporter after Thailand, banned exports for several months. Such moves may reassure those living in the exporting countries, but they are creating panic in importing countries, which must rely on what is then left of the world's exportable grain.[67]

Joachim von Braun, the head of the International Food Policy Research Institute in Washington, D.C., says, "World agriculture has entered a new, unsustainable and politically risky period."[68] There have been food riots in many parts of the world, including Austria, Hungary, Mexico, Namibia, Zimbabwe, Morocco, Yemen, Mauritania, Senegal, and Uzbekistan. Russia and Pakistan have introduced food rationing. Pakistani troops guard imported wheat. India has banned the export of rice, except for high-end basmati.[69]

The World Bank estimates that 33 countries are threatened by social unrest; experts at the International Food Aid Conference described a "perfect storm" of factors that have produced a rise in food costs. Between 2006 and 2009, the prices of wheat, rice, and corn tripled. Japan, China, and South Korea are now buying overseas farmland. China, Egypt, Vietnam, Egypt, Ethiopia, Haiti, Indonesia, the Philippines, Senegal, and India are cutting crop exports, driving global prices even higher. In 2008, riots took place in Cameroon, Peru, Guinea, Mauritania, Uzbekistan, and Yemen. Even where there is food to eat, people often don't have the money to buy it.[70]

The rising food prices and hunger are breaking down the social order in some countries.[71] In Haiti, in 2008, protesters chanting "We're hungry" forced the Prime Minister to resign; 24 people were killed in riots in Cameroon; Egypt's president ordered the army to start baking bread; and the Philippines made hoarding rice punishable by life imprisonment. This explosive situation threatens political stability.[72]

When the food bubble bursts, food prices will soar worldwide, threatening economic and political stability everywhere. For those living on the lower rungs of the global economic ladder, survival itself could be at stake.

Lester Brown
World on the Edge

Like many other countries, Haiti was subjected to trade liberalization and privatization in the mid-1980s by international financial institutions such as the World Bank and donor countries like the U.S. During this time, U.S. agribusinesses flooded the local market with massive quantities of cheap, subsidized staple foods with which Haitian peasants couldn't compete. After the large-scale imports had succeeded in paralyzing local production, prices skyrocketed. A kilo of imported rice is now worth an average day's salary in the Artibonite, a region once known as Haiti's "rice bowl."[73]

Brown says that to restore food security, we must "stabilize population, stop the use of grain to produce automotive fuel, stabilize climate, stabilize water tables and aquifers, protect cropland, and conserve soils."[74]

Additional Problems in Our Current Food System

Our current industrial agriculture looks at farming mechanistically, as an outdoor factory. Sustainable agriculture sees the farm as a living system that is approached holistically. Global economics and politics have allowed corporations to have the best farmland in the U.S. and then to destroy forests for more farmland. Our food system has led to America's obesity epidemic. We will never produce real health care until we address nutrition as a cause of disease.

Our current farming system is dependent on taxpayer subsidies to underwrite cheap grain, and cheap fuel allows food to be shipped around the world. But peak oil and the resulting price increase of fertilizers, pesticides, and transportation will bankrupt the American agricultural system as well as end the globalization of the food system. In fact, John Cavanagh and Jerry Mander, in *Alternatives to Economic Globalization*, say that "the undermining of small-scale, diversified, self-reliant, community-based agricultural systems and their replacement by corporate-run, export-oriented monocultures has been the

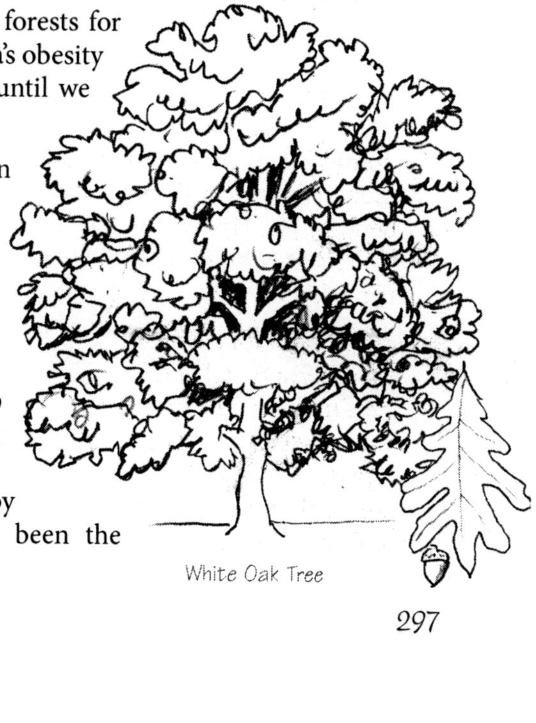

White Oak Tree

primary cause of landlessness, hunger, and food insecurity."[75] This system is maintained by the lobbying, advertising, and public relations efforts of agriculture corporations. These corporations also control the trade policies that make American agriculture highly profitable, but they undercut and drive out of business small farmers in developing counties.

Industrial agriculture corporations argue that corporate farms are more efficient and better able to feed a hungry world. Yet, Cavanagh and Mander say, these efficiencies are sustained only by large government subsidies.[76] Brian Halweil, in a *Worldwatch* magazine article, says, "The small farm doing traditional polyculture (raising more than one kind of crop, using different root depths or soil nutrients on the same piece of land) makes more intensive and efficient use of resources and can produce significantly more food overall per acre."[77]

In evaluating our industrial food system, we look only at output and profit and ignore all other consequences. We need to look at the food system as a whole. When we do, we find further problems that are being ignored but that make our current food system unsustainable.

> Agriculture is America's largest, most diffuse, and most anonymous water polluter.
>
> Paul Hawken, Amory Lovins, and L. Hunter Lovins
> *Natural Capitalism*

The Enormous Impact of Raising Livestock

Producing a pound of beef creates 11 times more emissions than a pound of chickens and 100 times more than a pound of carrots.[78] Livestock now uses 10 billion acres, which represents 30% of Earth's entire land surface and 70% of all agricultural land. This includes land used to grow the grain that is fed to livestock.[79] (More than 70% of the grain produced in the United States is fed to livestock, most of it to cattle.[80]) It takes 5 pounds of grain to make 1 pound of beef.

Raising livestock creates serious problems. Most reports say livestock alone creates 18% of global warming emissions, but an article in *Worldwatch* magazine says that livestock (cows, pigs, and chickens) accounts not for 18% but for 51% of greenhouse gas emissions.[81] Therefore, replacing livestock products with alternatives would have far more rapid effects on reducing greenhouse gas emissions than actions to replace fossil fuels with renewable energy. The article reviews the direct and indirect sources of greenhouse gas emissions from livestock. Some of these sources were underestimated, some were overlooked, and some are emissions sources that are already counted but that have been assigned to the wrong sectors.[82] Included are the emissions associated with the 70% of the grain produced in the United States that is fed to livestock, most of it to cattle.[83] Not eating meat has more of an impact on reducing emissions than any other choice you could make. Or, as Michael Pollen says, "Eat food. Not too much. Mostly plants."[84]

In addition, in the U.S., industrialized animal agriculture typically uses crowded, unsanitary, and stressful growing conditions that encourage disease. Although sick animals do need effective antibiotics, livestock producers use massive amounts to promote animal growth and make their business more efficient and profitable. Better animal housing and more attention to animal hygiene could avoid much of the disease problems and reduce antibiotic use.[85] Michael Pollan says we ended up with Confined Animal Feeding Operations (CAFOs) because of many factors. Subsidized grain monocultures led to animal monocultures because the agriculture subsidies allowed the animal factory farms to buy grain for less than it cost farmers to grow it. This allowed the separation of meat and dairy animals from farming.

CAFOs were also encouraged because the Food and Drug Administration (FDA) approved the routine use of antibiotics in feed (otherwise the animals could not survive in these conditions). Every year, farmers feed about 25 million pounds of antibiotics to livestock for growth promotion and disease prevention, almost 8 times the amount given to humans to treat disease. Farmers use antibiotics in the huge, concentrated livestock operations that dominate animal agriculture. Both the pharmaceutical industry and agribusiness fight to continue their use. Overuse of antibiotics has become a growing public health threat because of antibiotic resistance. Doctors overprescribe them, especially when patients demand them. But 70% of antibiotics in the U.S. are fed to cows, chickens, and hogs. More and more bacterial strains have become resistant to antibiotics. The Centers for Disease Control and Prevention and the World Health Organization have called for an end to the use of antibiotics to promote growth. The FDA has the authority to withdraw approvals of antibiotics because of resistance-related risks to human health, but they focus almost exclusively on approving new drugs.

An additional problem is that the government also does not require CAFOs to treat their wastes like any other industry or municipality must do.[86] This creates an enormous problem of animal waste, which is now a major source of pollution; originally, animal waste was a major source of fertilizers for crops on the farm.[87]

One-half of all fresh water worldwide is used for thirsty livestock. Producing 8 ounces of beef requires an unimaginable 25,000 liters of water, and the water necessary for 1 pound of steak equals the water consumption of the average household for a year.[88]

The Loss of Biodiversity

Industrial agriculture has resulted in an enormous loss of diversity in plants and animals. This loss of biodiversity threatens our food supply. Eventually, this system could cause a sudden, catastrophic collapse. Since 1900, 75% of the genetic diversity of crops has been lost.[89] Cary Fowler and Pat Mooney, two experts on sustainable agriculture, write, "If enough diversity is lost, the ability of crops to adapt and evolve will have been destroyed. We will not have to wait for the last wheat plant to shrivel up and die before wheat can be considered extinct. It will become extinct when it loses the ability to evolve and when neither its genetic defenses nor our chemicals are able to protect it. And that day might come quietly even as millions of acres of wheat blanket the earth."[90]

Currently, large-scale agriculture plants large plots of land with the

Araripe Manakin
On Ten Most Endangered List (ICF)

same crop year after year (referred to as monocropping). This approach has increased yield, but the lack of natural diversity deprives the soil of minerals and nutrients, so more and more chemical fertilizers must replace them. Artificial nitrogenous fertilizers also encourage lusher, sappier growth, which attracts insects, fungal growth, and decay-causing bacteria.[91] Widespread use of nitrogen fertilizers has been a major contributor to global warming, for when the fertilizers break down, they release nitrous oxide.

Industrial agriculture also eliminates hedgerows, where the pests' predators once lived. Hedgerows are removed because industrial agriculture requires larger fields to make room for bigger, more powerful equipment.

Those in industrial agriculture want their entire crop to look the same and taste the same. This specialization and standardization threatens the genetic diversity of plants and animals. In *Short Circuit*, Richard Douthwaite says that for years,

> Plant breeders have been selecting particular characteristics from among the multitude available in landraces and in the wild, then crossing and re-crossing their selections until they arrive at a 'pure' line, a strain that breeds true—that is, without any variability—from generation to generation." The industrial agriculture system is running a desperate race between plant breed and pests. Millions will starve if the breeders loose. It was only six years after introducing DDT that resistance began to appear.[92]

For thousands of years, instead of monocropping, farmers planted natural assortments of seeds adapted to local conditions (assortments called landraces). Seeds with little genetic diversity did not exist. These seeds had developed defenses over hundreds of years in agriculture and thousands of years in the wild. Although diseases or pests were always present, only vulnerable plants would succumb because there was no pressure on the pest to adapt or change. When thousands of acres of a pure line are planted, pests and diseases are forced to overcome the various resistances the line has had bred into it.[93] Few people are still planting landraces and only then in extremely remote areas.

> *In India, peasants grow over forty different crops on localities that have been cultivated for more than two thousand years without a drop in yields, yet have remained free of pests.*
>
> UNEP

The potato famine in Ireland is a warning about the loss of biodiversity. Agriculture around the world today is in the same position as Ireland was in the 1800s. Douthwaite said that from Ireland, within two months the blight had spread to every country in Europe: "Whole fields became blackened and stinking almost overnight. There was nothing anyone could do."[94] Between 1825 and 1849, there were 14 years of bad or disastrous crops, at least 8 involving famine. Perhaps 1 million people lost their lives, and another 1 million emigrated.[95] The loss of genetic varieties caused the potato famine in Ireland. Before the famine in 1845, Ireland had 12 varieties that came originally from the two introduced from the New World in the 1500s. The potato had left natural enemies behind, but eventually various pests made their way across the Atlantic. There were different problems; first, it was a dry rot, later leaf curl, then mold, and finally, blight.

Even today, Europe can only grow potatoes because they found a gene that confers blight resistance among thousands of potato varieties that peasants cultivated in the Andes and Mexico.[96] In 1908, another disease hit—potato wart disease. An inspector found two unaffected varieties and developed resistant potatoes from them. This was only possible because they could choose from a wide range of potato varieties.

When rice cultivation began to be done on an industrial scale, it promoted monoculture, using a few high-yield rice varieties. These seeds are not as hardy as local varieties and can only perform well under certain circumstances: they need a lot of fertilizer and pesticides and are water intensive. The small farmers from rural communities, who used local knowledge built up over centuries to generate new varieties of seeds that blended with the local environment, were marginalized.

The Asia-Pacific arm of the Pesticide Action Network has been warning about impending rice shortage for years. Clare Westwood, campaign coordinator for Pesticide Action Network's "Save Our Rice Campaign," says that we are in rice crisis and that food security and food sovereignty are being undermined. But they were ignored by the governments of the world's top 10 producers of the grain: China, India, Indonesia, Bangladesh, Vietnam, Thailand, Burma, the Philippines, and Japan. The only non-Asian major rice grower is Brazil.[97] In the Philippines, where thousands of traditional rice varieties were once cultivated, just two varieties accounted for 98% of all rice production in the mid-1980s.

A new threat to wheat was found in Uganda in 1999, called Ug99. It is a new strain of a type of fungus called stem rust. It could potentially wipe out 80% of worldwide wheat crops. It has already traveled to Iran.[98] The world's wheat scientists are working to find genes that will protect plants; however, incorporating them into crops takes 9 to 12 years, and that process has only just begun.[99] Scientists will need to replace almost all the commercial wheat grown in the world today.[100] Rick Ward, Coordinator of the Durable Rust Resistance in Wheat Project at Cornell University in Ithaca, N.Y., said, "A significant humanitarian crisis is inevitable."[101]

Genetic Engineering and Biopiracy

Genetic engineering is a technology in which scientists transfer genes from one species to another. This practice goes far beyond selective breeding or hybridization. For example, scientists have spliced genes from viruses, bacteria, and animals into food crops. Presently, two-thirds of processed foods are made using genetic engineering. Our laws do not require long-term testing for safety, so long-term effects are still unknown.[102] Sufficient evidence has already emerged to raise serious safety concerns about damage to health and the environment. Ecological concerns

Goliath Beetle

include the irreversible effects on the environment if these genes spread to other plants. A battle to stop genetic engineering is being fought across the world between those who champion farmers' rights to seeds, livelihood, and land and those who would privatize them. Global corporations now dominate farming and have destroyed the community way of life farming once sustained.

> Since the beginning of the last century, 75 percent of the genetic diversity of agricultural crops has been lost. This creeping homogeneity handicaps the ability of farmers everywhere to respond to pests, disease, and changes in climate.
>
> State of the World 2005

The reasons companies give for gene splicing is that the added gene offers resistance to insect pests or weed killers. Some modifications increase the size or speed of plant growth. In practice, the real reason for gene modification is that it allows a company to hold a patent on the altered seeds. A farmer cannot legally gather seeds from this year's crop to share or to plant the next year but must buy seed again each year.

This ability to patent seeds has had serious consequences. Through the WTO, the U.S. has succeeded in forcing other nations to adopt U.S.-style intellectual property laws. Life science corporations like Monsanto, Novartis, DuPont, Pioneer, and others have benefited enormously from the WTO's Trade-Related Intellectual Property Rights (TRIP) agreement, which allows companies to patent the genetic makeup of plants and seeds. The interpretation of this concept now goes far beyond protection of technology, products, or inventions that the corporations have developed internally. According to Andrew Kimbrell, of the International Center for Technological Assessment, "Corporations are now scouring the globe seeking valuable plant, animal, and human genes that they can claim as their own private property, as if they invented them. Corporations have already received thousands of gene patents. They are now able to patent whole life forms and own them."[103]

When global biotech and pharmaceutical companies aggressively patent the intellectual commons, they prevent its common use. Global corporations claim that small communities should not "lock up" this valuable genetic material but that the whole world should have access to it. However, corporations do not pay for the "valuable genetic material" they obtain from the community. When they get patents on the material, they argue that for "the benefit of humanity" they need monopoly rights to recoup their research investment.[104]

Critics strongly oppose trade agreements that allow patents on indigenous knowledge or that have unreasonable patenting requirements that interfere with the provision of affordable medicine.[105] Indigenous farming communities in India have developed seed varieties over centuries and have always freely shared these with one another. However, these farmers must now pay fees to use these seeds, which are now "owned" by corporations.[106] Monsanto actually hires detectives to locate farmers who might be engaging in such "theft" and brings them to court.[107] Vandana Shiva, an activist from India, says,

> Living organisms and life forms that are self-creating were thus redefined as if they were machines and artifacts made and invented by the patentee, despite the fact that the corporations made only minor modifications in their labs—just enough to be redefined as "inventions." Intellectual property rights laws and patents then give the patent holder a monopolistic right to prevent others from making, using, or selling these "inventions" even if the seeds were originally developed by local farmers. Seed saving by farmers had thus been redefined from a sacred duty to a criminal offense: stealing corporate "property."[108]

Shiva offered the example of the U.S. government granting a patent for the antibiotic properties of karela, jamun, and brinfal to two nonresident Indians, Onkar S. Tomer and Kripanath Borah, and their colleague Peter Gloniski. The use of these substances for control of diabetes has been everyday knowledge and practice in India for ages. Now a corporation owns it, and the Indian people can no longer use it without paying for it.[109]

Activists in the developing world call this patenting process biopiracy. Shiva says that this piracy is easier in India because a great deal of indigenous knowledge about plants and seeds has been systematically documented.[110] Two-thirds of the people in the Southern Hemisphere depend on free access to biodiversity of plants and seeds for their livelihoods and needs.[111] Latin American and Indian farmers, indigenous people, and peasant communities have protested in the streets by the millions against the WTO's TRIP agreement.[112] However, the U.S. media rarely—if ever—reports on these intense conflicts.

Over time, the consequences of the new TRIP agreements on diversity will be severe. No one will be allowed to produce or reproduce patented agricultural, medicinal, or animal products freely. The livelihoods of small producers will be eroded, and the poor will be prevented from using their own resources and knowledge to meet basic needs of health and nutrition.[113] Farmers in developing countries must now purchase seeds, fertilizers, pesticides, and herbicides from a foreign company. Farmers are driven off land that has provided food for their communities for generations, making communities dependent on imports, which increases hunger and migration.

Food First, together with the Pesticide Action Network, has summarized the criticisms of genetically engineered food.[114]

- Since 1999, research shows that genetically engineered crops have done more harm than good. Gains in yield or pest resistance were not as significant as promised. Transgenic contamination has cost the U.S. an estimated $12 billion in product recalls, lost sales, and farm subsidies.

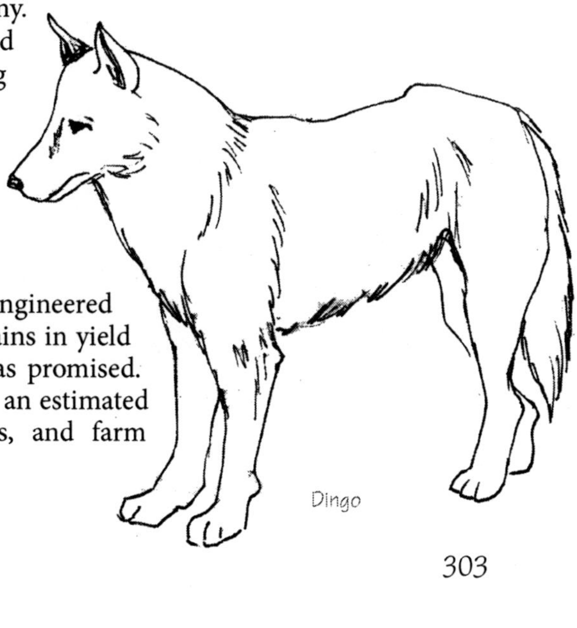

Dingo

- ❖ Wind-blown pollen is a major source of extensive transgenic contamination that cannot be contained or reversed. The Bt proteins in 25% of the genetically engineered crops can be potent allergens and kill benign and helpful insects as well as pests.
- ❖ Other food safety concerns include the effects of glufosinate, ammonium, and glyphosate, which are herbicides used on herbicide-tolerant genetically engineered crops. These products are linked to birth defects, spontaneous late-term abortion, and some cancers; they are toxic to earthworms, butterflies, clam and oyster larvae, some fish, especially rainbow trout, frogs, and mammals.
- ❖ Disease viruses and bacteria are the materials used in most genetic engineering. The genetic engineering process itself can result in creation of viruses and bacteria that cause disease epidemics. (In 2001 one genetic engineering experiment "accidentally" created a killer mouse virus.) Too many aspects and effects of genetic engineered foods have not been proven to be safe. Agricultural corporations have suppressed and misrepresented scientific evidence of potential dangers of genetic engineering. Needed experiments were either not begun, were not completed, or were invalid and then misrepresented. However, more than sufficient evidence has emerged to raise serious safety concerns about irreversible damage to health and the environment.[115]

Claims are made that genetically modified seeds are wildly successful. But it is difficult to know whether that is true because the agricultural companies restrict what scientists can say about the seeds. These corporations claim that to protect a company's intellectual property, anyone buying these seeds (including scientists) must sign an agreement that limits what can be done with them. For a decade, their user agreements have explicitly forbidden the use of the seeds for any independent research. In *Scientific American*, the editors say that "under the threat of litigation, scientists cannot test a seed to explore the different conditions under which it thrives or fails. They cannot compare seeds from one company against those from another company. And perhaps most important, they cannot examine whether the genetically modified crops lead to unintended environmental side effects." The research that is published must be approved by these companies first.[116]

> The dangers of monoculture agriculture, like those of global warming, will never be reflected in free-market prices before it's too late to act.
>
> Michael Schuman
> *Going Local*

Elson J. Shields, an entomologist at Cornell University, in a letter to an official at the EPA (the body tasked with regulating the environmental consequences of genetically modified crops), says, "Selective denials and permissions [are made] based on industry perceptions of how 'friendly' or 'hostile' a particular scientist may be toward [seed-enhancement] technology."[117] A group of 24 corn insect scientists has submitted a statement to the EPA protesting that "as a result of restricted access, no truly independent research can be legally conducted on many critical questions regarding the [genetic engineering] technology."[118]

No genetically modified crops have led to dramatically higher yields, comparable to the doubling or tripling of wheat and rice yields that took place during the green revolution. Nor do they seem likely to do so, simply because conventional plant-breeding techniques have already tapped most of the potential for raising crop yields.[119]

For years, the biotechnology industry has claimed that its genetically engineered crops will produce higher yields. But *Failure to Yield*, a Union of Concerned Scientists report by Doug Gurian-Sherman, says that despite 20 years of research and 13 years of commercialization, genetic engineering has failed. The Union of Concerned Scientists report concluded that genetically engineered herbicide-tolerant soybeans and herbicide-tolerant corn has not increased yields. Insect-resistant corn, meanwhile, has improved yields only marginally. The increase in yields for both crops over the past 13 years, the report found, was largely the result of traditional breeding or improvements in agricultural practices. Traditional breeding greatly outperforms genetic engineering.

Creating Self-reliant, Organic, Local Food Economies

Global warming and peak oil will force our food system to change. All over the world, people are challenging agro-industry and calling for "food sovereignty." The alternative to industrial farming is sustainable farming, which includes developing local food supplies, organic farming, urban farming, and green roofs. Sustainable agriculture approaches the farm holistically, as a living system, in which farmers rely on crop diversity to replenish soils and combat pests. In addition, local agriculture is much less likely to use monoculture, which Michael Pollan calls "the original sin from which almost every other problem of our food system flows."[120]

Sustainable Farming Is Done without Pesticides and Fertilizers

The EPA approved and registered many pesticides long before research established their links to cancer and other diseases. Now, the EPA considers 60% of all herbicides, 90% of all fungicides, and 30% of all insecticides to be carcinogenic. Pesticides can cause birth defects, nerve damage, and genetic mutation.[121] A National Cancer Institute study found that farmers exposed to herbicides had six times the normal risk of contracting cancer. Pesticide poisonings in California among farm workers have increased by an average of 14% a year since 1973, doubling between 1975 and 1985. Field workers suffer the highest rates of illness. Developing nations poorly regulate pesticide use, so farm workers are at even greater risk. Pesticides poison an estimated 1 million people annually.[122]

Sustainable agriculture is organic. As long as pesticides have not sterilized the soil, it teems with life. Good soils can improve yields and produce robust crops less vulnerable to pests. Incorporation of animal waste (manure), living plants, or plant debris (compost) will maintain and enhance soil quality.

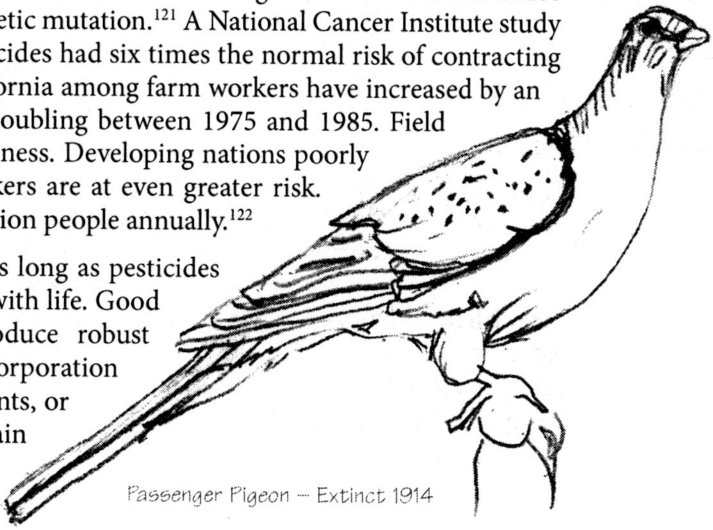

Passenger Pigeon — Extinct 1914

The widespread use of nitrogen fertilizers has contributed substantially to greenhouse gas emissions. When fertilizers break down, they create nitrogen oxide—the most potent greenhouse gas. These fertilizers also pour into the Mississippi watershed and create a dead zone at the mouth of the Mississippi River. Fertilizers have created 400 dead zones around the world. In the water, the nitrogen produces more algae, which sucks all the oxygen out of the water, killing marine life.

American feed takes so much energy to grow. . . . counting fuel for farm machinery and for making fertilizers and pesticides . . . that it might as well be a petroleum byproduct.

Alan Thein Durning
Worldwatch

Approaching agriculture as an ecosystem rather than a factory offers new methods for pest control. A new technology called integrated pest management (IPM) begins with accurate diagnosis of pest problems and then chooses from a range of responses. These include introducing sterile males, biocontrol agents like ladybugs, or, as a last resort, reduced-risk chemical pesticides. Biointensive IPM relies as much as possible on biological rather than chemical measures. IPM emphasizes pest prevention through crop rotation, microbial control of root pathogens, and release of beneficial organisms that prey on the pests. It minimizes soil erosion through crop choices, cover-cropping, and low-till methods. It emphasizes crop diversity of both species and varieties, which provides inherent resilience to pests, disease, and weather extremes.

Many birds, insects, and spiders are natural predators of agricultural pests. But intensive use of chemical pesticides indiscriminately kills birds, bats, and other pest predators. Managing farms so that they harbor populations of pest predators is an effective pest-control technique.

Crop rotation can also help in pest control because many pests prefer specific crops. Growing the same crop year after year guarantees pests a food supply, so populations increase. Rotation is a key solution to permanent and effective pest control because it reduces pest pressure on all the crops in the rotation by breaking the pest reproductive cycles. In rotations, farmers can also plant crops that replenish plant nutrients, reducing the need for chemical fertilizers. Corn grown in a field previously used to grow soybeans needs less added nitrogen to produce high yields. Although they are not grown to sell, cover crops like hairy vetch, clover, or oats prevent soil erosion, suppress weeds, and enhance soil quality. Using appropriate cover crops can reduce the need for chemical herbicides, insecticides, and fertilizers.

Phasing out pesticide use will create cleaner water, thriving populations of fish and birds, fewer industrial and transportation accidents, soil conservation, soil health, and more wildlife habitat; rural streams and rivers would again run clean enough for people to swim in and for fish, birds, and other wildlife to flourish in.[123]

Threats to Pollination

Pesticides may also be poisoning bees. It is estimated that one-third of everything we eat depends on honeybee pollination.[124] The honeybee pollinates 90 commercial crops worldwide, including most fruits and vegetables (including apples, oranges, strawberries, onions and carrots) nuts, sunflowers and oil-seed rape, coffee, soya beans, clovers, like alfalfa, which is used for cattle feed, and even cotton.[125]

In May 2010, according to the annual survey by the Apiary Inspectors of America and the U.S. government's Agricultural Research Service, for the fourth year in a row, more than one-third of bee colonies in the U.S. had failed to survive the winter. The decline of the estimated 2.4 million beehives began in 2006, when a phenomenon called colony collapse disorder led to the disappearance of hundreds of thousands of colonies. Since then, more than 3 million colonies in the U.S. and billions of honeybees worldwide have died. The collapse in the global honeybee population is a major threat to crops.[126]

Potential causes of the collapse range from parasites, such as the bloodsucking varroa mite, viral and bacterial infections, and pesticides and poor nutrition, stemming from intensive farming methods. U.S. scientists have found 121 different pesticides in samples of bees, wax, and pollen, which suggests that pesticides are a key problem. Jeffery Pettis, of the Agricultural Research Service, says, "We believe that some subtle interactions between nutrition, pesticide exposure and other stressors are converging to kill colonies."[127] Bernard Vallat, Director-General for the World Organization for Animal Health, warned, "Bees contribute to global food security, and their extinction would represent a terrible biological disaster."[128]

Our Food System and Health Care Costs

We will never reduce health care costs unless we first address the food system. We have become dependent on cheap, mass-produced food filled with sugar and fat, causing an epidemic of obesity. Corn is used for refined sugars like high fructose corn syrup, and soy is used to make hydrogenated oil, the fast-food oil. Local food reduces energy needs and also maximizes nutrition.

Obesity and lack of nutrition are particularly a problem for the poor. There is a lack of grocery stores in the inner city, although there are plenty of fast-food chains. In addition, the most processed food is also the cheapest. Many of our health issues begin with our diets: too much meat, too much processed food, and too many toxins. The pesticides and fertilizers put on our food pollute our rivers and land, as well as our food. Health care must begin with prevention, and most of that involves diet and nutrition.

The health risks of our food system include not just lower nutritional values. In addition, our society has created about 100,000 industrial chemicals. Many of these chemicals end

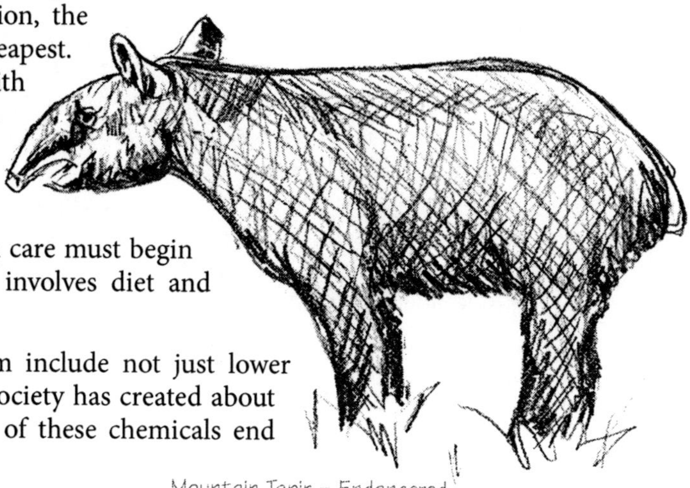

Mountain Tapir – Endangered

up in the food chain one way or another. There are also many issues with meat and poultry. One study showed that perhaps as many as 70% of the chickens grown in this country are routinely given a form of arsenic in their feed because it makes them grow faster. This practice was never approved as safe in Europe. An added problem is that antibiotics are used heavily in raising farm animals. Any time bacteria are exposed to antibiotics, it tends to make them more resistant to these drugs, which is true whether the antibiotics are used in hospitals or on farms. So, the scientific consensus now is that with the overuse and misuse of antibiotics in the farm setting, we're helping to create drug-resistant superbugs that affect human health.

Sustainable Farming Is Local and Regional

Local food, as opposed to organic, implies a new economy as well as a new agriculture—new social and economic relationships as well as new ecological ones. It's a lot more complicated than just food.[129] Farms produce a lot more than food; they also produce a kind of landscape and community.[130] In reclaiming local food production, we are returning to community. Chris Hedges, in *Food Is Power and the Powerful Are Poisoning Us*, says that local food "brings us back to the values that sustain community. It is a return to the recognition of the fragility, interconnectedness and sacredness of all living systems and our dependence on each other. It turns back to an ethic that can save us."[131]

> Every time we buy food, our choices shape farms and nature and communities somewhere in the world. It's the most political decision we make, and we make it every day.
>
> Christian Aid
> *Fighting Food Shortages: Hungry for Change*

One way to create local food networks is with Community Supported Agriculture (CSA) (in Canada, it is called Community Shared Agriculture). CSA connects local farmers with local consumers, develops a regional food supply and a strong local economy, and maintains a sense of community.[132] CSA is a partnership between a farm and a community—a direct link between production and consumption of food. Supporters purchase a share of the season's harvest in the spring, giving the farmer cash at the beginning of the growing season to cover a farm's yearly operating budget.

The farm, in turn, provides, to the best of its ability, a healthy supply of seasonal fresh produce throughout the growing season, generally delivered weekly. In buying a membership, people share the risk with the farmer. CSA began 30 years ago in Japan, when a group of women concerned about the increase in food imports and decrease in the farming population initiated a direct growing and purchasing relationship between their group and local farms. This arrangement, called "teikei" in Japanese, translates to "putting the farmers' face on food." Europe and then the U.S. adapted this plan and called it CSA.[133] A typical food item grown locally travels 200 miles instead of 1,300.

CSA has many advantages:

- ❖ Direct marketing gives farmers and growers the fairest return on their products.

- ❖ Food dollars stay in the local community to maintain and establish regional food production.
- ❖ Closer communication encourages cooperation among farmers and opens up a dialogue between farmers and customers.
- ❖ A "guaranteed market" means farmers can invest their time in doing the best job they can, rather than looking for buyers.
- ❖ Small farms producing a wide variety of crops support the biodiversity of a given area and the diversity of agriculture.
- ❖ CSA creates a sense of social responsibility and stewardship of local land.
- ❖ CSA puts "the farmers' face on food" and increases understanding of how, where, and by whom our food is grown.[134]

The community food security (CFS) movement simultaneously addresses issues of ecology, environmentalists, and community development groups. The CFS Coalition backs projects like a farmers' market in Milwaukee that accepts food stamps and a food bank–sponsored 12-acre farm near Washington, D.C., that provides fresh food for hundreds of food pantries, free meal programs, and farm stands in the city.[135]

The number of CSA farms in the U.S. is increasing, with most of the new ones being small and often run by women. Many of these new farms cater to local markets. The number of farmers' markets nearly tripled between 1994 and mid-2009. Many now take food stamps, giving low-income people access to fresh produce. In places, year-round farmers' markets are developing that only carry locally produced foods such as fruit, vegetables, meat, milk, cheese, and eggs. Many schools now have gardens that provide fresh produce for school lunches. California is the leader, with 6,000 school gardens.[136]

Urban Agriculture. Another way to create local agriculture is with urban agriculture, which can alleviate poverty, reuse waste resources, generate jobs and income, protect the environment, offer adequate food and health resources, make productive use of open-space management, and create micro-enterprises.[137] The effort for a sustainable local food system for urban areas includes a number of related interests: urban agriculture, community gardens, CFS, Local Food Policy Councils, and rooftop agriculture. Urban farming includes vegetables grown on roofs and patios, market gardens on vacant plots, and fish harvested from tanks and sewage lagoons.[138]

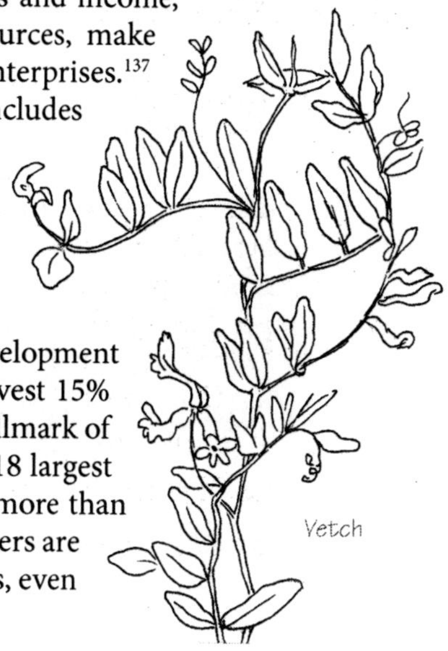

In the first global survey of urban agriculture, the U.N. Development Programme estimates that 800 million urban farmers today harvest 15% of the world's food supply. High-intensity urban farming was a hallmark of early civilizations. Even into the 1980s, urban farmers in China's 18 largest cities met more than 90% of their cities' vegetable demand and more than one-half of their meat and poultry demand. However, these numbers are lower now because of intensive building. With protective measures, even dense urban islands can raise much of their own food. Hong Kong produces two-thirds of the poultry and close to one-half of the vegetable it consumes, and Singapore, which licenses nearly 10,000 farmers, is self-sufficient in meat and produces one-fourth of its own vegetables. From the 1850s through the early 1900s, intensive horticulture covered one-sixth of Paris. The crops, sustained by manure from the city's horse stables, produced so many fresh vegetables that they exported the surplus.[139]

> In Saudi Arabia, which was totally dependent on a now-depleted fossil aquifer for its wheat self-sufficiency, production is in a freefall. From 2007 to 2010, Saudi wheat production fell by more than two thirds.
>
> Lester Brown
> *The Great Food Crisis of 2011*

The most serious objection to urban agriculture is its potential threat to public health. But rather than eliminate urban farming in order to solve this problem, cities can address the problem by promoting safe practices, supplying compost and treated sewage, mandating organic farming methods, and prohibiting the chemical fertilizers and pesticides that contaminate water and air.[140]

Urban organic waste and wastewater can nourish city farms and gardens, thereby reducing the problems and costs of waste management. For instance, aquatic plants, such as duckweed and water hyacinth, grown in sewage lagoons provide a food crop for livestock and purified water.[141] Communities can also create short-term gardens in vacant lots and long term gardens for farming in parkland, floodplains, and steep slopes and wetlands that are not safe for building.[142]

The Netherlands is leading the rebirth of urban agriculture in Europe. It is densely populated yet maintains a vibrant green core in its major cities. In Germany, 80,000 Berliners tend community gardens, and another 16,000 are on waiting lists.[143]

Over the past 20 years, New York City has opened 1,000 community gardens on public land and 18 public markets in which to sell the food. Boston and Philadelphia have even more gardens per capita. Several cities, such as Chattanooga and Hartford, and states, such as Massachusetts and Oregon, have developed comprehensive urban food policies. Land in Detroit is now 40% vacant; Pittsburgh has more than 37,000 empty lots, creating plenty of potential.[144]

Green Roofs. One area of urban agriculture is green roofs. The Fairmount Waterfront Hotel in Vancouver grows herbs, flowers, and vegetables on its accessible roof, saving its kitchen an estimated $30,000 a year in food costs.[145]

Local Food Policy Councils. To try to bring many issues together, many localities are starting Local Food Policy Councils to coordinate various local efforts around food. In the *Worldwatch* report titled *Home Grown: The Case for Local Food in a Global Market*, Brian Halweil states that these councils often emerge out of informal coalitions of local politicians, hunger activists, environmentalists, sustainable agriculture advocates, and community development interests. Some issues they deal with are

- distributing coupons to low-income households for use at farmers' markets,
- improving public transportation to food outlets,

Issues in Sustainable Agriculture

Sustainable Agriculture addresses the problems that large-scale agribusiness has created: topsoil depletion, groundwater contamination, high use of pesticides, the decline of family farms, continued neglect of the living and working conditions of farm laborers, increasing costs of production, and the disintegration of the economic and social conditions in rural communities.

Community-Supported Agriculture (CSA) involves members who participate by buying shares in a local, organic farm. The farmer delivers a box of produce per share every week to convenient drop-off sites during the growing season.

Urban Agriculture is a movement to use small plots of unused urban land to grow food. It is an activity that contributes to a solid community economic base through micro-enterprises.

Food Security or the Community Food Security Movement is a global issue of providing sufficient food for the world's growing population without creating widespread environmental damage. Initially, the focus was on productivity and profitability; now, it is also about sustainability, which considers the long-term.

- creating a grocery delivery service for homebound elderly people,
- helping broker business between local farmers and school food programs, food co-ops, and hospitals, and
- helping develop a kitchen incubator for new food businesses.[146]

In Vancouver, the group Farm Folk/City Folk coordinates food delivery and farmers' markets. They have started a rooftop garden project and opened a healthy café in inner-city areas of Vancouver. The group holds an annual "Feast of Fields" harvest festival, which features local foods. Farm Folk/City Folk has also converted some large city parks that were to be developed, into working farms.

Future food security has also been affected by trade rules. For 30 years, industrialized agricultural nations in Europe and North America have dumped heavily subsidized foods on poor countries, devastating their local food systems. The agriculture section of trade treaties favors large-scale agriculture. The U.S. and the European Union give huge subsidies to their agricultural sectors, which depresses global prices and harms farming communities worldwide. Rich countries have won exemptions for the types of subsidies they use while still prohibiting the types of subsidies (generally tariffs) used by developing countries.[147]

Chipmunk

The WTO says its goal in agriculture is to eliminate tariffs on agricultural imports and to promote "free trade" in agriculture. However, high tariffs on agricultural products in many developing countries serve to protect their own local farmers when there is a crisis in global market prices and to help guarantee food security through local food production. Once their agriculture is run by "free trade," their farmers are forced out of business, and the country becomes permanently dependent on food from the outside, making their food security vulnerable to any unforeseen changes. Now, a few large corporations increasingly control the global food supply, which harms family farmer income, biodiversity, environmental sustainability, and food security.

Suburbia is where the developer bulldozes out the trees, then names the streets after them.

Bill Vaughan

The important next step is to democratize better quality food for people. In creating a more local and organic food system, it is important that the poor be able to access and afford it.[148] Michael Pollan says that the government supports unhealthy food and does very little to support healthy food, so it is more expensive. The incentives we have make processed food, fast food, and junk food very cheap. We subsidize high-fructose corn syrup and hydrogenated corn oil, but we do not subsidize organic food.[149]

Grocery stores are rarely located in the inner city; therefore, people there have to get on a bus and take a long ride to get to a source of fresh produce. So, they buy food at small convenience stores or eat at fast-food places. We have to figure out a way of getting supermarkets and farmers markets into the inner cities.[150] Pollan says that when you use the Women, Infants, and Children program or food stamps to give people farmers' market vouchers, the farmers' markets will show up. One of the best things we could do would be to build 12-month farmers' markets in inner cities. This would be a wonderful way to make farmers' markets a central part of these communities, which would drive economic development and local agriculture. Pollan says that wherever there is a permanently located farmers' market, economic redevelopment follows close behind.[151]

The other way to democratize the food movement is through the public school system. If schools can buy local, fresh food, we could improve the health of the students, the health of the local economy, and the performance of the students. Pollan says it would take about $1 per student per day to really change the food system.[152]

Government policies created our current food system, and new policies can change it. The federal government has enormous power to determine how food is grown. Science needs to again be publicly funded and in the hands of land-grant universities committed to conducting research in the public interest, rather than in the hands of the private commercial interests who hijacked the research agenda and privatized its technologies, leaving us with only for-profit solutions.[153]

In addition, current government subsidies discourage farmers from growing healthful, fresh food. For one thing, farmers are prohibited from growing "specialty crops," meaning fruits and vegetables (rules pushed by California and Florida produce growers). We need to change what is subsidized. The government should use subsidies to encourage growing as many different crops, including animals, as possible, because the greater the diversity of crops on a farm, the less need there is for both fertilizers and pesticides.

> Asphalt is the land's last crop.
>
> Rupert Cutler
> Environmentalist

In addition, subsidies should also pay for the number of days of the year fields are green. If farmers planted a cover crop after the fall harvest, they would significantly reduce their need for fertilizer while also cutting down on soil erosion, but they don't do this because right now fertilizers are cheaper.[154] Compost could be another substitute for fertilizers. Applying compost to fields improves the fertility of the soil and its ability to hold water and therefore withstand drought. It could be made mandatory that the food wasted by consumers, retailers, wholesalers, and institutions be included in municipal composting of food and yard waste and that this compost be distributed to area farmers.[155]

Industrial agriculture corporations argue that corporate farms are more efficient and better able to feed a hungry world. Yet Cavanagh and Mander say, "All the evidence consistently shows the contrary; industrial farming's so-called efficiencies are sustained only by large government subsidies, and it causes more hunger than it solves."[156] Brian Halweil also says the rationale that large, corporate farms are more efficient or productive is a myth: "While a large monoculture operation may produce more output per acre of that crop than a small farm does, the small farm doing traditional polyculture (raising more than one kind of crop, using different root depths or soil nutrients on the same piece of land) makes more intensive and efficient use of resources and can produce significantly more food overall per acre."[157]

Creating Millions of New Jobs and Renewing Rural America

Making these changes will require educating a new generation of farmers, who will need to be more highly skilled. These farms will be small-scale family farms growing crops organically. These changes will generate tens of millions of new jobs and will create economic renewal of rural areas and small towns, for they will generate the need for all kinds of goods and services. They will allow much of the new economy to shift to areas where and at a scale at which relationships and community come naturally. Farmers can earn rent from windmills put in their fields. Additional farm income could come from making paper from plants and creating small-scale local paper mills. Synthetics could be made from plants rather than petroleum—again with local refinery plants. This would also greatly reduce the toxins that are poisoning our planet. These new products for farmers are discussed in the next chapter.

Striped Hyena

Chapter 14

Sustainability in Our Use of Materials

In our economy, the production, processing, and disposal of materials waste both energy and materials. Therefore, we must drastically decrease our use of materials. There are many different approaches, but all of them involve shifting from linear thinking to cyclical, systems thinking.

Our materialistic economy creates great environmental harm. The U.S. produces billions of tons of waste, primarily because industries, retail firms, governments, and consumers have no incentive to use natural resources sustainably. The *State of the World 2004* report on consumption says, "From the standpoint of global justice and equality, the solution cannot be a system of consumer apartheid that upholds western binge habits but denies the poor a decent standard of living. Instead, the rich need to curb their outsized material appetites. Rough calculations suggest that in order to accommodate the imperatives of environmental protection and social equity, the rich nations may need to cut their total use of materials by as much as 90% over the next few decades."[1]

However, reducing the use of materials by 90% does not mean we need to have 90% less "stuff," only that the stuff we do have needs to be made using 90% fewer materials. The throwaway economy is now coming up against Earth's geological limits, as Lester Brown describes:

> Aside from running out of landfills near cities, the world is also fast running out of the cheap oil that is used to manufacture and transport throwaway products. Perhaps more fundamentally, there is not enough readily accessible lead, tin, copper, iron ore, or bauxite to sustain the throwaway economy beyond another generation or two. Assuming an annual 2-percent growth in extraction, U.S. Geological Survey data on economically recoverable reserves show the world has 17 years of reserves remaining for lead, 19 years for tin, 25 years for copper, 54 years for iron ore, and 68 years for bauxite.[2]

There are three ways currently used to reduce the use of energy and materials in industry. First, industry can accomplish steep reductions by using a process called Zero Waste in which manufacturers design the production system rather than the product, so that everything is reused in a closed loop. Second, industry can sell a product as a service that is leased and maintained by the company producing it. This creates incentives to produce goods for durability. Third, government can mandate that anything that cannot be reused or recycled must be biodegradable.

Sustainable Materials Extraction

Reducing the use of materials begins with mining. The majority of environmental destruction happens even *before* materials are sold and used. Even though leaking landfills are a problem, the major impact of our materials economy on the environment comes from the quantity and toxicity of mining. The damage done and the wastes created by mining far exceed the damage done by consumers disposing of goods. Part of the reason is that we are "liquidating" natural resources at their *extraction* cost rather than at their *replacement* value.

> Materials use is a proxy indicator for environmental impact: the greater the tonnage of virgin materials extracted, processed, consumed, and disposed of, the greater an economy's environmental footprint.
>
> Gary Gardner
> Vital Trends

The concept of sustainable materials extraction has two parts. It means using only non-toxic materials that are abundant and can be safely returned to the Earth (soil, sand, gravel, iron, hydrogen, silicone, titanium, aluminum, etc.). It also means that if materials are scarce, toxic, difficult to safely extract, or difficult to safely sequester—like uranium, mercury, cadmium, or platinum—they are used only in tightly controlled loops that do not leak into the biosphere. In this concept, whatever industry extracts must be done in careful, environmentally sustainable ways, with the impacts and costs included in the price of the finished product.

Industry, including the production of plastics, fertilizers, steel, cement, and paper, accounts for more than 30% of world energy consumption. There is a vast worldwide potential for cutting carbon emissions by reducing materials use. This begins with the major metals—steel, aluminum, and copper—where recycling requires only a fraction of the energy needed to produce these metals from virgin ore. It is estimated that there is 40 million tons of copper in landfills.[3]

Manufacturing Based on Systems Thinking

Our materials economy needs to be regenerative and not deplete the natural world. Our current linear model ("take—make—waste") needs to be redesigned into a circular model of closed loops that emulates nature. The GrassRoots Recycling Network says that for every single can of discarded products and packaging by individuals, industry produces 71 garbage cans' worth of industrial waste. Industrial waste includes waste from mining, clear-cutting timber, oil and gas drilling, and refining and manufacturing raw materials.[4] Currently, tax subsidies and incentives favor wasteful extraction industries and distort accurate market prices.

In the 1990s, Schmidt-Bleek, founder of the Factor Ten Institute in France, showed that resource productivity could rise by a factor of 10 with existing technology and management, given the right policy incentives. In 2002, American architect William McDonough and German chemist Michael

Duskylory

rt coauthored *Cradle to Cradle: Remaking the Way We Make Things*. They ued that waste and pollution could be avoided entirely. "Pollution," they said, "is mbol of design failure."

> o Waste is a design philosophy that can help communities achieve a local economy that operates efficiently, sustains good jobs, and provides a measure of self-sufficiency.
>
> Peter Montague
> Environmental Research Foundation

The primary way to eliminate waste is called Zero Waste. It is also called Life-Cycle Assessment, or a "cradle-to-cradle" analysis. Zero Waste is a philosophy and design principle that uses a whole-systems approach to designing products and packaging for reuse and recycling. Manufacturing using Zero Waste means creating materials in a way that eliminates all wastes. Zero Waste eliminates the entire concept of waste and instead thinks of waste as a "residual product" or a "potential resource."

In Zero Waste, discards are not garbage in need of disposal but become valuable resources and alternatives to mining additional resources. What is now thrown away represents jobs, financial opportunities, and raw materials for new products. This process results in reduced costs, increased profits, and reduced environmental impacts.

A Zero Waste strategy improves on "cleaner production" and "pollution prevention" strategies. It begins by defining the goal as not "just a little less" waste and pollution but as creating *no* waste and *no* pollution. It is a vision that leads to innovative structural responses.

Zero Waste puts the responsibility for materials' entering the waste stream with the manufacturer, up front, at the beginning of a product's life, rather than with the taxpayers, who pay for clean-up after production, use, and disposal of goods. The goal of Zero Waste includes zero solid waste, zero hazardous waste, zero toxic emissions, zero material waste, zero energy waste, and zero waste of human resources. Using a systems view, it optimizes the total materials cycle, from raw materials through manufacturing, use, and final disposal.

Zero Waste creates a ripple effect when it merges with a full range of related concepts.

- *Design for the Environment* is a new discipline initiated by designers that ensures that all costs—including environmental costs—are internalized at the design stage.
- *Design for Disassembly* is a design discipline aimed at ensuring that products are designed for ease of disassembly so that the parts can be reintegrated into new models and materials can be recycled.[9]
- *Dynamic Modularity* products are made in modules so that only some modules need to be replaced to lengthen product life, for example, the 'skin' of a product.
- *Reverse Logistics* is a process in which retail chains use their distribution systems in reverse to have all broken and unsold merchandise returned to central locations for repair, reuse, or disassembly for recycling. Retailers report huge economic savings from reverse logistics. Manufacturers also get faster feedback about product failures, which helps in redesign.[10]

Examples of Zero Waste

Xerox: In 1990, Xerox embarked on an Asset Recycle Management initiative. While the company had previously done some remanufacturing, this program led Xerox to design its products from the very beginning with remanufacturing in mind and to make every part reusable or recyclable. As a result, 70%–90% of the equipment (measured by weight) returned to Xerox at the end of its life can be rebuilt.

Fetzer: This wine maker is converting to organic agriculture and aims to achieve Zero Waste. By reducing, reusing, and recycling, the company has already cut waste by 93%. They compost (mostly grape seeds and cork) and buy recycled products (even the roof on their building is recycled). Fetzer's buildings use solar power, which supplies enough energy to fill and cap 1.2 million bottles of wine each year.

Interface: The world's largest carpet manufacturer has set the goal of becoming the world's first truly sustainable company by rethinking the way their products are made, transported, maintained, and recovered. For example, Interface offers the "Evergreen Lease," an innovative carpet rental and service contract. They use stronger, more durable carpet fibers, recover worn carpet fibers for recycling, and service carpets on a section-by-section (rather than whole-carpet) basis.

BioCorp: This company offers cups, plates, and cutlery made from fully biodegradable cornstarch and from natural polymers that can be composted. They also offer biodegradable leaf-collection bags that can be composted along with their contents.

Kodak: In 1990, Kodak began a "Take-Back" program to reuse and recycle disposable cameras. Components are reused up to 10 times in new cameras and then are recycled into new components. Kodak pays photofinishers for used cameras, providing a financial incentive for retailers to collect them. The program has saved 20,000 tons of waste; however, 24% of the cameras are never recycled by the purchaser and are most likely in landfills.

- *Cleaner Production* is an efficiency concept used mainly by businesses to reduce the impacts of production on the environment.
- *Remanufacturing* means returning a used, worn-out product to as close to new as possible. The product is completely disassembled, cleaned, inspected, remachined, reassembled, and tested to ensure functional quality. Remanufacturing is ecological and economical recycling and has a substantial effect on energy, as well as on material use. The *State of the World 2004* report says, "Remanufacturing operations—particularly in areas like motor vehicle components—save at least 11 million barrels of oil each year, an amount of electricity equal to that generated by five nuclear power plants and a volume of raw materials that would fill 155,000 railroad cars annually."[11]
- *De-materialization* is the name Paul Hawken, in *Natural Capitalism*, calls the process of using fewer materials to create the same service.

Golden Frog– Endangered

- *Industrial Ecology* is the objective, multidisciplinary study of industrial and economic systems and their linkages with fundamental natural systems. It incorporates, among other things, research involving energy supply and use, new materials, new technologies and technological systems, basic sciences, economics, law, management, and social sciences.[12]

> Keep it in the crust of the Earth
> Don't make nasty stuff
> Don't pave paradise
> Be thrifty and fair.
>
> The Natural Step

One way to create new designs is by studying nature. Janine Benyus, in *Biomimicry: Innovation Inspired by Nature,* tries to learn from nature rather than focusing on what we can extract from nature. She asks, "How might we apply biological designs, processes, and laws to the design of human systems? Biomimicry is a new way of viewing and valuing nature. After 3.8 billion years of research and development, failures are fossils, and what surrounds us is the secret to survival. We can use nature to design our systems, and current learning is exploding—biological knowledge doubles every five years."[13] She says that after years of evolution, nature learned what works. Therefore, Benyus says, "'Doing it nature's way' has the potential to change the way we grow food, make materials, harness energy, heal ourselves, store information, and conduct business. In each case, nature would be 'model, measure, and mentor.'"[14] Benyus gives many examples, such as how solar-powered transpiration in trees can silently move tons of water up hundreds of feet, how mangroves desalinate water, and how termites thermoregulate their shelters through structural design. And we can learn "how to harness energy like a leaf, grow food like a prairie, build ceramics like an abalone, self-medicate like a chimp, compute like a cell, and run a business like a hickory forest."[15] Biomimicry uses an ecological standard to judge the "rightness" of our innovations.

The Natural Step Framework

The Natural Step is a not-for-profit organization that provides education, advisory work, and research related to sustainable development. They created The Natural Step Framework, a comprehensive model for planning in complex systems, to help organizations and communities define principles of sustainability based on our system – Earth.

They coined the term "backcast" (as opposed to "forecast") to describe a method that begins with defining a future point of success and developing the most effective steps to reach that goal. "Forecasting" plans usually reinforce our existing systems and structures by taking them as a given. Changing the word "forecast" to "backcast" helps us to re-examine and reconsider whether our current systems and structures must change to reach our goal. Green options appear to cost more in up-front investment but they are much cheaper over their full life span when full-cost pricing is used. Daily decisions about sustainability can often seem like we are simply trading one problem for another. An overarching framework to help us evaluate our decisions give us the true pros and cons in those decisions.

The Natural Step Framework

The Natural Step developed four conditions that define characteristics of a sustainable society. These conditions are rooted in science and are used for strategic planning toward sustainability.

The Four System Conditions ...	*Reworded as The Four Principles of Sustainability*
In a sustainable society, nature is not subject to systematically increasing:	To become a sustainable society we must:
1. concentrations of substances from the Earth's crust	1. eliminate our contribution to the progressive buildup of substances extracted from the Earth's crust (for example, heavy metals and fossil fuels)
2. concentrations of substances produced by society	2. eliminate our contribution to the progressive buildup of chemicals and compounds produced by society (for example, dioxins, PCBs, and DDT)
3. degradation by physical means	3. eliminate our contribution to the progressive physical degradation and destruction of nature and natural processes (for example, over-harvesting forests and paving over critical wildlife habitat)
4. and, in that society, people are not subject to conditions that systemically undermine their capacity to meet their needs	4. and, eliminate our contribution to conditions that undermine people's capacity to meet their basic human needs (for example, unsafe working conditions and not enough pay to live on)

Extended Producer Responsibility

Blister Beetle

Another tool to reduce the use of materials by 90% is Extended Producer Responsibility. This concept focuses on recycling after the product is used. The goal is to redesign products in a way that facilitates reuse, recycling, and recovery. With this concept, manufacturers are responsible for all the costs of collecting and recycling a product. This change would ultimately encourage reparable products made from fewer, more easily separated recycled materials.

Producer Responsibility or Take Back laws, where the manufacturer must take their product or packaging back for remanufacture or reuse, have now been enacted in 29 countries. Manufacturers have to include new labeling to facilitate the reuse/return process. To accomplish this, the Conservation Economy website states,

> Just as plastic bottles carry a recycling code from 1-7 indicating their chemical composition, before too long VCRs, steel beams, windows, desks, boilers, or pipe connections will announce what they are made of, how they may be safely disassembled, and where they should be returned. Embedded chips, miniature text, web links, or other information media can be used to signal this information. One of the largest Swedish construction companies is already stamping every building component with this information. As even mundane objects like toasters acquire the intelligence of computer chips, there is no reason why they can't carry information about their remanufacture.[16]

One way to accept producer responsibility is by having manufacturers sell a product as a *service*. Instead of selling as many goods as possible, manufacturers can provide a desired service—leasing or renting their products to consumers—and thereby take responsibility for a product for its entire life. With this approach, Paul Hawken says, "The goal is selling results rather than selling equipment, performance, and satisfaction rather than motors, fans, plastics, or condensers. In a *service economy*, the product is a means, not an end."[17] When manufacturers provide services rather than goods, it changes their long- and short-term incentives. They design a product as a continuous flow of service and completely reclaim materials for use in the next generation of the same or another product. Any toxic substances in the product can be kept tightly controlled and prevented from leaking into ecosystems.

> If [United States] homeowners reduced their use of toxic chemicals including pesticides by 10%, 2 million kilograms of toxic chemicals would be removed from the environment each year. If manufacturing firms reduced their toxic releases by 10%, 700 million kilograms would be removed.
>
> Biodiversity 911: Saving the Earth
> WWF and the Worldwatch Institute

As an example, Carrier, the world's largest manufacturer of air-conditioning equipment, now offers "thermal comfort" contracts for buildings. Carrier maintains the desired comfort level through a combination of energy-efficient building retrofits, new equipment, and improved control and management. Or a washer and dryer could be leased and cleaned by the manufacturer every 6 months or once a year, with any worn parts replaced. When the product wears out, it is replaced with a new one, and the old one is recycled by the manufacturer. The consumer pays a monthly fee for the service, and the manufacturer has a steady income.

Shifting to providing a service reduces what Jeremy Rifkin calls "our near fanatical commitment to private property."[18] This shift to selling a service instead of a product is affecting many different industries. Instead of selling a CD, some music companies like Rhapsody charge a monthly subscription fee for unlimited access to their music. This allows them to keep an ongoing relationship with their customers rather than selling CDs one at a time. Encyclopedia Britannica still sells their 32-volume set, but the encyclopedia is also on the Web and continually updated, and people can pay a subscription to access the information for a given time period.[19]

Rifkin says that in these networks, "providers and users replace sellers and buyers, and access to the use of goods in extended time segments substitutes for the physical exchange of the goods."[20] In pure networks, "property still exists, but it stays with the producer and is accessed in time segments by the user. Subscriptions, memberships, rentals, time-shares, retainers, leases, and licensing agreements become the new medium of exchange."[21]

Anything Not Reusable Must Be Biodegradable

Anything that cannot be contained in a closed loop or recycled and reused must be biodegradable and nontoxic. A good description of non-toxic production is given in an *Atlantic Monthly* article called "The NEXT Industrial Revolution." The authors, William McDonough and Michael Braungart, who wrote *Cradle to Cradle*, present a new way to design industrial production. They say, "Real change would not be regulating the release of particles, but attempting to eliminate dangerous emissions altogether—by design."[22] In such a case, "regulations can be seen as signals of design failure."[23]

Waste is too expensive; it's cheaper to do the right thing.

Paul Hawken

Approximately 100,000 synthetic chemicals are now on the market, with 1,000 new chemicals added every year. Chemical companies are not required to tell the EPA how their compounds are used or to monitor where their products end up in the environment. Approximately 99% of chemicals sold are not subject to safety testing, and scientists have never studied the dangers of exposure to these chemicals in combination.[24] Therefore, we have no idea how much our current system is poisoning people and the environment. Scientists measuring the threat of toxic substances initially misunderstood the risk. They assumed that if they could rule out cancer caused by large quantities of a chemical, people would be protected from any threat from smaller amounts. However, they did not look at the effect of pollutants and synthetic chemicals on embryonic development in the womb. Embryonic development is driven by hormones, which work at a concentration of one-tenth of a trillionth of a gram. Fetal development is extremely sensitive to any variations in hormone signals. For a fetus to develop in accordance with its genetic blueprint, the right hormone message has to arrive in the right amount at the right place at the right time. When chemicals interfere with these messages, they are called endocrine disruptors.

WWF senior scientist Dr. Theo Colborn has a background in pharmacology, a field that has grown into toxicology and

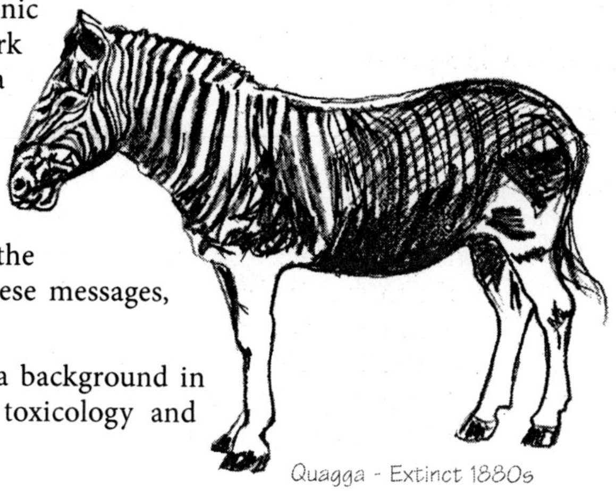

Quagga - Extinct 1880s

studies chemicals produced and released into the environment. As she studied the toxic effects of chemicals, she plotted on a spreadsheet all the health effects and chemicals being seen in animals from hundreds of studies. This synthesis resulted in her book, *Our Stolen Future,* which demonstrates that many human-made chemicals interfere with the body's hormones even after birth. Endocrine disruptors include any chemical that interferes with hormones, such as thyroid, cortisol, insulin, or growth hormones. Endocrine systems control body growth, organ development, metabolism, and regular body processes, such as kidney function, body temperature, and calcium regulation. The effects may not appear until puberty or later. *Our Stolen Future* lists the four most serious aspects of endocrine disruption:

Some effects are seen at *very* low exposures;

- ❖ Dose-response is usually understood as a linear relationship. [The greater the dose, the greater the effect.] For some endocrine-disrupting chemicals, however, effects may disappear at higher levels or vary qualitatively, or they may appear at levels below the "no observable adverse effect level." This means that high-dose testing, the standard procedure of risk assessment, cannot reveal the health risks posed by endocrine disrupters.
- ❖ In some circumstances, there is *no* threshold level without effect.
- ❖ Exposures to contaminants take place in mixtures, and these mixtures can interact additively, synergistically, or not at all.[25]

Currently, the government uses a policy of "risk assessment" with chemicals; this policy assumes that science and governments can and must provide definitive proof of harm before protective action is taken against harmful chemicals. However, Colborn says that identifying proof of harm is very difficult because harms rarely have a single cause; causes and outcomes are often multiple; periods for effects are often long; timing of exposure is often crucial; there are no unexposed "control" populations; and many complicating factors are unidentified. All of these complications make scientific standards of certainty impossible to attain. The use of risk assessment measures creates a loophole in regulations, giving the benefit of the doubt to products—even those that are likely to have harmful side effects. In addition, the WTO names as "illegal barriers" to trade any preventive measures used to stop a product or practice and has ordered these measures stopped.

The burden of proof that a product is safe should rest with the company producing it. Because we are now learning how difficult it is to know what levels of contamination, if any, are safe, we need to use a *precautionary principle* rather than risk assessment. Germany and Sweden have precautionary principle laws, and other countries are considering doing so. Without the adoption of a precautionary principle, citizens lose the right to decide what risks to themselves or the natural environment they are willing to accept.[26] The precautionary principle also brings democracy and transparency to decisions that affect health and the environment.[27]

One reason why chemicals create so much harm is that so many products are made from petroleum. After World War II, petrochemicals began to dominate the synthetics market. By 1990, for every industrial product except paper, petroleum had replaced starch, vegetable oil, and cellulose—the three components of plant matter that had been used to create synthetics. These petrochemicals have created a highly toxic environment. But there are alternatives. Instead of producing everything from petrochemicals, we need to shift to a "carbohydrate economy." *Anything made from a hydrocarbon (petroleum) can be made from a carbohydrate (a plant).* Substituting biochemicals can be a permanent solution to regulatory compliance; it reduces both upstream (manufacturing) and downstream (waste) pollution. Using benign materials makes products safer to produce, use, and eventually compost. When all costs are considered, biochemicals are competitive with petrochemicals, especially if subsidies go to biodegradable products rather than petrochemical products.

The Environmental Defense Fund has set up a huge interactive site that enables anyone in the US to learn about what pollutants are being released into the air, water and soil of any community, and by whom—all by entering a zip code, or clicking on their maps.

Environmental Defense Fund
Scorecard
http://www.scorecard.org/

Replacing petrochemicals with biochemicals will create new industrial markets for plant matter. Plant matter, unlike petroleum, is bulky and expensive to transport; therefore, biochemical refineries need to be smaller and closer to their raw material source than petrochemical refineries. This encourages local and regional processing, preferably owned by the farmers who can supply adequate amounts of biomass. Farming has a long tradition of cooperative ownership, primarily service and marketing co-ops, but there are also processing and production cooperatives as well. A major expansion of the use of plant matter for nonfood and feed purposes could result in thousands of new small processing and manufacturing facilities in rural America. A proliferation of modestly scaled production enterprises will encourage innovation and cost reduction while promoting healthy communities and strong local economies. Biochemical production can be combined with growing plants for paper and creating local paper mills. The two would bring a strong economic base to rural areas and small towns. However, using plants to make biochemicals means we need to stop using plants for biofuels or for livestock. With the various methods used to change manufacturing, we really can actually reduce our use of materials by 90%; all it takes is a new mindset and new goals.

Special Case: Restore Sinks for Carbon in Soils and Forests

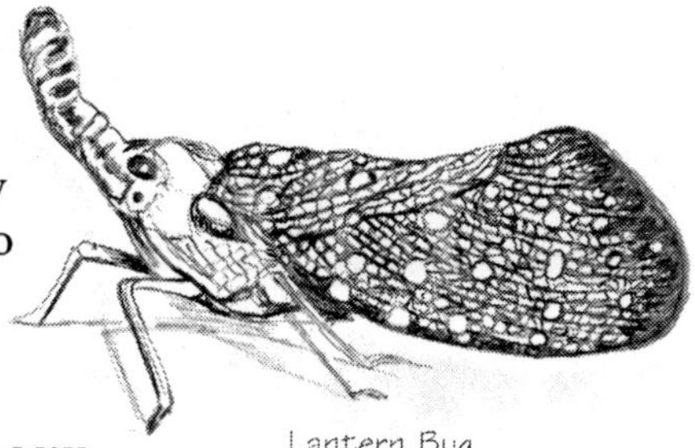

Lantern Bug

A special case for materials is the use of trees. We not only need to stop the increasing rate of emissions, we also need to drastically reduce what is now in the atmosphere. There are companies promoting exotic, very expensive, and potentially disastrous ideas like sequestering CO_2 under the ocean floor. We seem to always look for the most high-tech and expensive way to do things—because they can make companies a lot of money.

There are much easier ways to remove greenhouse gases from the atmosphere, such as sequestering emissions in soil. The other major way is to reduce logging and plant more trees.

Consequences of Deforestation

Removing forests (and their natural functions) causes many serious problems. Through photosynthesis, trees remove CO_2 from the air, produce oxygen, and store carbon as wood. The international community needs to find ways to help developing countries in the tropics slow and halt tropical deforestation. Tropical deforestation is responsible for approximately 20% of total human-caused CO_2 emissions each year.[28]

- *Loss of Trees Makes Global Warming Worse.* Through photosynthesis, trees remove CO_2 from the air, produce oxygen, and store carbon as wood. One ton of carbon in wood or forest biomass represents 3.67 tons of atmospheric CO_2 recycled. We are creating warming not only by putting more CO_2 into the air, but also by getting rid of trees that absorb and remove carbon from the air.
- *Impact on Ecosystems.* Forests preserve water, soils, plants, and wildlife. Their destruction aggravates droughts, soil erosion, and pollution of watercourses, and it causes extensive flooding and increased pest populations as a result of the ecological imbalance.
- *Loss of Species.* Tropical forests contain at least one-half of Earth's species, so their loss causes a dramatic loss of biodiversity. Clearing and destructive logging of forests is the single greatest cause of species' extinction worldwide.
- *Harm to Water.* Forests are natural dams that catch rainwater in their canopies and in leaves and litter on the forest floor, retaining and purifying rainwater. Forest logging allows rapid runoff and destroys the ability of the soil to absorb water.

Ban Deforestation and Start to Plant Trees

Banning deforestation is essential to reducing global warming. To start, we need to eliminate our timber subsidies. The U.S. has given away to logging firms timber worth half a billion dollars in our national forests. They then have built roads for them at taxpayers' expense, at a cost of nearly 1 billion dollars. These logging subsidies keep the price of wood artificially low and make it harder for non-wood alternatives to compete. To change this, the price of wood and paper needs to reflect the true cost of destroying a tree so that alternatives to deforestation can compete financially.

> From now on, the question of whether we get our protein from animals or plants has direct implications for how much more of the world's remaining forest we have to raze.
>
> Mark Lynas
> *Worldwatch*

A major part of the problem is the trade in illegally extracted timber. Legal supplies of wood fiber fall short of demand by up to 40 million cubic meters per year. Illegal logging fills the gap—accounting for almost 70% of wood supply, meaning that illegal logging *exceeds* the volume of legal logging. The WWF says the trade in illegally extracted timber is a multi-million dollar industry that goes on in over 70 countries in all types of forests, even in national forests.[29] Since illegal logging activities tend to be concentrated in forests rich in plant and animal diversity, environmental costs are also high.[30] Illegal logging threatens 65% of the 200 areas with high biological diversity worldwide.

To fight illegal logging, all domestic production and imports of wood products should come from responsibly managed forests that meet rigorous environmental and social standards. It is important to use only wood certified by the Forest Stewardship Council (FSC)—the only label recognized as providing "ecologically sound" timber. This certification stamp means the wood is from well-managed, environmentally sensitive logging operations and is not the result of monoculture plantations, clear-cutting, violations of indigenous land claims, or environmental hazards. Because of these other considerations, relying on wood labeled "second growth" is not enough. The FSC label is the only guarantee that the wood purchased is environmentally sound.

Along with reduced logging (legal or illegal), we also need to have a massive tree-planting campaign. In late 2006, UNEP announced plans for a worldwide effort to plant 1 billion trees in one year, but by mid-2008, more than 2 billion trees had been planted in more than 150 countries. The campaign now aims to facilitate the planting of 7 billion trees.[31]

Producing Paper from Tree-Free Sources

At least 40% of the trees that timber corporations log (4 billion trees) are used to make paper. One solution to cutting down trees for paper is to use *tree-free pulp* for papermaking. Treeless pulp is not a new concept. Before the 1900s, cutting down forests to make paper was rare. For example, in China, over 85% of paper pulp comes from non-wood raw materials, predominantly straw. In India, approximately 55% of papermaking pulp is from non-wood sources, with about one-half of that from agricultural residues.[32]

Treeless paper can be made using either using "on-purpose" crops (kenaf or hemp) or agricultural residues (the wastes left over from crops). *Agricultural residues* are what are left in the field after the harvest. They come from wheat, rice, corn, rye, banana, cotton, or flax. About one-half of the residual straw is left in the field and plowed back into the ground; even so, a significant amount of straw remains.

Until recently, many farmers disposed of agricultural wastes by burning or depositing them in a landfill. Burning agricultural waste causes air pollution, soil erosion, and a decrease in soil biological activity.[33] Turning these residues instead into paper is an example of using one industry's waste as the necessary raw material for another. Using agricultural residues does not require any additional land use, and there are a lot of them.

These "waste" fibers blend well with stronger non-wood fibers such as kenaf and industrial hemp. Agripulp can also replace wood in products such as building and insulating materials, particleboard, and various grades of paper and paperboard. Agriboard Industries claim that local farmers stand to

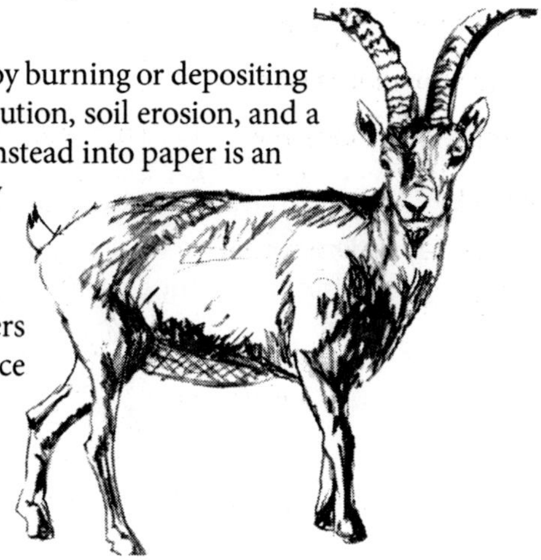

Pyrenean Ibex – Extinct 2000

make more from selling the wheat straw to them than from selling the actual crop.[34] The increased interest in making paper out of agricultural wastes means that farmers can reap a "second harvest" from grain plantings.

On-purpose crops are crops that are specifically grown to make paper. The two best plant fibers for paper are kenaf and hemp. For more than 40 years, the Agricultural Research Service has investigated more than 500 plants to find the best fiber for pulping paper. While hemp has historically been the most recognized plant for tree-free paper, the Agricultural Research Service has identified another plant, kenaf, as a viable fiber to produce tree-free paper.

God has cared for these trees, saved them from drought, disease, avalanches, and a thousand tempests and floods. But he cannot save them from fools.

John Muir

Kenaf is a crop grown specifically for paper and can replace or be rotated into existing crops. It matures in 5 months at 12 to 18 feet high and yields 6 to 10 tons of dry fiber per acre (3 to 5 times the yield for paper from growing southern pine for paper). Kenaf's core fiber can be used for such products as particleboard, potting soil, horse bedding, and oil and chemical absorbents. Moreover, kenaf is ideal for rotation with legumes, corn, sorghum, and other food crops because it helps to maintain soil quality, maximizes yields, and increases bug and disease resistance.[35]

Kenaf has only a very limited need for pesticides, herbicides, or fertilizers, and it is low in lignin content (a "glue-like" substance), compared with wood; therefore, it needs far fewer chemicals and less energy to process into paper. Because of its lighter color, kenaf can be bleached using chlorine-free agents, such as ozone, hydrogen peroxide, and oxygen. Creating totally chlorine-free paper eliminates the creation of the most deadly of pollutants—dioxins._

Hemp. Paper can also be made from industrial hemp. Hemp is the world's most versatile fiber. Industrial hemp produces the longest, strongest natural fiber in the plant kingdom. Almost any product that can be made from wood, cotton, or petroleum (including plastics) can be made from hemp. There are more than 25,000 known uses. The greatest future market for industrial hemp products is predicted to be automobile parts (industrial hemp is an excellent, cost-efficient replacement for fiberglass parts). The crop's current largest market is foods.

Hemp is related to the marijuana plant, and even though it does not have psychoactive levels of THC, the Drug Enforcement Administration will not allow industrial hemp to be grown. There is no federal statute outlawing the cultivation of hemp, just the Drug Enforcement Administration's insistence that hemp is an illegal drug. Hemp was originally banned with marijuana in 1937. During World War II, the ban was lifted and the U.S. Department of Agriculture encouraged farmers across the country to "Grow Hemp for Victory." But after World War II, the restrictions were reinstated.[36] Hemp is lawfully grown in 32 nations (in Canada and most European and Asian countries); in the EU, it is a subsidized crop. Hemp is easily distinguishable in fields from marijuana. Marijuana plants are harvested for their flowers and leaves and so are

Hemp Facts:

- Hemp is the oldest cultivated fiber plant in the world. The first Gutenberg Bible was printed on hemp paper. Christopher Columbus's sails and ropes were made from hemp. The first drafts of the Declaration of Independence were printed on hemp paper. Ben Franklin owned a mill that made hemp paper. The first American flag was made out of hemp. George Washington and Thomas Jefferson grew it. In 1794, Washington said, "Make the most of the hemp seed. Sow it everywhere." Rembrandt and Van Gogh painted on hemp canvas.
- In 1941, Henry Ford built a plastic car made of fiber from hemp and wheat straw. "Why use up the forests which were centuries in the making and the mines which required ages to lay down, if we can get the equivalent of forest and mineral products in the annual growth of the hemp fields?" —Henry Ford
- The original Levi's jeans, made for Sierra Nevada gold rushers, were made of rugged hemp sailcloth. A current vintage line includes 40% hemp. Hemp can be imported for use in products made here—it just cannot be grown here.
- "I feel the industrial hemp crop could very easily be the soybean crop of the new millennium," Jeffrey W. Gain, U.S. Department of Agriculture.
- Hemp particle board may be up to 2 times stronger than wood particleboard and holds nails better.
- Hemp is an excellent rotation crop: it crowds out weeds and its deep roots break up hard pan soils. Hemp uses no pesticides and crowds out weeds without herbicides.
- Hemp could replace cotton because:
  - One acre of hemp can produce as much usable fiber as 2 acres of cotton.
  - Hemp fabric requires fewer chemicals than cotton and is stronger and longer lasting. Hemp does not require pesticides, while 40% of the pesticides in agriculture are used on cotton.
  - Cotton has a drinking problem: extensive water subsidies. Hemp requires less water than cotton and grows in cooler climates.

http://www.ecomall.com/biz/hemp.htm

short and bushy. Industrial hemp is harvested for its stalk before flowering occurs and so is tall and straight, with leaves at the top of the stalk. Marijuana is grown widely spaced to maximize leaves. Hemp is grown in tightly spaced rows to maximize stalk. Marijuana could not be illicitly grown in the middle of a field of industrial hemp because crossbreeding between the two plants quickly eliminates the THC content in the marijuana seeds.

Hemp typically reaches 6 to 12 feet in its 4-month growing period, producing approximately 3 to 6 tons of dry fiber per acre annually—nearly twice as much as southern pine (kenaf is 3 to 5 times). It grows well in a variety of climates and soil types. It does not need pesticides or herbicides. Hemp

African Wild Dog – Endangered

paper is longer lasting than wood pulp paper, is stronger and acid-free, and can be processed without chlorine.

A World Health Organization 1997 study estimates that 80 percent of cancer cases in the 21st century will be attributable to the environment. Breast cancer rates have increased by more than 50 percent in the past three decades, while prostate and testicular cancer rates have almost doubled.

World Health Organization

Were U.S. hemp cultivation to become legal, only 1% of the nation's farmland would be required to achieve paper self-sufficiency. Like kenaf, hemp production would also give farmers new sources of income.

Creating this new paper industry will require consumer demand for using less wood fiber and support systems for agricultural non-wood products. Recycled paper was rare and expensive until some major publishing corporations committed to using it. Now, we must commit to tree-free paper or 100% recycled post-consumer waste paper.

However, creating paper from plants increases transportation costs. Pulping agricultural residues is most efficient and economical at small, local mills. With state-of-the-art technology, smaller, local mills can improve regional economies by creating jobs at a number of facilities rather than centralizing them at huge mills.[37] Therefore, small and medium-sized paper mills will need to be built in rural areas near the farms that supply the residues, again with farmer-owned co-ops owning the mills.

Rural areas can then be revitalized by three new sources of income: local small-scale chlorine-free paper mills; small factories for making plastics from plants rather than from petroleum; and renting farmland for windmills.

So how will this end? Our current global economy is making it all but impossible to address global warming. In discussing these required changes in *The Great Turning*, David Korten asks,

> By what name will future generations know our time? Will they speak in anger and frustration of the time of the Great Unraveling, when profligate consumption exceeded Earth's capacity to sustain and led to an accelerating wave of collapsing environmental systems, violent competition for what remained of the planet's resources, and a dramatic dieback of the human population? Alternatively, will they look back in joyful celebration on the time of the Great Turning, when their forebears embraced the higher-order potential of their human nature, turned crisis into opportunity, and learned to live in creative partnership with one another and Earth?[38]

California Condor – Endangered

Part V

Conclusion

Zebra Family

Chapter 15

Facilitating Self-Organization for Social Transformation

How do we create the critical mass in public opinion we need to shift the global warming conversation from an emphasis on voluntary steps by individuals to a focus on massive social change? It is possible. The only barrier to change is our political will, and change requires education. However, educating adults in our culture is very difficult because large, multinational corporations own the media—that is, television and newspapers do not want this kind of economic conversation to take place.

We have to find a way to respond to global warming that moves beyond the lists of "10 Things You Can Do to Help Solve Global Warming." Focusing on changes individuals can do is important because behavioral change can raise people's consciousness and give us a sense of how much we are interconnected--that the choices of each of us affect all of us. But changes by individuals are most important if they change the national political will.

> Human history becomes more and more a race between education and catastrophe.
>
> H. G. Wells
> *The Outline of History*

The 2007 IPCC Working Group III report says, "Some corporations, local and regional authorities, NGOs and civil groups are adopting a wide variety of voluntary actions. These voluntary actions may limit GHG emissions, stimulate innovative policies, and encourage the deployment of new technologies. But on their own, they generally have limited impact on the national or regional level emissions." If the focus on efforts by individuals is what "tackling global warming" means, it is wholly inadequate, misrepresents the problem, and is not fair to all the hard work and energy people are putting into working on global warming. It gives the false impression that individuals voluntarily changing light bulbs "addresses global warming." It's too late for that; we've lost the past 20 years, and now a global crash program is the only thing that will work. Individuals, even individual cities, can only address conservation. Gelbspan says, "Climate change will not be solved through energy efficiency. It requires an energy revolution."

Cities alone cannot do it. The mayor of Seattle and others began a campaign when the Kyoto Protocol went into effect on February 16, 2005, and the U.S. did not join. They challenged mayors to commit their cities to the same targets as Kyoto had. As of early 2011, more than 1,000 cities had joined, and this campaign is now the U.S. Conference of Mayors Climate Protection Agreement. Yet cities find reducing their energy use to

7% below the 1990 level to be extremely difficult. Even whole cities have only a limited ability to affect their use of energy, and all of it is generally in conservation.

The issue is scale. Asking people to change light bulbs to compact fluorescent bulbs will never approach the effect of creating a national and global ban on the production of incandescent bulbs. Driving a little less or in a car that gets better gas mileage cannot begin address the problem: we need to mandate a large raise in CAFÉ gas mileage standards, put a tax on carbon, and produce only electric or hybrid cars.

Asking people to turn their thermostats down a few degrees in winter is not the same as mandating that all new construction be 50% more efficient than it is now (as recommended by the American Institute of Architects), eventually making buildings carbon neutral. In addition, to encourage the use of mass transit, we need to make all public transportation free, just as the roads have been "free" (that is, paid for by taxes). However, asking people to take mass transit more often is not the same as stopping all highway construction and all airport construction and using the money for a crash program in mass transit across the nation. It's not the same to encourage people to know whether their energy company uses some renewable energy as it is to move all subsidies now going to oil, coal, and nuclear and give them instead to renewable energy.

The fact that few people realize that we need to reduce CO_2 emissions by 80% by 2020 is no accident. The media does not report the truth. You have to hunt for this information, and most of it will only be on the Internet. Understanding our climate crisis requires that we begin to learn as fast as possible. In addition, as we do, we need to create a political movement that understands the issues and demands adequate change.

Remember that continuing as we are is not an option because of what is physically happening to Earth and the inevitability of nature's response. The U.S. government is paralyzed (Republicans and Democrats). Therefore, change on the necessary scale will never happen unless the people demand it, but people cannot demand something they know little about. However, ideas can spread like wildfire once they get started. I think the best place to create this education is in churches and other religious institutions because global warming is, above all, a question of values and morality.

The climate crisis is a type of problem human beings have never faced before, and it is going to require a type of activism never seen before. People should do what they can in their own lives, just in principle, but when advocating these responses, the real goal should be education about the real scale and urgency of the problem and the creation of a new political movement. Eban Goodstein, in

Florida Panther

Worldwatch magazine, says, "Global warming will not be solved by lifestyle changes or any shift to simple living, desirable as those changes might be. We need to pledge our lives, our creative energies, and our broad ranging collective talents to creating a political movement powerful enough, in a few short years, to launch the new clean energy markets and global forest preservation efforts that will mark the beginning of the end of the fossil fuels era."

Creating a political movement for a new sustainable world is primarily a function of transformative learning, particularly changing our world view. Ecology is not only about *what* to do, it is also about *how* to do it. How can we get enough people, soon enough, to learn enough? And how do we do that democratically? We need to put in place processes that will help people learn and work together. As we work to create a new world, these changes will cascade throughout our social systems. In other words, consciousness is contagious. So is hope.

Learning in Democratic Theory

However, creating sufficient adult learning is difficult in our political system. Democratic theory assumes people are entitled to rights because of their capacity to reason, which is what distinguishes them as adults. The theory assumes adults have already formed their preferences, and only children need to learn; therefore, it ignores how adults form their preferences in the first place, or how they can change them over time. Political discussions assume people already know what they need to know, and so we just take a vote or conduct a poll. Activists generally help people to agree to actions that they want to take, but what if the real issues and choices are outside the realm of what they know?

> The adaptive demands of our societies require leadership that takes responsibility without waiting for revelation or request. One may lead perhaps with no more than a question in hand.
>
> Ronald Heifetz
> *Leadership Without Easy Answers*

Samuel Bowles and Herbert Gintis, in *Democracy and Capitalism*, review the creation of new choices in adults and say that democratic theory "ignores the choices determining how individuals are to develop their preferences, their capacities for social participation and their abilities to make informed decisions. [Democratic theory] claims that the marketplace and the ballot box allow people to get what they want. But [it] is silent on how people might get to be what they want to be, and how they might get to want what they want to want."[1]

Instead of any adult learning or real dialogue, we get fragmentary and misleading information from marketing, advertising, and 30-second sound bites. Because of this blind spot in our political theory, we do not even notice the absence of real collective learning, or serious dialogue about serious problems. Ecology is a complex argument for profound changes in social thought and practice, in economics, values, and morality. Because our world requires this level of radical change in our thinking, we need to develop methods to create rapid social learning.

Facing the Emotional Challenges of Learning for Change

Perhaps human beings are simply incapable, emotionally and politically, of sufficient change. In *Leadership without Easy Answers,* Ronald Heifetz, a psychiatrist who teaches oat the Kennedy School of Government at Harvard, gives three reasons why people fail to adapt, even when faced with a threat to their survival.

First, people can only face and adjust to threats they can see. So far, most people do not know how serious our ecological crises really are. The failure to change can be the result of a lack of information or simply denial. Heifetz says, "A society may operate without increasing levels of stress, quite oblivious to the bankruptcy that lies ahead."[2]

Second, people can also fail to adapt because the nature of the challenge is so huge and so swift that it exceeds the culture's adaptive capability—such as a major plague or war. Heifetz states, "Innumerable human tribes and organizations have disappeared with the onslaught of disease, environmental challenge, invasion, or competition because they could not develop the ability or find the means to adjust appropriately."[3]

> When I care to be powerful – to use my strength in the service of my vision, then it becomes less and less important whether I am afraid.
>
> Audré Lorde

Third, people can also fail to adapt because the necessary changes cause them too much distress. Unlike other animals, we have the technical ability to anticipate a meteor, a coming ice age, or global warming but can lack the psychological capacity to adapt appropriately. Heifetz says, "Living with long periods of disorientation and distress can make people cruel; they block their own empathy, compassion, and flexibility in a desperate desire for order. People need time to develop resilience and learn to tolerate the high pressures that tough issues generate."[4] He says people may react by using coping mechanisms like blaming authority, scapegoating, externalizing the enemy, denying the problem, or finding a distracting issue rather than facing and taking responsibility for a complex challenge.[5] In other words, we often choose denial.

It is still possible that we will fail to adapt sufficiently. However, with adequate education, we can at least avoid a future in which we fail to survive because we never saw the crisis coming. Even so, Heifetz says that leaders must find strategies to counteract a search for scapegoats, enemies, or quick technical fixes. Effective leaders will look for counterproductive patterns of denial and redirect attention to real issues. Open communication is vital because people's denial of the real issues is not conscious or deliberate, just an automatic fear reaction to help them get back to what feels "normal."

Hippopotamus

The focus, Heifetz says, needs to be on an *educative strategy*, although he concedes, "Telling people to face pain can feel like cruelty, particularly when one has no answer to give, only questions, perspectives, and some hard facts."[6] Therefore, a key question, he says, is, "How can one counteract the expected work avoidance and help people learn despite resistance?"[7]

Reframing Issues

All discussions need to begin with a definition of the problem. Any definition of global warming that does not require fundamental whole-systems change is inaccurate. Within our current social systems, adequate solutions are simply not available—and people intuitively know it. So either they try not to think about global warming or they feel overwhelmed and defeated. The changes we need seem impossible because we are using old models of change and are thinking in mechanistic, linear cause-and-effect terms. In that way of thinking, any enormous change must have an enormous cause, such as a revolution. However, revolution—the idea that coercive power is necessary for radical change—is a concept of mechanistic thinking. In systems thinking, large changes can be the result of small changes in the system or its inputs.

> It is the debate itself, and debate alone, that gives rise to the desire for usable information. In the absence of democratic exchange, most people have no incentive to master the knowledge that would make them capable citizens.
>
> Christopher Lasch
> *Revolt of the Elites and the Betrayal of Democracy*

In fact, a new understanding of politics and democracy has been developing in many areas at the same time: ecology, feminism, and peace. Activists in each of these areas have been rethinking these problems and the required changes and have come to very similar conclusions. These three issues share the same worldview—and often the same members. People in the peace movement saw that disarmament is insufficient and came to believe that we can only achieve peace when the entire system changes. Environmentalists found that fixing one environmental crisis after another was inadequate, that ecological health would not be possible until the values and beliefs inherent in the present system were changed. People in the women's movement saw that although women wanted equal opportunities, "equality" within a hierarchical system would only become another trap. Observers and participants have referred to these movements collectively as "the new social movements." Thomas Rochon, in *Mobilizing for Peace,* says the shared ideas and value systems of these new social movements have expanded to encompass a shared vision of solutions. He says these new social movements are:

> important primarily because of their commitment to radically democratic political procedures based on interpersonal solidarities that are fundamentally antithetical to large organizations and to political legitimacy based on appeals to material interests..... Rejecting the present organization of authority and material production, new social movements are said to give primacy to issues of lifestyle and social relations. Harmony and complementarily are emphasized as the proper basis

> for relations between people, and between man and nature. The role of the new social movements is to substitute new habits of thought for the old.[8]

Values researchers Paul Ray and Sherry Anderson, in *Cultural Creatives,* also discuss the similarities in an even broader range of issues:

> Unlike the social movements of 1880 to 1930—the Wobblies, the fascists, the communists, the socialists, and so on—those involved in the post-1960s movements are not trying to take over the government. Nor are they primarily concerned with "more for us" issues, like wages and benefits, for example. Rather these movements are *reframing issues* in a way that changes how people understand the world. . . .
>
> Dr. Martin Luther King, Jr., for example, didn't say, "It's time the Blacks got theirs." He said, "This is about freedom, and justice, and dignity, and the Constitution, and who we are as an American people." Rachel Carson didn't advocate NIMBYism—"keep pollution out of my back yard." She said, "This is about the death of nature." Betty Friedan didn't just say, "It's time that women got through the glass ceiling." She asked, "Who are we as human beings?" The alternative health care movement isn't about getting insurance coverage for chiropractic care. It's about the real meaning of health. . . . What happens when somebody gets involved in a half dozen of these issues and has their world reframed six times? Their entire worldview changes.[9]

Ray and Anderson say there are about 50 million people they call "cultural creatives" in America. This group represents people who have been "involved in or cared intensely about three to six social movements: very strong environmentalism, the condition of the whole planet, civil rights, peace, social justice, new spiritualities, organic food, holistic health and the like."[10] However, they say it takes a decade for people to change the way they live in order to align it with their values and beliefs; therefore, there are many more people still in transition. And the 50 million people who share the same values all feel they are alone because there are no "media mir The media either ignores these opinions (the media assumes they a an aberration) or they ridicule them. Yet Ray and Anderson say rea change is possible and it will emerge from conversations about wh is possible and what kind of world we want. They say, "The hallma of this profound culture shift is going to be reinventing practic every institution of society from the ground up. And that not only possible it is rather likely."[11]

We need to reinvent practically all of society so that we do not continue to destroy our planet. This involves reinventing economics, politics, and even religion. Part III focused on the content of these new creations. This chapter focuses on the method by which we can create th transformation.

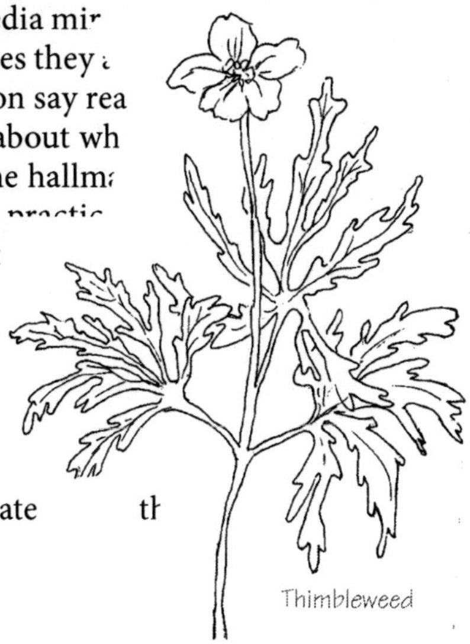

Thimbleweed

Creating Social Transformation

Transforming society begins with transforming ourselves. We can't succeed if we are acting out of fear or with grim determination. We need to be serious, but we can also work with a sense of joy. Fundamentally, the issue is about who we are as human beings; therefore, I see the entire effort as a spiritual journey. A profound shift in planetary awareness is coming, and a spiritual awakening is already beginning to happen.

We need to reconnect to our inner world and to each other. Groups, especially religious groups, can provide a great deal of mutual support. After getting ourselves well grounded, creating change in complex systems requires three steps:

- We need to *use systems thinking* to understand problems and solutions.
- Then, we must *develop the learning,* required for creating a shared vision of the new worldview.
- And last, we need to *increase connections and networks* between people and between groups working on change so that the learning is widely shared until we reach a critical mass and our culture crosses a threshold and jumps to a new way of organizing society. This process of learning in community creates a form of deliberative democracy.

Thinking in Systems Terms

Systems thinking offers a language to restructure how we think; it is about not fragmenting the world, but about seeing it as one interconnected whole. Jay Forrester, a computer engineer at MIT, first applied complex system ideas to social systems. Linear models are extremely unreliable as predictors, so Forrester's group developed models that shifted the emphasis away from prediction. Instead, "the modeler is not seeking to control the complex system by quantifying it and mastering its causality, (s)he wants to increase her 'intuitions' about how the system works so (s)he can interact with it more harmoniously."[12] That is, in systems thinking, we focus on the process—on dynamics of interactions—rather than focus on the outcome or on final goals.

> The quality of our knowledge depends on the quality of our questions. Questions are alive. Questions that matter to us create motion.
>
> Tom Atlee
> Co-Intelligence Institute

Peter Senge, in his book on systems thinking, *The Fifth Discipline: The Art and Practice of the Learning Organization*, says we need to begin by telling the truth about current reality and then explore the alternatives. Senge's idea for creating collective learning is to build *shared visions*, which are basically the answer to the question, What do we want to create?[13]

The gap between the vision we see and the reality that exists can generates a "creative tension." Senge says we can use this tension to generate energy for change.[14] But this work of examining our own thinking and building shared visions needs to be done with others.

Because causation is created by multiple inputs and outputs that go in all directions, total systems change—the tipping point—can be reached without warning. It is called threshold behavior. We have no way of knowing how close we are either to runaway global warming or to massive social transformation. Malcolm Gladwell, in *The Tipping Point,* describes this threshold behavior in societies: "The Tipping Point is that magic moment when an idea, trend, or social behavior crosses a threshold, tips, and spreads like wildfire."

Transformative Learning

The primary barrier to change is the lack of information about alternatives, particularly an alternative worldview. Once we become comfortable with systems thinking, we can focus on the two agents of change: new information and new connections. This learning-based approach means we do not need to try to force change through fear and guilt. This learning is done in community, and change will evolve as people begin to understand the nature of the problem and the range of solutions and then choose which topic they want to work on.

Hierarchy is antithetical to dialogue.

David Bohm

Margaret Wheatly studies how organizations can change through self-organization, and she says that in learning, "The essential lesson is this: When people understand the forces creating the adverse conditions of their life, and how they might change those forces, they become eager and rapid learners. They are capable of learning sophisticated skills that far surpass traditional assumptions about their intellectual capacity. And they learn these skills faster than anyone would have thought possible."[15]

As communities begin to act on the new knowledge, a direct democracy process is created. Thomas Prugh wrote *The Local Politics of Global Sustainability* with Robert Costanza and Herman Daly. In an essay based on this book, he says they found similar lessons to Wheatly:

> First, direct democracy engages people, despite the extra demands it makes on their time and energies, because their decisions count. Second, a robust tradition of civic engagement can be created and maintained for a very long time; people can be educated into citizenship. Third, ordinary people can grasp complex ethical and environmental issues and contribute thoughtfully to social decisions about them. Finally, people with varying interests and viewpoints can come together as political creatures and will a common environmental future. These virtues are exactly what the challenge of sustainability requires of us. Direct democracy would give us both better eyes to see the environmental problems we face and superior political means to address them.[16]

But at the beginning, Prugh says, "We have a bootstrap problem: direct democracy develops engagement, but

Cardinal

people must be engaged to move toward direct democracy."[17]

Transformative learning requires that we understand how much the problems are not just "out there," but how much they result from how we think. We generally believe that the changes we need to make are in our outer world, not in our inner world.

In creating this shared vision of a sustainable world, Senge asks us to consider these questions:

- Why do we confront learning opportunities with fear rather than wonder?
- Why do we derive our self-esteem from knowing as opposed to learning?
- Why do we criticize before we even understand?
- Why do we create controlling bureaucracies when we attempt to form visionary enterprises?
- Why do we persist in fragmentation and piecemeal analysis as the world becomes more and more connected?[18]

Democracy ought to be a method of learning, but we have no structured process to use. Many groups are developing new processes and methods to help people connect with each other to create collective learning and develop collective wisdom. We know it is possible to have conversations that can create new choices, where creativity and wisdom happen among ordinary people who make decisions jointly. People long for these conversations in small groups, in large organizations, and throughout the country.

More than 100 different processes can help create true democracies to address our ecological crises.[19] Two of these methods are the World Café and the Open Source Technology.

The World Café

The first process is the World Café. The World Café is a method for getting people together to create collective wisdom. Consultants Juanita Brown and David Isaacs created this method for dialogue in community out of conversations and experimentation. It applies principles of dynamic networks and living systems to access a source of deeper creativity and shared knowledge.[20] The purpose of the World Café is to connect people who would not otherwise be connected and to dissolve the barriers that separate us. This method enhances collaborative thinking about critical issues by linking small- and large-group conversations.

> In tribal council: "We just keep talking until there's nothing left but the obvious truth."
>
> Oren Lyons, faithkeeper of the Turtle Clan of the Onandaga Iroquois

Through collaborative thinking, the World Café helps people clarify values and increases their adaptive capacity. It creates the "holding environment" where we collectively work through our fear, uncertainty, and disorientation, and it creates trust within a community and increases community resilience so we can address difficult issues together.

Brown describes the World Café: "We have always brought forth the future through webs of personal commitments born through human conversation. By choosing to focus on questions worth asking, we are changing the conversation. As we change the conversation, we are beginning to change our common future. Through mutual listening and a spirit of discovery, a certain type of *magic* appears—the magic of a new collective intelligence arising from the individual minds present in the conversation."[21]

We need to become comfortable with group processes that enable self-organization, support individual and collective meaning making, and help us recognize the increases in intelligence that group conversations generate. The practice of the World Café is based on a set of working assumptions:

- The future is born in webs of human conversation.
- Compelling questions encourage collective learning.
- Networks are the underlying pattern of living systems.
- Human systems—organizations, families, communities—are living systems.
- Intelligence emerges as the system connects to itself in diverse and creative ways.
- *We collectively have all the wisdom and resources we need.*[22]

The basic principles are to clarify the context, create hospitable space, and explore questions that matter; connect diverse perspectives; encourage each person's contribution; listen together for patterns, insights, and deeper questions; and share collective discoveries.

- suspend judgment as best you can
- listen…with respect
- seek to understand rather than persuade
- invite and honor diversity of opinion
- speak what has personal heart and meaning
- go for honesty and depth…without going on and on.[23]

Begin by explaining the café structure. People gather at round, low tables to discuss the questions at hand. Provide markers and paper tablecloths where people can record any ideas, insights, or questions that have emerged. (Tables are coffee table height rather than dining table height. This helps keep the atmosphere casual.) *After the first round* of 30–45 minutes at each table, each table decides who will remain at the table for the whole session, as host. Everyone else randomly moves to new tables. When everyone is in their new places, the home table host shares the key ideas and questions that emerged from that table's earlier discussion. Then, others

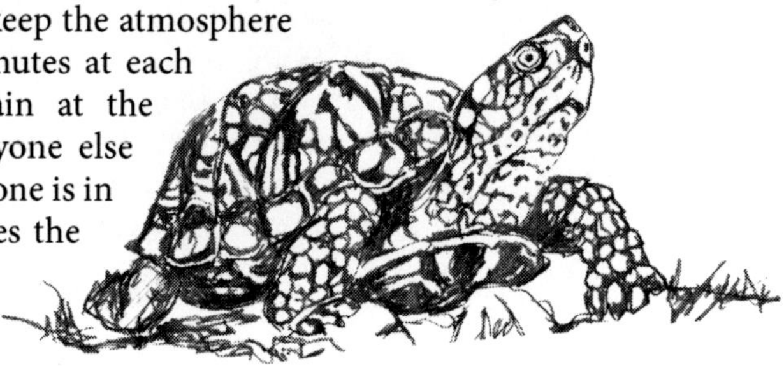

Box Turtle

share what occurred at their original tables and continue the conversation. *After the second round,* everyone returns to their home tables to compare notes with their original companions. After this round, most people in the room will have heard the ideas generated by the others in the Café.

Explore questions that matter. Well-crafted strategic questions define intention, focus energy, and direct attention toward what really counts. Focus on what is possible in the future rather than what is wrong in the present.

Focus on questions rather than answers. Questions such as

- What is our common good?
- What do we really care about?
- What am I here to do?
- What do you do when you don't know what to do?
- What beliefs and values do we hold most dear in this time of uncertainty and change?
- What do we really mean by "security"?
- What common purpose might grow out of the meaning shared by this community?
- How can this common purpose be translated into effective action?
- What is the natural evolution of learning in this community?
- What questions do we have that, if answered, would take us to the next level?
- What would have enough juice to get us committed?
- What are we in service to?
- How do we translate this dialogue into practice?
- What is our emerging identity?[24]

For more on this, see *The World Café: Shaping Our Futures through Conversations That Matter.*

Open Space Technology

Harrison Owen developed a second process of group learning to facilitate learning specific content that he called Open Space Technology. Open Space defines issues and clarifies which issues are just technical and which involve new thinking (for example, new thinking in the new worldview). This process began as a response to the common observation that in traditional conferences coffee breaks are often the best part. Owen believes the phenomenon of self-organization lies at the heart of Open Space and says, "There is no such thing as a non-self-organizing system, at least in the natural world, which would include us."[25]

The process begins by creating a powerful theme statement for an interested and committed group (such as creating a sustainable world that will be an effective response to global warming). Then, in parallel working sessions around a central theme, participants create and manage their own agenda. Owen says that the key ingredients for deep creative learning are real freedom and real responsibility. These meetings do not have keynote speakers, scheduled workshops, or panel discussions. People create their own conference on the spot.

To listen, one has to live with doubt.

Ronald Heifetz
Leadership without Easy Answers

It begins with anyone who wants to create a discussion on a sub-issue going to the microphone, announcing their workshop, picking a time and place, and then posting it on the wall in big letters. Within a short time, everyone has spoken who wants to, and the workshops are posted. In what Owen calls "the village market place," everyone at the meeting then mills around and signs up for the discussions in which they want to participate. The first meetings begin immediately. The conference self-organizes: no one is in control. Four principles and one law govern the entire process. The principles state:

1) *Whoever comes is the right people,* which reminds people in the small groups that getting something done is not a matter of having 100,000 people and the chairman of the board. The fundamental requirement is people who care to do something. And by showing up, that essential care is demonstrated.
2) *Whatever happens is the only thing that could have,* which keeps people focused on the here and now, and eliminates all of the could-have-beens, should-have-beens, or might-have-beens. What is the only thing there is at the moment.
3) *Whenever it starts is the right time,* which alerts people to the fact that inspired performance and genuine creativity rarely, if ever, pay attention to the clock. They happen (or not) when they happen.
4) Lastly, *When it's over it's over.* In a word, don't waste time. Do what you have to do, and when it's done, move on to something more useful.

The *Law of Two Feet* states, "If at any time you find yourself in any situation where you are neither learning nor contributing—use your two feet and move to some place more to your liking. Such a place might be another group, or even outside into the sunshine. No matter what, don't sit there feeling miserable."[26]

Groups of 20 to 1,000 can create this process, and they can create one-day workshops, three-day conferences, or a regular weekly staff meeting. Owen says Open Space works best when the work to be done is complex, the people and ideas involved are diverse, the passion for resolution (and potential for conflict) is high, and the time to get it done was yesterday.

The process has an apparent lack of structure and welcomes surprises, yet the Open Space

Squirrel Monkeys

meetings are actually very structured, and this structure supports, rather than blocks, the best work. Owen says the only way to ruin an Open Space gathering is to attempt to control it. Participants find that control is not only unavailable, but also unnecessary. Emergent order appears in Open Space, while the process creates the conditions for self-organization.

Stop being lazy about thinking.

Maya Angelou

In addition to these two examples, there are hundreds of other deliberative democracy methods.[27] These include Listening Circles, Fishbowl, Consensus Process, Open Dialogue, Technology Panels (in Denmark), and Study Circles. The website www.democracyinnovations.org lists these and other innovative practices and ideas, along with their web addresses. These processes can create new solutions as well as the needed emotional support to address global warming. By creating cooperative communities and using participatory democracy processes and new technologies, the chances of creating transformative change are much greater than people now realize.

Increasing Connections and Building Relationships

In addition to creating new information and learning, we also need to increase connections between people and groups involved in this learning. By increasing the amount and kind of connections, the new information can remain fluid and affect the whole system. Then, the networked communities can self-organize. Each community can become a model for others. Every single effort eventually affects every other one. By becoming aware of each other's efforts, we magnify the power of each person, and we can draw on each other for support, connections, feedback, and new information.

Creating a sustainable world is a very complex problem; in dealing with it, we will need a lot of experimentation. We must be willing to try many more things than can be expected to work. When everything is shared with as many groups as possible, things that haven't worked in one area may work in another, and groups can share the reasons. Many groups may find things that other groups have not yet tried. All this growing knowledge belongs to everyone. In addition, learning happens on many levels—both consciously and unconsciously. The collective unconscious can pull a lot of others along.

These increased connections help in learning and self-organizing; they also help people deal with the emotional challenges this effort creates. But distress is a symptom, not a problem, and we can help each other deal with it. In working together, we can help each other have the emotional capacity to tolerate the uncertainty, frustration, and fear that this level of change will generate. This is a particular problem because so many people have a worldview and morality system that requires maintaining order and control.

In creating these new connections, it is important to realize how different social change can be now that we have the Internet, cell phones, and text messaging. These new technologies make certain kinds of human interactions possible that were never possible before. This results in new social practices, which creates new kinds of relationships, enterprises, and communities, along with new ways to be citizens.

The Internet has been a truly new form of communication in both form and function. Technology is a type of thinking, and we can use it to create the changes we need. The Internet and cell phones with text messaging are transforming activism. Both can flood the social system with new information and new feedback, allowing the system to learn and self-organize in completely new ways. New communications technologies can strengthen neighborhood assemblies, create televised town meetings, and even create a national civic communications cooperative.

All of these are a means with which to develop stronger citizen participation. This allows everyone, potentially, to have equal access to information and promotes the citizen education required to respond to global warming. Cell phones now make wireless Internet connections possible. E-mail on the Internet must wait for the recipient to look at their e-mail. E-mail and text messaging on cell phones goes immediately to people wherever they are and can be sent to many people at the same time. Text messaging may have special significance in developing countries, where reliable, affordable Internet connections are lacking in many regions and phone calls are expensive. Text messaging is an economical and effective way to spread messages quickly and have them reach many members of the population who don't have access to other types of technology.

Howard Rheingold, who studies the social implications of technology, says cell phones have created a phenomenon called "swarming," a social practice that amplifies human cooperation and teaches people that they can coordinate instantly without a leader. In his book, *Smart Mobs: The Next Social Revolution*, Rheingold discusses how new technology allows people to act in concert. In fact, he says, the "practice of sending short instant text messages triggered youth revolutions in Tokyo and Helsinki, and a political revolution in Manila." Rheingold explains:

> On January 20th, 2001, President Joseph Estrada of the Philippines became the first head of state in history to lose power to a smart mob. More than one million Manila residents, mobilized and coordinated by waves of text messages, assembled at the site of the 1986 "People Power" peaceful demonstrations that had toppled the Marcos regime. Tens of thousands of Filipinos converged on Epifanio de los Santas Avenue, known as "Edsa," within an hour of the first text message volleys. Over four days, more than a million citizens showed up.[29]

Websites are already being developed for people to share what they are doing. These sites can be organized geographically, so everyone in a city can know what

Ruby-throated Hummingbird

people are doing by topic. There can also be sites organized by topic from many geographical areas. These sites can report on progress and problems, increasing knowledge and a sense of being in this together. Thousands of other people can know every change, at no additional cost. One problem this creates, of course, is information overload, so inventing new ways to organize sites and groups of sites will be important. One process that can be helpful for sharing information on the web is software for collaboration called Wiki (the name "Wiki" is Hawaiian for "quick"). This software, available for free on the Internet, allows users to create and edit web page content using a web browser. This "open editing" encourages democratic use of the web and promotes content composition by any user.

Social Change in Complex Systems

By creating cooperative communities and using participatory democracy processes with new technologies, the chances of creating transformative change are much greater than people now realize. Remember that complex systems self-organize because of tight interconnectedness and close relationships between all the parts. In any complex system, self-organization emerges spontaneously as the network "learns" how to function in coordinated and seemingly goal-directed ways without a central coordinating authority. There are many ways to get an intuitive sense of how transformative change can happen.

In thinking about social change, remember that

1) Social self-organizing is a collective experience, created by interactions and relationships between the parts.
2) Self-organization in societies is the result of information flooding and cascading through a system. Therefore, changing social systems requires enormous amounts of information and learning. It also requires close interconnections, and the creation of the Internet now makes this possible in ways that were unavailable 20 years ago.
3) The future will be uncertain and unpredictable and will require an ability to tolerate ambiguity and uncertainty. However, this lack of predictability also allows for greater creativity and innovation. Therefore, we need to evaluate proposals for change not on their ability to control and predict the outcome, but on their level of creativity and connectivity.
4) Learning requires living on the "edge of chaos," another term for the critical state, which means that the confusion generated by our efforts is a natural part of the process.

Rules for Reaching Consensus

by Steven Saint and James R. Lawson

Voting is a means by which we choose one alternative from several. Consensus, on the other hand, is a process of synthesizing many diverse elements together.

Voting is a win or lose model, in which people are more often concerned with the numbers it takes to "win" than with the issue itself.

Voting does not take into account individual feelings or needs. In essence, it is a quantitative, rather than qualitative, method of decision-making.

How to Foster Consensus

- Create unity
- Value diversity
- Involve everybody
- Look for the third way
- Promote imagination, creativity, innovation
- Don't be afraid of compromise
- Use internal rewards
- Promote the value of learning
- Encourage self-reliance
- Encourage empathetic listening

Robert's Rules of Order was designed to manage competition.
Rules for Reaching Consensus is designed to manage cooperation.

http://www.hbo3.com/consensus/rules.html

Red Kangaroo

Chapter 16

A Shared Moral Vision Can Transform the World

Human beings are now an endangered species, and the threat is of our own making. But a shared moral vision can transform and heal the world. Only a profound shift in consciousness to a new worldview can reverse global warming. All over the world, consciousness is already changing. The most powerful force for change occurs when an idea that everyone has taken for granted is no longer seen as legitimate. Ideas matter. Ideas can have tremendous force. In addition, consciousness is causative, and new consciousness is contagious. But can human beings accept the reality and urgency of global warming when they are given every chance to deny it, and are human beings, psychologically and politically, capable of creating transformative social change?

> Not to dream boldly may turn out to be simply irresponsible.
>
> George Leonard

To repeat: the science is definitive. Fully 98% of scientists believe that global warming is unequivocal, catastrophic, and extremely urgent. Because of two decades of delays, all the slack in the system is now gone, and we have only a few years to change course and stop global warming. We are close to creating runaway warming or sufficient warming to eventually melt most of the ice on Earth. The last time that happened, sea levels were 83 feet higher than they are today. Our survival now depends on our ability to awaken from this trance of denial and accept the reality of global warming.

The first crucial step is for us to become aware of our own thinking and thereby make our unconscious assumptions about people and the world conscious. We need to understand the global consequences that our automatic ways of thinking and our limited cultural worldview create.

Definition of the Problem

Initially, we need to ignore the question of how to make these changes and focus on clarifying the definition of the problem. I have said that political arguments are rarely over solutions—they are over the definition of the problem. In the case of global warming, there are two parts to an adequate definition: 1) addressing global warming requires reinventing society, especially economics; and 2) knowing that the consequences of inaction are catastrophic and time is short. An adequate definition is needed to break through denial. If politics is the art of the possible, then by definition, politics is incapable of radical change, for power always resides with the status quo.

Systems Thinking

By using the human capacity for conscious change and by creating the capacity for rapid social learning, we can create the systemic solutions our interconnected world requires. Any chance we have of creating transformative change must come from people in conversation with each other, in communities, organizations, and over the Internet. That is, first, we must build relationships. Using systems thinking, we need to emphasize interdependence and facilitate self-organization so that new order can "emerge."

> There is nothing more powerful than an idea whose time has come.
>
> Victor Hugo

Our current woldview is resulting in the destruction of Earth. The changes we need are available in an alternative to our current worldview and its social systems. This woldview is based on complex systems science, and therefore, it sees reality whole and all our social systems as interconnected. This new worldview needs an encompassing term, the term I use is "ecological social theory."

Ecological Political Theory: Restore Community

In the *current worldview,* (primarily conservaive) individuals are thought to be selfish, ruthless, competitive, and materialistic. People are assumed to relate in a hierarchical, authoritarian, coercive manner, which is regarded as the only "natural" or "realistic" way to see the world. Judgment and punishment are emphasized. There is a belief in the special righteousness of those in power, and care is taken to protect current patterns of privilege. To be rational is to be cynical—not to be cynical is naïve. We focus on rights, with no obligations or accountability to the whole. Freedom is seen as the right to be free of obligations to others. The emphasis is on the individual or individual families. In particular, an individual's situation is always due only to his own hard work and self-discipline. Therefore, the wealthy are wealthy because they deserve to be, not because the system maximizes additional wealth for the wealthy. The poor or unemployed must be unworthy, not self-disciplined, or even lazy. This makes it easy to ignore the poor.

The *new worldview* emphasizes the *common* good, not special interests. Instead of patriarchal "family values," the ecological view emphasizes cooperative community values. The new worldview assumes we are meant to take care of each other ("love your neighbor as yourself") and recognizes that because of our interconnectedness, the self is created in community. This worldview emphasizes that new patterns of power could be collegial and democratic and can focus on partnership rather than dominance and coercion. It includes an understanding of human nature as kind, generous, and cooperative—if given the chance. It celebrates "the feminine" in the sense of compassion, nurturing, mutuality, and a cooperative understanding of power.

This alternative goes beyond seeing "the environment" as only one

Rufous Hummingbird

progressive issue among many. Instead, it places progressivism under the larger umbrella of ecology, so that all political and economic issues flow from there. The social message is progressive, but the economic message goes far beyond what progressivism is today. Progressives do a poor job of defining any progressive economic vision.

Ecological Economic Theory: Putting Earth and People First

In our current worldview, all economic choices are made to maximize profit for the wealthy while people and the environment are ignored and harmed. By using "cost-benefit" analysis and labeling the economy "value-free," we allow all moral considerations to be seen as irrelevant and unnecessary in making economic choices. Wall Street and corporations are in control, and anyone who dares to question their natural right to rule is ridiculed as "socialist." This worldview assumes that, if left free of any restraints, the Market will always make the best choices, but we don't aks best for whom or over what time frame. Because there are no economic indicators that measure or acknowledge the destruction this worldview causes, we ignore the fact that we are destroying our planet. Adequately addressing global warming cannot happen within this current worldview. For effective change, we need communities of strong individuals who, for the common good, agree to accept Earth-imposed limits.

Ecological economics takes economics out of the realm of opinion and grounds it in the concrete requirements of preserving Earth and life. Ecological economics emphasizes the whole and includes the cost of all previously unacknowledged consequences of our behaviors. It assumes that economies must function within nature-imposed limits and that perpetual economic growth cannot continue on a finite planet. This alternative economic worldview puts people before profit and Main Street before Wall Street. This in turn, creates an emphasis on eradicating poverty, to put people first, and also as a requirement for reducing population growth. It emphasizes local and regional economies and the free sharing of information.

> A problem solver tries to make something go away. A creator tries to bring something new into being.
>
> Peter Senge

In this new worldview, we could create a new money system in which new money was created by the government rather than private banks. This would create a sense of abundance. As British commentator Ron Morrison said, "Modern monetary reform is about displacing the current paradigm of 'what can be afforded' with 'what we have the capacity to undertake.'"[1] Once the fear of scarcity is over, we can experience people as essentially good and compassionate. We can offer an alternative to greed.

Create an Economy that Can Reduce Emissions by 80% by 2020

We must reduce greenhouse gas emissions by 80% by 2020, and doing so will require a new economy. We can start by ignoring the economy that is there now and build a new economy next to it. The jobs that have been lost are never coming back; to create new jobs, we need new industries. Sustainable industries could revitalize the economy. These efforts can become mutually reinforcing and would include profound changes.

The new economy will:

- Create local or regional economies to reduce transportation needs, increase stability, and keep jobs at home.,
- Emphasize small and medium-sized businesses, along with regional duplication. Business schools will teach a new kind of consulting in which consultants help hundreds of small local businesses start the same type of company for their own local markets, allowing for duplication and sharing of information. Companies can create business plans together for human-scale, decentralized businesses. On the Internet, they can be connected either by topic or geographically.

> The choice will be made by our generation, but it will affect life on earth for all generations to come.
>
> Lester Brown

- Emphasize efficiency, conservation, and renewable energy, along with creating a national unified smart grid. Create decentralized micropower.
- Restructure taxes to create honest markets. Move from taxing work to taxing environmentally destructive activities. This will create true prices for the cost of goods. We also need to shift subsidies—stop subsidizing coal and oil, and start subsidizing renewable energy; stop subsidizing road construction, and start subsidizing rail and mass transit systems.
- Build mass transit and high-speed rail. When we begin to close coal-fired power plants, coal miners will build subway systems. Create plug-in hybrid cars that recharge with electricity created from renewable sources.
- Facilitate open information and sharing. This includes the free transfer of technology to developing countries. We need to forgive the debts of developing countries so that they, too, can shift to local economies.
- Increase carbon sinks: shift to organic gardening so that soil can recover and absorb massive amounts of carbon, ban deforestation worldwide, plant billions of trees worldwide, and use tree-free paper.
- Design manufacturing systems for zero waste; lease and maintain products instead of selling them.
- Retrofit buildings. The American Institute of Architects says that even now, we can build new buildings and renovate buildings so that they use 50% less energy; by 2030, all buildings designed can be carbon neutral.
- Revitalize rural America and small towns. Farmers can rent their land to wind energy companies, create local refineries for safe chemicals made from plants rather than petroleum, and build small-scale paper mills for crops of hemp and kenaf for tree-free paper. To free up land for these new crops, we need to stop feeding corn to livestock and either become vegetarians or eat meat from grass-fed cattle.

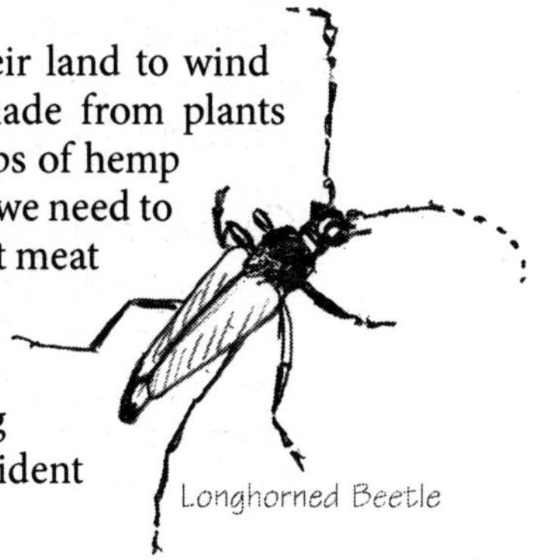

Longhorned Beetle

As I finish this book in August 2011, Congress is choosing an austerity program in the midst of a major depression—limiting demand for goods even further. Our political system, including President

Obama, seems locked into an economic view that has been proven false. Our economy may be headed for chaos. As dangerous as that will be, it is probably the only way to create profound change. Profound structural change is never gradual or incremental. The current structure becomes more and more brittle until it breaks. This chaos allows for rapid restructuring, and can jump a system outside the original container until suddenly "the emperor has no clothes" and the old system is over. Then we suddenly see a new reality and the new structure can solidify and become permanent.

Where there is no vision, the people perish.

Proverbs 29:18

Keep in mind that although no one can control the final outcome of this change, it is in the nature of chaos that small actions can have a great effect on the outcome: the butterfly effect. Whatever the process and outcome, respond by creating local economies, local renewable energy, and local organic food, along with a strong sense of community and compassion, for ourselves and others. I hope people can use this book as a resource and for support.

Restore the Natural World

The oil spill in the Gulf made our impact on the oceans, wetlands, and many species very visible. However, we are also destroying many other ecosystems: the world's forests, rivers, land, and of course, the atmosphere. Therefore, we need to undo the damage already done to Earth. This would require that we

- Plant trees to reduce flooding and conserve soil
- Plant trees to sequester carbon
- Protect topsoil on cropland
- Restore rangelands
- Restore fisheries
- Protect biological diversity
- Stabilize water tables

The U.S. Moral Leadership and Ending War

The U.S. can no longer lead by dominating the world economically and militarily, but we can regain our role of moral leadership, even after a decade of disastrous wars and advocating torture. For 50 years, the U.S. led the world in the fight for human rights, and overnight, we became the worst offender—all because we got scared. We should praise the courage of our soldiers, but I think people greatly underestimate how much courage everyday life requires, particularly the courage to choose wisely even in the midst of profound uncertainty. That is, not allowing fear to turn us to war.

When we are afraid we demand an artificial certainty so that we can feel that we are in control—even if control is impossible. We back politicians who promise control—a promise that necessarily requires violence. Even if war always causes more problems than it solves, it makes us feel powerful and decisive, but assuming war solves anything is naïve. The money we spend to make us feel powerful should be spent on ending poverty, allowing us to change the world dynamics that lead people to terrorism. We can't completely stop terrorism any more than we will completely stop other types of crime, but we can mitigate the conditions that create it.

World War III is coming, but the enemy will be global warming. It is said that we are always fighting the last war. Our military is still emphasizing weapon systems to fight the old Soviet Union. And if China ever decides to control the U.S., they could do it with their economy (as we have done) and would not need bombs. World War III will be global and our very survival is at stake. But bombs will be useless. As Walt Kelly and Pogo said, "We have meet the enemy and he is us." This war will be won only if we see that everyone is up against a common enemy: concepts of dominance, ruthless competition, "natural" hierarchies, and winner-take-all.

And we might not pull it off. There's no guarantee that this tremendous shift will kick in before our life support systems unravel irretrievably. Actually, that very fact that there's no guarantee of success is what will draw forth our greatest courage and creativity.

Joanna Macy

Social Change Issues Spiritually Defined

The social and economic transformation required for survival is entirely possible, if we act as a global community. But action means we each need to be centered and courageous. Courage and confidence will be difficult because we live in a world in which feedback from the material world is amplified and feedback from the soul is dampened.

To deal with uncertainty and chaos we must be grounded in a spiritual practice to keep us centered, whether this practice is prayer, mysticism, yoga or painting. Logic may tell us to give up, but logic alone is brittle and limited. When grounded in the Spirit, all things are possible.

But the transition will be chaotic because that is how systems shift to new levels—they transition through chaos. Keeping grounded amid chaos is what faith is for: faith gives us resilience. Spirituality is not the answer; it is the *process* that allows us to create the answer and make the world new in spite of chaos and uncertainty. You need not believe in God, because being grounded in the Spirit means that you are centered in process and able to let go of the outcome. Thought creates reality. So naming your intention is important. Prayer is one way to name your intention.

The threats before us create an unprecedented opportunity to create the world anew. The effort will unleash our innate human capacities for cooperation and creativity.

Focus on the Process, Not the Outcome

The task before us is overwhelming, but to address fear and doubt—trust the process and ignore the end goal. The first step is to create a consensus on the reality of global warming. Then choose your own vision of possibility and ask: What does the world soul communicate to my soul? How can I be a citizen in today's world? Identify your path in this process of transformation. Organizations can help people identify their path

Lily

in this process of transformation. It doesn't matter where you start, as long as you find your path exciting.

I realize that people are overwhelmed in their own lives because our economy requires so much intense effort. A local economy would leave time for leisure and community involvement. But until then, facing global change is particularly suited to seniors, and senior population is growing. Addressing global warming can bring focus and meaning to people who have great wisdom, experience, balanced judgment, and who inherently understand the long term. We need wise elders who can put energy and commitment into creating a positive future.

Jeremy Rifkin asks whether there may be an underlying meaning to the human saga. That is, does human history have a purpose? Perhaps God is cautioning us—because of our past behavior, we are at the end of the line: we have only one more chance to learn to turn away from war, cutthroat competition, radical individualism, and materialism. We either put the planet and each other first, or the consequences of our current lifestyles will be the extinction of millions of species, and one of them will be humans. We will have also destroyed God's Creation; there can't possibly be a greater sin.

To change course, we need to believe in people and ourselves, to believe in possibility and potential. We can do this. To stop the threats to our world, we need to find within ourselves a deep sense of purpose; humans must decide what they are and what they will become.

But we are running out of time.

Blue-footed Boobies Mating Dance

Acknowledgments

Most of all I want to thank my husband, Jim. After our youngest began kindergarten, I began to look for a job as a librarian. Even though we really needed a second income, he told me I needed to go back to school to study ecology. I would never have done that on my own. His support made all the difference. He also frequently read the manuscript and helped with the organization of the book.

I am grateful for the support of my whole family. My sister, Morgan Brooke, helped edit many versions and designed the book, cover art, and illustrations. My stepdaughter Heidi Orcutt-Gachiri copyedited the book. My son, Jeff, helped to edit various versions of the book. Thanks also to my daughter, Arianne, for her steadfast support and love; to Heidi and Joe for my new grandson, James (who turned one in April 2011), and to stepdaughter Holly and Dan for my magical twin granddaughters, who turned four in April.

I want to thank Dr. David Rhoads, New Testament Professor at Lutheran School of Theology at Chicago. His belief in this book and his support and encouragement were invaluable. His commitment to the environment has made LSTC a center for education on ecology and religion. I co-taught two courses on ecology and environment and want to thank the students for their feedback on reading my book. I also want to thank Steve Perkins, Senior Vice President, Center for Neighborhood Technology for years of support and encouragement.

Bibliography

Abram, David. *The Spell of the Sensuous: Perception and Language in a More-Than-Human World.* New York City, NY: Random House, Vintage Books, 1997.

Barlow, Maude and Tony Clarke. *Blue Gold: The Fight to Stop the Corporate Theft of the World's Water.* Darby, PA: Diane Pub Co., 2002.

Berry, Thomas. *The Great Work: Our Way into the Future.* New York City, NY: Harmony/Bell Tower, 2000.

Borg, Marcus. *The Emerging Christian Way,* Chapter 1 in Michael Schwartzentruber, ed., *The Emerging Christian Way: Thoughts, Stories, and Wisdom for a Faith of Transformation.* Incline Village, NV: CopperHouse. 2006.

Borg, Marcus J. *Jesus A New Vision: Spirit, Culture, and the Life of Discipleship*. San Francisco, CA:HarperSanFrancisco, 1987.

----------*Meeting Jesus Again for the First Time: The Historical Jesus and the Heart of Contemporary Faith.* San Francisco, CA: HarperSanFrancisco, 1994.

----------*The God We never Knew: Beyond Dogmatic Religion To A More Authentic Contemporary Faith.* New York City, NY: HarperOne, 1998.

Bowles, Samuel and Herbert Gintis. *Democracy and Capitalism: Property, Community and the Contradictions of Modern Social Thought*. New York City, NY: Basic Books, Inc., 1986.

Bowen, Mark. *Censoring Science: Inside the Political Attack on Dr. James Hansen and the Truth of Global Warming.* New York City, NY: Dutton Adult, 2007.

Briggs, John and F. David Peat. *Turbulent Mirror: An Illustrated Guide to Chaos Theory and the Science of Wholeness.* New York City, NY: Harper & Row Publishers, 1989.

Brown, Juanita, and David Isaacs. *The World Café: Shaping Our Futures through Conversations That Matter.* San Francisco, CA: Berrett-Koehler Publishers, 2005.

Brown, Lester R. *Plan B:4.0 Mobilizing to Save Civilization*. New York City, NY: W. W. Norton & Company, 2009. There is a slide show for *Plan B 3.0* http://www.earth-policy.org/index.php?/books/pb3/presentation, And for *Plan B 4.0* at http://www.earth-policy.org/books/pb4/pb4_presentation

Brown, Ellen. *Web of Debt*. 3rd ed. Baton Rouge, LA: Third Millennium Press, 2008.

Buchanan, Mark. *Ubiquity: Why Catastrophes Happen*. New York City, NY: Broadway, 2002.

Capra, Fritjof. *The Turning Point: Science, Society, and the Rising Culture.* New York City, NY: Simon and Schuster, 1982.

Cavanagh, John and Jerry Mander. *Alternatives to Economic Globalization: Another World Is Possible.* San Francisco, CA: Berrett-Koehler Publishers, 2002. A 2nd ed. is available.

Condren, Mary. *The Serpent and the Goddess.* San Francisco, CA: Harper SanFrancisco, 1989.

Daly, Herman E. and John B. Cobb, Jr. *For the Common Good: Redirecting the Economy toward Community, the Environment, and a Sustainable Future*. Boston, MA: Beacon Press. 1989.

Daly, Herman. *Beyond Growth: The Economics of Sustainable Development*. Boston, MA: Beacon Press, 1996.

----------. "The Steady-State Economy: Postmodern Alternatives to Growthmania" in *Spirituality and Society: Postmodern Visions,* David Ray Griffin, ed., Albany, NY: State University of New York Press, 1988.

----------. *Steady-State Economics,* First published in 1977, There is also *Steady-State Economics* with New Essays. Washington D.C.: Island Press, 1991. http://www.worldpolicy.org/globalrights/environment/daly.html

Douthwaite, Richard. *Short Circuit: Strengthening Local Economies for Security in an Unstable World.* Dublin, Ireland: Lilliput Press, 1996. Also at http://www.feasta.org/documents/shortcircuit/index.html?sc4/share.html

Drucker, Peter. *Post-Capitalist Society*. New York City, NY: HarperBusiness, 1993.

Durning, Alan. *How Much is Enough?* New York City, NY: W.W. Norton Company, 1992.

Fox, Matthew. *The Coming of the Cosmic Christ.* New York City, NY: HarperOne, 1988.

Fox-Genovese, Elizabeth. *Feminism without Illusions: A Critique of Individualism.* Chapel Hill, NC: The University of North Carolina Press, 1991.

Gablik, Suzi. *The Reenchantment of Art*. New York City, NY: Thames and Hudson, 1991.

Gelbspan, Ross. *Boiling Point: How Politicians, Big Oil and Coal, Journalists, and Activists Have Fueled a Climate Crisis--And What We Can Do to Avert Disaster.* New York City, NY: Basic Books, 2005.

Gladwell, Malcolm. *The Tipping Point: How Little Things Can Make a Big Difference.* Boston, MA: Little, Brown, 2000.

Gleick, James. *Chaos: Making a New Science.* New York City, NY: Viking, 1987.

Griffin, David Ray, ed. *Spirituality and Society: Postmodern Visions.* Albany, NY: State University of New York Press, 1988.

Harmon, Willis and John Hormann. *Creative Work: The Constructive Role of Business in a Transforming Society.* Indianapolis, IN: Knowledge Systems, Inc. 1990.

Hawken, Paul, Amory Lovins and L. Hunter Lovins. *Natural Capitalism: Creating the Next Industrial Revolution.* Boston, MA: Little Brown and Co., 1999.

Hartmann, Thom. *Unequal Protection: the Rise of Corporate Dominance and the Theft of Human Rights.* Emmaus, PA: Rodale Books, 2002.

Heifetz, Ronald. *Leadership without Easy Answers.* Boston, MA: Harvard University Press, 1998.

Hertz, Noreen. *The Silent Takeover: Global Capitalism and the Death of Democracy.* New York City, NY: Simon & Schuster, 2002.

Kauffman, Stuart. *At Home in the Universe: The Search for Laws of Self-Organization and Complexity.* New York City, NY: Oxford University Press, 1995.

Kelly, Margorie. *The Divine Right of Capital: Dethroning the Corporate Aristocracy.* San Francisco, CA: Berrett-Koehler Publishers, 2003.

Kennedy, Jr. Robert F. *Crimes against Nature: How George W. Bush and his Corporate Pals Are Plundering the Country and Hijacking Our Democracy.* New York City, NY: Harper Perennial, 2005.

Kohn, Alfie, *No Contest: the Case against Competition.* 2nd ed. New York City, NY: Houghton Mifflin, 1992.

Korten, David C. *Agenda for a New Economy: From Phantom Wealth to Real Wealth.* 2nd ed. San Francisco, CA: Berrett-Koehler Publishers, 2009. A second edition is available.

---------- *The Great Turning: from Empire to Earth Community.* San Francisco, CA: Berrett-Koehler Publishers, 2006.

---------- *The Post-Corporate World.* San Francisco, CA: Berrett-Koehler Publishers, 1999.

Lakoff, George. *Don't think of an Elephant: Know Your Values and Frame the Debate.* White River Junction, VT: Chelsea Green Publishing, 2006.

----------*Moral Politics: What Conservatives Know that Liberals Don't.* Chicago, IL: University of Chicago Press, 1996.

----------*The Political Mind: Why You Can't Understand 21st Century Politics with an 18th Century Brain.* New York City, NY: Viking, 2008.

--------- *Thinking Points: Communicating Our American Values and Vision.* New York City, NY: Farrar, Straus and Giroux, 2006.

--------*Whose Freedom.* New York City, NY: Farrar, Straus and Giroux, 2006.

Lietaer, Bernard. *The Future of Money: Creating New Wealth, Work and a Wiser World.* London, England: Century, 2001.

Mead, Loren B. *The Once and Future Church: Reinventing the Congregation for a New Mission Frontier.* Washington, D.C.: The Alban Institute, 1991.

Merchant, Carol. *The Death of Nature: Women, Ecology and the Scientific Revolution.* New York City, NY: Harper & Row, 1980.

Mooney, Chris. *Storm World, Hurricanes, Politics and the Battle over Global Warming.* New York City, NY: Mariner Books, 2008.

Moore, Thomas. *Care of the Soul: A Guide for Cultivating Depth and Sacredness in Everyday Life.* New York City, NY: HarperPerennial, 1992.

National Academy of Science. *Abrupt Climate Change: Inevitable Surprises.* Washington D.C.: National Academies Press, 2002. See http://www.nap.edu/books/0309074347/html/.

Nelson, Robert H. *Economics as Religion.* University Park, PA: The Pennsylvania State University Press, 2001.

Owen, Harrison. *Open Space Technology: A User's Guide.* 3rd ed. San Francisco, CA: Berrett-Koehler Publishers, 2008.

Page, Cristina, *How the Pro-choice Movement Saved America: Freedom, Politics and the War on Sex.* New York City, NY: Basic Books, 2006.

Pagels, Elaine. *The Origin of Satan.* New York City, NY: Random House, 1995.

Pieris, Aloysius. *An Asian Theology of Liberation.* Maryknoll, NY: Orbis Books, 1988.

Rochon, Thomas. *Mobilizing for Peace: The Antinuclear Movements in Western Europe*. Princeton N.J.: Princeton University Press, 1988.

Ray, Paul, and Sherry Anderson. *The Cultural Creatives: How 50 Million People Are Changing the World*. New York City, NY: Three Rivers Press 2001.

Rheingold, Howard. *Smart Mobs: the Next Social Revolution*. New York City, NY: Basic Books, 2003.

Ruether, Rosemary Radford. *Sexism and God-Talk: Toward a Feminist Theology*. New York City, NY: Beacon Press 1993.

Rubin, Jeff. *Why Your World is about to Get a Whole Lot Smaller: Oil and the End of Globalization*. New York City, NY: Random House, 2009.

Rifkin, Jeremy. *The Empathic Civilization: the Race to Global Consciousness in a World in Crisis*. New York City, NY: Jeremy P. Tarcher/Penguin, 2009.

Rokeach, Milton. *The Open and Closed Mind: Investigations into the Nature of Belief Systems and Personality Systems*. New York City, NY: Basic Books, 1960.

Schmookler, Andrew Bard. *The Illusion of Choice: How the Market Economy Shapes Our Destiny*. Albany, NY: State University of New York Press, 1993.

Senge, Peter M. *The Fifth Discipline: The Art and Practice of the Learning Organization*. New York City, NY: Doubleday/Currency, 1990.

--------- *Presence: An Exploration of Profound Change in People, Organizations, and Society*. New York City, NY: Crown Books 2005.

Shlain, Leonard. *The Alphabet versus the Goddess: the Conflict between Word and Image*. New York City, NY: Penguin, 1999.

Shuman, Michael. *Going Local: Creating Self-Reliant Communities in a Global Age*. New York City, NY: Routledge, 2000.

Tarnas, Richard. *The Passion of the Western Mind*. New York City, NY: Balantine Books, 1991.

Union of Concerned Scientists (UCS) *Smoke, Mirrors & Hot Air: How Exxon Mobile Uses Big Tobacco's Tactics to Manufacture Uncertainty on Climate Science*. 2007. http://www.ucsusa.org/assets/documents/global_warming/exxon_report.pdf.

Union of Concerned Scientists, *Atmosphere of Pressure: Political Interference in Federal Climate Science*. See http://www.ucsusa.org/scientific_integrity/interference/atmosphere-of-pressure.html.

Wheatley, Margaret J. *Leadership and the New Science : Discovering Order in a Chaotic World*. 2nd edition. San Francisco, CA: Berrett-Koehler Pub. 1999.

----------*Turning To One Another: Simple Conversations to Restore Hope to the Future*. San Fransisco, CA: Berrett-Koehler Publishers, 2002.

White, Lynn. "The Historical Roots of Our Ecological Crisis," pp. 8-30; and "Continuing the Conversation" pp. 55-64, in *Western Man & Environmental Ethics*, Ian Barbour, ed., Reading, MA: Addison-Westly.1973.

Endnotes

Introduction

1 Dr. Nafeez Mosaddeq Ahmed, "The Real ClimateGate: Getting Over the Nonexistent 'Email Scandal,'" (Part 1) November 13, 2010, http://www.truth-out.org/the-real-climategate-getting-over-nonexistent-email-scandal-part-165012.

2 Ibid. The recent poll is at http://www.rasmussenreports.com/public_content/politics/current_events/environment_energy/energy_update.

3 Discussed further at the end of Chapter 1.

4 Geoffrey Lean, "Global Warming Approaching Point of No Return, Warns Leading Climate Expert" Independent/UK, January 23, 2005, http://news.independent.co.uk/world/environment/story.jsp?story=603752.

5 Elisabeth Rosenthal, "U.N. Report Describes Risks of Inaction on Climate Change," New York Times, November 17, 2007.

6 James Hansen, "G-8 Failure Reflects U.S. Failure on Climate Change," July 9, 2009, http://www.huffingtonpost.com/dr-james-hansen/g-8-failure-reflects-us-f_b_228597.html.

7 Quoted in Ross Gelbspan, "Straddling Solutions and Survivalism, a Talk at Mass. Climate Action Network Conference, November 19, 2006, http://www.heatisonline.org.

8 Ross Gelbspan, Boiling Point, p. 144. Italics added.

9 Joe Romm, "Science: CO2 levels haven't been this high for 15 million years, when it was 5° to 10°F warmer and seas were 75 to 120 feet higher," http://climateprogress.org/2009/10/18/science-co2-levels-havent-been-this-high-for-15-million-years-when-it-was-5%c2%b0-to-10%c2%b0f-warmer-and-seas-were-75-to-120-feet-higher-we-have-shown-that-this-dramatic-rise-in-sea-level-i.

10 Climate Change 2007: Synthesis Report, Summary for Policymakers http://www.ipcc.ch/pdf/assessment-report/ar4/syr/ar4_syr_spm.pdf. . Quoted in Joe Romm, Hadley Center: "Catastrophic" 5-7°C warming by 2100 on current emissions path," December 21, 2008, http://climateprogress.org/2008/12/21/hadley-study-warns-of-catastrophic-5%c2%b0c-warming-by-2100-on-current-emissions-path.

11 Joe Romm, Hadley Center: "Catastrophic" 5-7°C warming by 2100 on current emissions path," December 21, 2008, http://climateprogress.org/2008/12/21/hadley-study-warns-of-catastrophic-5%c2%b0c-warming-by-2100-on-current-emissions-path.

12 Bill McKibben, "Copenhagen: Too Hot to Handle," Mother Jones, October 23, 2009, http://www.motherjones.com/environment/2009/11/copenhagen-too-hot-handle.

13 For both, see http://www.3nov.com.

14 Gelbspan, Boiling Point, p. 138.

15 Ibid.

16 Ibid. p. 146. Italics added.

17 Lester Brown, Plan B 3.0, p. 280.

18 Ibid, p. 280. Italics added.

19 Lester Brown, Janet Larsen, Jonathan G. Dorn, and Frances Moore, Time for Plan B: Cutting Carbon Emissions 80 Percent by 2020. The full report is available in pdf without endnotes http://www.earthpolicy.org/Books/PB3/80by2020.pdf, or with endnotes http://www.earthpolicy.org/Books/PB3/80by2020notes.pdf.

20 Michael Pollan, "Farmer in Chief," New York Times Magazine, October 12, 2008, http://www.nytimes.com/2008/10/12/magazine/12policy-t.html.

Chapter 1

1. Woods Hole Oceanographic Institute, http://www.whoi.edu/institutes/occi/currenttopics/abruptclimate_15misconceptions.html.

2. "Consensus Emerges Earth is Warming—Now What?" *Washington Post*, November 12, 1997.

3. See http://earth-policy.org/Indicators/indicator8.htm.

4. To get an estimate, Fahrenheit is roughly ½ of centigrade. Specifically, one degree Centigrade equals 1.8 degrees Fahrenheit. In degrees Fahrenheit, the freezing point of water is 32, in degrees Celsius it is 0. In degrees Fahrenheit, the boiling point of water is 212; in degrees Celsius, it is 100. To convert Fahrenheit to Celsius (Centigrade), subtract 32 and divide by 1.8. To convert Celsius (Centigrade) to Fahrenheit, multiply by 1.8 and add 32.

5. NASA, "2009: Second Warmest Year on Record; End of Warmest Decade" January 21, 2010 http://www.nasa.gov/topics/earth/features/temp-analysis-2009.html.

6. "NASA: First half of 2010 breaks the thermometer — despite 'recent minimum of solar irradiance,'" http://climateprogress.org/2010/07/10/nasa-hottest-year-solar-minimum/.

7. Chart is at http://climateprogress.org/wp-content/uploads/2010/10/TSI.gif. Article is at http://www.columbia.edu/~jeh1/mailings/2010/20101001_SummerTemperatures.pdf.

8. GISS Surface Temperature Analysis, http://data.giss.nasa.gov/gistemp/graphs/.

9. IPCC, Frequently Asked Questions 9.2. Can the Warming of the 20th Century be Explained by Natural Variability? http://www.ipcc.ch/publications_and_data/ar4/wg1/en/faq-9-2.html.

10. National Oceanic and Atmospheric Administration, "State of the Climate Global Analysis Annual 2010" http://www.ncdc.noaa.gov/sotc/global/2010/13.

11. IPCC. See http://www.ipcc.ch. IPCC Fourth Assessment Report (AR4) is at http://www.ipcc.ch/index.htm. There is a full report, a Summary for Policymakers, and a Technical

Summary for all three areas: Working Group I Report, "The Physical Science Basis," http://www.ipcc.ch/ipccreports/ar4-wg1.htm. Working Group II Report, "Impacts, Adaptation and Vulnerability," http://www.ipcc.ch/ipccreports/ar4-wg2.htm. Working Group III Report, "Mitigation of Climate Change," http://www.ipcc.ch/ipccreports/ar4-wg3.htm. Climate Change 2007: Synthesis Report, Summary for Policymakers, http://www.ipcc.ch/pdf/assessment-report/ar4/syr/ar4_syr_spm.pdf.

12. IPCC. See http://www.ipcc.ch/.

13. IPCC, *Working Group I Summary for Policy Makers*, p. 5. See http://www.ipcc.ch/SPM2feb07.pdf.

14. Ibid, p. 10. See http://www.ipcc.ch/SPM2feb07.pdf.

15. Ibid, p. 2. See http://www.ipcc.ch/SPM2feb07.pdf.

16. Ibid, p. 2. See http://www.ipcc.ch/SPM2feb07.pdf.

17. IPCC, *Working Group II Summary for Policy Makers*, pp. 8–9. See http://www.ipcc.ch/SPM13apr07.pdf.

18. George Monbiot, "Giving up on two degrees," *The Guardian*, May 1, 2007, http://www.monbiot.com/archives/2007/05/01/1058/

19. IPCC. *Working Group III Summary for Policy Makers*, p. 23. See http://www.ipcc.ch/SPM040507.pdf.

20. Climate Change 2007: Synthesis Report, Summary for Policymakers, http://www.ipcc.ch/pdf/assessment-report/ar4/syr/ar4_syr_spm.pdf.

21. James Kanter and Andrew C. Revkin, "Last-minute wrangling on global warming report," *International Herald Tribune*, February 1, 2007, http://www.iht.com/articles/2007/02/01/news/warm.php?page=2.

22. Joe Romm, "Has the IPCC rendered itself irrelevant?" Climate Progress, http://climateprogress.org/2009/04/15/ipcc-2014-fifth-assessment-irrelevant/.

23. Robert Collier, "China About to Pass US as World's Top Generator of Greenhouse Gases," *San Francisco Chronicle*, March 5, 2007, http://www.commondreams.org/headlines07/0305-02.htm.

24. Naomi Klein, "Copenhagen's failure belongs to Obama" *Guardian*, December 21, 2009, http://www.guardian.co.uk/commentisfree/cif-green/2009/dec/21/copenhagen-failure-obama-climate-change.

25. Kate Sheppard, "Copenhagen Decoded" *Mother Jones,* December 21, 2009, http://motherjones.com/politics/2009/12/copenhagen-decoded.

26. David Corn and Kate Sheppard, "Obama's Copenhagen Deal" *Mother Jones,* December 18, 2009, http://motherjones.com/environment/2009/12/deal-or-no-deal-time-copenhagen.

27. Sheppard, "Copenhagen Decoded."

28. Final text of the Copenhagen Accord is at http://unfccc.int/files/meetings/cop_15/application/pdf/cop15_cph_auv.pdf. For other information, see http://unfccc.int/2860.php.

29. Emma Hayward, "Future of Global Climate Talks Remain Unclear," *Al-Jazeera/English*, December 30, 2009, http://www.commondreams.org/video/2009/12/30-0

30. Sheppard, "Copenhagen Decoded."

31. James Hansen, "G-8 Failure Reflects U.S. Failure on Climate Change," July 9, 2009,

32. http://www.huffingtonpost.com/dr-james-hansen/g-8-failure-reflects-us-f_b_228597.html.

33. Ibid.

34. Ted Glick, "Climate: Time Is Short," *Truthout*, April 25, 2007, http://www.truthout.org/docs_2006/042507N.shtml.

35. Paul Hawken, Amory Lovins, and L. Hunter Lovins, *Natural Capitalism*, p. 205.

36. Fred Pearce, "Peat Bogs Harbour Carbon Time Bomb," *NewScientist*, July 10–16, 2004, p. 9, http://www.newscientist.com/article/mg18324551.500-peat-bogs-harbour-carbon-time-bomb.html

37. Ibid.

38. Ibid.

39. See "Major Biomes of the World" www.runet.edu/~swoodwar/CLASSES/GEOG235/biomes/main.html.

40. Dmitry Solovyov, "Unusual warming signal: Prehistoric dung," *Reuters News Service*, September 17, 2007, http://www.msnbc.msn.com/id/20819312/from/ET/.

41. Ibid.

42. All of these are by Joe Romm at his blog, www.climateProgress.org. The drying of the Northern peatlands (bogs, moors, and mires). "For peat's sake: A point of no return as alarming as the tundra feedback," October 13, 2008, http://climateprogress.org/2008/10/13/for-peats-sake-a-point-of-no-return-as-alarming-as-the-tundra-feedback. "The defrosting of the permafrost Tundra, Part 2: The point of no return" *May 23, 2008* http://climateprogress.org/2008/05/23/tundra-part-2-the-point-of-no-return. "Tundra 3: Forests and fires foster feedbacks," *May 29, 2008* http://climateprogress.org/2008/05/29/tundra-3-forests-and-fires-further-foster-feedbacks. "Breaking News — Tundra 4: Permafrost loss linked to Arctic sea ice loss," *June 12, 2008.* http://climateprogress.org/2008/06/12/breaking-news-tundra-4-permafrost-loss-linked-to-arctic-sea-ice-loss. "Wetlands destruction — another climate feedback ," July 23, 2008, http://climateprogress.org/2008/07/23/wetland-destruction-another-amplifying-feedback. "Decelerating growth in tropical forest trees — thanks to accelerating carbon dioxide," January 21, 2008, http://climateprogress.org/2008/01/21/decelerating-growth-in-tropical-forest-trees-thanks-to-accelerating-carbon-dioxide. "Wildfires and Climate-Driven forest destruction by pests: "Global warming and the California wildfires," October 24, 2007. http://climateprogress.org/2007/10/24/global-warming-and-the-california-wildfires and" "Climate-Driven Pest Devours N. American Forests," August 1, 2007, http://climateprogress.org/2007/08/01/climate-driven-pest-devours-n-american-forests. "The desertification-global warming feedback," September 12, 2007, http://climateprogress.org/2007/09/12/the-desertification-global-warming-feedback. "Big news: The ocean carbon sink is saturating," October 22, 2007, http://climateprogress.org/2007/10/22/big-news-the-ocean-carbon-sink-is-saturating.

43. See http://www.ghgonline.org/co2sinkocean.htm.

44. Stephen Leahy, "Oceans Passing Critical CO2 Threshold,"

November 25, 2008, by Inter Press Service, http://www.commondreams.org/headline/2008/11/25-6. Published in the Proceedings of the National Academy of Sciences.

45. Ibid.

46. "How do we know that recent CO_2 increases are due to human activities?" See http://www.realclimate.org/index.php?p=87.

47. See http://www.koshland-science-museum.org/exhibitgcc/carbon02.jsp.

48. See "Methane." http://www.epa.gov/methane/.

49. Matthew Hornbach, Demian Saffer, and Steve Holbrook. "Critically pressured free-gas reservoirs below gas-hydrate provinces," *Nature*, January 8, 2004. http://www.nature.com/nature/journal/v427/n6970/abs/nature02172.html, or http://faculty.gg.uwyo.edu/holbrook/papers/Hornbach_2004_Nature_hydrates.pdf.

50. Ibid.

51. Ibid.

52. IPCC. *Working Group I Summary for Policy Makers*, p. 34. See http://www.ipcc.ch/SPM2feb07.pdf.

53. Ibid, p. 12.

54. Fred Pearce, "Harbingers of doom?" *New Scientist*, July 24, 2004, p. 44. See http://www.newscientist.com/article/mg18324575.600-harbingers-of-doom.html.

55. Ibid.

56. Ibid.

57. Ibid.

58. Ibid.

59. National Academy of Sciences. *Abrupt Climate Change*, http://www.nap.edu/catalog/10136.html. The book can be read for free online.

60. Ibid.

61. Ibid, p. 14.

62. Ibid, p. 155.

63. IPCC. *Working Group II Summary for Policy Makers*, p. 7. See http://www.ipcc.ch/SPM13apr07.pdf.

64. Bill McKibben, "Copenhagen: Too Hot to Handle "*Mother Jones*, October 23, 2009, http://www.motherjones.com/environment/2009/11/copenhagen-too-hot-handle.

65. "Kelpie Wilson Interviews Ross Gelbspan," *Truthout*, February 24, 2005, http://www.truthout.org/issues_05/E022405Y.shtml.

66. Sierra Club. See http://www.sierraclub.org/globalwarming/health/disease.asp.

67. Ibid.

68. Ibid.

69. See Janet Larsen, "Record Heat Wave in Europe Takes 35,000 Lives: Far Greater Losses May Lie Ahead," http://www.earth-policy.org/Updates/Update29_printable.htm.

70. "Setting the Record Straight: More than 52,000 Europeans Died from Heat in Summer 2003," *Update*, July 28, 2006, http://www.earth-policy.org/Updates/2006/Update56.htm. (See http://www.earth-policy.org/Updates/2006/Update56_data.htm for the supporting data.)

71. See http://yosemite.epa.gov/OAR/globalwarming.nsf/content/ImpactsHealth.html.

72. NOVA "Evolution" http://www.pbs.org/wgbh/nova/evolution.

73. Virginia Morell, "The Sixth Extinction" *National Geographic*, February 1999, http://www.nationalgeographic.com/ngm/9902/fngm/ .

74. "Climate change threatens a million species with extinction" January 7, 2004, http://www.leeds.ac.uk/media/current/extinction.htm and C. D. Thomas et al., 2004, Extinction risk from climate change, *Nature*, vol. 427, pp. 145–148, http://www.nature.com/nature/journal/v427/n6970/full/nature02121.html or http://eprints.whiterose.ac.uk/83/1/thomascd1.pdf.

75. Richard E. Leakey, *The Sixth Extinction: Patterns of Life and the Future of Humankind,* New York, N.Y., Anchor Books, 1996.

76. Morell, "The Sixth Extinction," and Leakey, *The Sixth Extinction.*

77. Jack Miles, "Global Requiem: The Apocalyptic Moment in Religion, Science, and Art." A keynote address at the fiftieth anniversary Cross Currents Consultation in New York City, http://www.crosscurrents.org/milesrequiem.htm.

78. C.D. Harvell et al. "Emerging Marine Diseases—Climate Links and Anthropogenic Factors," *Science*, September 22, 1999, http://www.sciencemag.org/cgi/content/abstract/285/5433/1505.

79. Worldwatch Institute. *Vital Signs 2001*, p. 92.

80. "Oceans' Acidity Worries Experts," *Atlanta Journal-Constitution*, September 25, 2003, http://www.heatisonline.org/contentserver/objecthandlers/index.cfm?id=4475&method=full. (See also Jeff Hecht, "Alarm over Acidifying Oceans," *New Scientist*, September 24, 2003, http://www.newscientist.com/article/dn4196-alarm-over-acidifying-oceans.html.)

81. Gelbspan, *Boiling Point*, pp. 65–66. Quoting Richard Sadler and Geoffrey Lean, "North Sea Faces Collapse of Its Ecosystem, Fish Stocks and Sea Bird Numbers Plummet as Soaring Water Temperatures Kill Off Vital Plankton," *Independent*, October 19, 2003, http://www.well.com/~davidu/northsea.html.

82. Ross Gelbspan, "Straddling Solutions and Survivalism," lecture at Massachusetts Climate Action Network Conference, November 19, 2006. See http://www.heatisonline.org/contentserver/objecthandlers/index.cfm?id=6161&method=full. Droughts can be self-reinforcing. Moisture in the upper soil evaporates when the air is hot. Changing water to vapor takes enormous energy, that is, as the water in the soil dries out and turns into vapor, this process absorbs enormous amounts of energy that otherwise would have contributed to higher air temperatures so it plays a significant role in moderating surface temperature changes. However, when the upper soil dries out, meaning that all the water in the soil has evaporated, further solar heating will no

longer be used to change the water to vapor. At that point, all the heat goes entirely into heating the surface air and land. As a result, once all the water in the soil is evaporated, heat waves and droughts begin to "feed" on themselves and become more intense as they persist.

83. Alister Doyle, "Global Warming: Pakistan Floods, Russia Heat Match Climate Trends" August 9, 2010 by Reuters . http://www.commondreams.org/headline/2010/08/09-4.

84. Ibid.

85. Janet Larsen, "Hurricane Damages Soar to New Levels," *Update*, August 29, 2006. See http://www.earth-policy.org/Updates/2006/Update58.htm. (See http://www.earthpolicy.org/Updates/2006/Update58_data.htm for the supporting data.)

86. Ross Gelbspan, "Why we need to worry about global warming, now," *San Jose Mercury News*, April 9, 2006, http://www.heatisonline.org/contentserver/objecthandlers/index.cfm?id=5896&method=full.

87. Ibid.

88. Chris Flavin, "A Dying Planet," July 24, 2006. See http://www.worldwatch.org/node/4457, or http://www.tompaine.com/articles/2006/07/24/a_dying_planet.php. Also, the entire August 2006 issue of *Worldwatch* magazine is related to Hurricane Katrina.

89. Larsen, "Hurricane Damages Soar to New Levels."

90. Kevin E. Trenberth, "The Extreme Weather Events of 1997 and 1998," *Consequences* 5 (1) 1999, pp. 3–15. See http://www.gcrio.org/CONSEQUENCES/vol5no1/extreme.html.

91. "Coral Reef Exposes Worst El Ninos Ever Are Now," *Reuters News Service*, January 28, 2001.

92. Kevin E. Trenberth, "The Extreme Weather Events of 1997 and 1998." http://www.ucsusa.org/global_warming/science_and_impacts/impacts/early-warning-signs-of-global-4.html

93. Gelbspan, "Straddling Solutions and Survivalism."

94. James Hansen et al., "Climate Change and Trace Gases," *Philosophical Transactions of the Royal Society – A*, Vol. 365, pp. 1925–1954. See http://pubs.giss.nasa.gov/docs/2007/2007_Hansen_etal_2.pdf. (Also see Monbiot, George. "Global Warming: A Sudden Change of State," *The Guardian*, July 3, 2007 [http://www.commondreams.org/archive/2007/07/03/2269.

95. James Hansen, "Why We Can't Wait," *The Nation*, May 7, 2007. See http://www.thenation.com/doc/20070507/hansen.

96. NASA Goddard Institute for Space Studies. See http://www.giss.nasa.gov/.

97. Jim Hansen, "Climate Change: On the Edge," *Independent*, February 17, 2006, http://www.commondreams.org/views06/0217-22.htm.

98. Steve Connor, "Sea Levels Likely to Rise Much Faster Than Was Predicted," *Independent*, February 17, 2006. http://www.commondreams.org/headlines06/0217-06.htm.

99. Hansen, "Climate Change: On the Edge."

100. Ibid.

101. Ibid.

102. James Hansen et al., "Climate Change and Trace Gases."

103. Steve Connor. "Global Warming 'Past the Point of No Return,'" *Independent*, September 16, 2005. See http://www.commondreams.org/headlines05/0916-09.htm.

104. Joe Romm, "Science: CO2 levels haven't been this high for 15 million years, when it was 5° to 10°F warmer and seas were 75 to 120 feet higher," http://climateprogress.org/2009/10/18/science-co2-levels-havent-been-this-high-for-15-million-years-when-it-was-5%c2%b0-to-10%c2%b0f-warmer-and-seas-were-75-to-120-feet-higher-we-have-shown-that-this-dramatic-rise-in-sea-level-i/.

105. "Arctic's Biggest Ice Shelf, a Sentinel of Climate Change, Cracks Apart" *The Los Angeles Times*, Sept. 23, 2003. http://www.heatisonline.org/contentserver/objecthandlers/index.cfm?id=4442&method=full.

106. Ibid.

107. Ibid.

108. See http://igloo.gsfc.nasa.gov/wais and http://www.glacier.rice.edu/land/5_antarcticicesheetparts.html; http://www.nsf.gov/.

109. Ibid.

110. Ibid.

111. Nicholas D. Kristof, "The Big Melt," Op-Ed, *New York Times*, August 17, 2007, and Revkin, Andrew C. "Scientists Report Severe Retreat of Arctic Ice," *New York Times* September 21, 2007. (See National Snow and Ice Data Center, Boulder, CO http://www.nsidc.org for research findings.)

112. Real Climate. "Arctic sea ice watch" http://www.realclimate.org/index.php/archives/2007/08/arctic-sea-ice-watch/ see also "Sea ice minimum forecasts" http://www.realclimate.org/index.php/archives/2009/07/summer-sea-ice-round-up/ .

113. Joe Romm, "Met Office's Richard Betts incorrectly asserts "the dramatic decline in Arctic sea ice in 2007 … was then partly reversed in 2008 and 2009." Climate Progress http://climateprogress.org/2010/01/12/sciencerichard-betts-andy-revkin-dramatic-decline-in-arctic-sea-ice-in-2007-was-not-reversed-in-2008-2009-recovery/.

114. David M. Lawrence et al., 2008. Accelerated Arctic land warming and permafrost degradation during rapid sea ice loss. Geophysical Research Letters, http://www.cgd.ucar.edu/ccr/dlawren/publications/lawrence.grl.submit.2008.pdf. Quoted in George Monbiot, "One Shot Left—The latest science suggests that preventing runaway climate change means total decarbonisation," November 25, 2008, *The Guardian/UK*. See http://www.monbiot.com/archives/2008/11/25/one-shot-left/.

115. United Nations Environment Project, 4 June 2007, "Melting Ice— a Hot Topic? Press Release, http://www.unep.org/Documents.Multilingual/Default.asp?DocumentID=512&ArticleID=5599&l=en. Quoted in George Monbiot "One Shot Left," at http://www.monbiot.com/archives/2008/11/25/one-shot-left/.

116. Kurt M. Cuffey and Shawn J. Marshall. "Substantial Contribution to Sea-Level Rise during the Last Interglacial from the Greenland Ice Sheet," *Nature*, April 6, 2000, http://

www.nature.com/nature/journal/v404/n6778/abs/404591a0.html.

117. Steve Connor, "Sea Levels Likely to Rise Much Faster Than Was Predicted," *Independent*, February 17, 2006. See http://www.commondreams.org/headlines06/0217-06.htm.

118. David Perlman, "Greenland's Ice Cap is Melting at a Frighteningly Fast Rate," *San Francisco Chronicle*, August 11, 2006. See http://www.commondreams.org/headlines06/0811-06.htm. (The article was written based upon information published in *Science* by scientists at the University of Texas at Austin—Chen, Tapley, and Wilson.)

119. Hansen, "Climate Change: On the Edge."

120. Perlman, "Greenland's Ice Cap is Melting at a Frighteningly Fast Rate."

121. Gelbspan, *Boiling Point*, p. 20.

122. Perlman, "Greenland's Ice Cap is Melting at a Frighteningly Fast Rate."

123. Robert A. Bindschadler and Charles R. Bentley. "On Thin Ice?" *Scientific American*, December 2002. http://www.scientificamerican.com/article.cfm?id=on-thin-ice .

124. World Wildlife Fund. "Going, Going, Gone: Climate Change and Global Glacier Decline." See http://www.panda.org/downloads/climate_change/glacierspaper.pdf. (From information presented by J. Oerlemans et al., "Modelling the response of glaciers to climate warming," *Climate Dynamics* 14:267–274, http://www.springerlink.com/content/xru0rmjk6xw47rwr/.)

125. Juan Forero, "As Andean Glaciers Shrink, Water Worries Grow," *New York Times*, November 24, 2002, http://www.nytimes.com/2002/11/24/world/as-andean-glaciers-shrink-water-worries-grow.html.

126. World Wildlife Fund, "Going, Going, Gone."

127. Anthony A. Arendt et al. "Rapid Wastage of Alaska Glaciers and Their Contribution to Rising Sea Level" *Science*, July 19, 2002, http://www.nasa.gov/pdf/121649main_Arendt.pdf.

128. Ibid.

129. See "Retreating Glaciers Spur Alaskan Earthquakes" 08/02/2004, http://www.nasa.gov/centers/goddard/news/topstory/2004/0715glacierquakes.html.

130. World Wildlife Fund, "Going, Going, Gone: Climate Change and Global Glacier Decline." See http://www.panda.org/downloads/climate_change/glacierspaper.pdf. (From information presented by C. K. Jain, "A hydro-chemical study of a mountainous watershed: the Ganga, India," *Water Research* 36(5) (2001), pp. 1262–1274; http://news.mongabay.com/2005/0907-wwf.html, and also by P. S. Singh et al., "Snow and glacier contribution in the Ganga River at Deoprayag," Technical report, CS(AR)132. National Institute of Hydrology, Roorkee, India, 1993–1994.)

131. World Wildlife Fund, "Going, Going, Gone: Climate Change and Global Glacier Decline." See http://www.panda.org/downloads/climate_change/glacierspaper.pdf. (From information presented by L. G. Thompson et al. "Kilimanjaro ice core records: evidence of Holocene climate change in tropical Africa," *Science*, October 18, 2002, http://www.sciencemag.org/content/298/5593/589.abstract or http://www.ncbi.nlm.nih.gov/pmc/articles/PMC2771743/.)

132. World Wildlife Fund. "Going, Going, Gone: Climate Change and Global Glacier Decline." See http://www.panda.org/downloads/climate_change/glacierspaper.pdf. (From information presented by M. H. P. Hall and D. B. Fagre, "Modeled climate-induced glacier change in Glacier National Park, 1850-2100," *BioScience* 53(2) (2003), pp. 131–140.)

133. Robin McKie, "Decades of devastation ahead as global warming melts the Alps," *The Observer,* Sunday, July 20, 2003. http://www.guardian.co.uk/environment/2003/jul/20/climatechangeenvironment.geology.

134. Elizabeth Rosenthal, "As the Climate Changes, Bits of England's Coast Crumble," *New York Times*, May 4, 2007, http://www.nytimes.com/2007/05/04/world/europe/04erode.html.

135. See Alliance of Small Island States, at http://www.sidsnet.org/aosis/, and Small Islands Developing States Network, at http://www.sidsnet.org.

136. Gelbspan, *Boiling Point*, p. 172. (Quoting from "Tuvalu: Global Warming's First Causality," *Honolulu Star Bulletin*, August 19, 2001.)

137. Lester R. Brown, "Rising Sea Level Forcing Evacuation of Island Country" *Update 2*, Earth Policy Institute, November 15, 2001. http://www.earth-policy.org/Updates/Update2.htm.

138. Ibid.

139. Maude Barlow and Tony Clarke, *Blue Gold: The Fight to Stop the Corporate Theft of the World's Water,* , p. 42.

140. Ibid. p. 24.

141. Ibid, p. 25.

142. Maude Barlow and Tony Clarke, "Who Owns Water?" The Nation, August 15, 2002. http://www.thenation.com/doc/20020902/barlow.

143. Don Hinrichsen "Appropriating the Water," *Worldwatch Magazine*, January-February 2003, p. 15. http://www.worldwatch.org/node/527.

144. Brown, *Plan B*, p. 163.

145. Lester R. Brown, "World Grain Harvest Falling Short by 54 Million Tons—Water Shortages Contributing to Shortfall," Earth Policy Institute, November 23, 2001. http://www.earthpolicy.org/Indicators/indicator6.htm.

146. Vital Climate Graphics provides maps at http://www.grida.no/climate/vital/32.htm.

147. William H. Calvin, "The great climate flip-flop," *The Atlantic Monthly*, January 1998. (Also see http://WilliamCalvin.com/1990s/1998AtlanticClimate.htm.)

148. Ibid.

149. Peter Schwartz and Doug Randall, *An Abrupt Climate Change Scenario and Its Implications for United States National Security*, October 2003, p. 1. http://www.greenpeace.org/international/press/reports/an-abrupt-climate-change-scena

150. Gelbspan, *Boiling Point*, p. 90.

151. IPCC. *Climate Change Impacts, Adaptation and Vulnerability*. For this summary from WG II, see http://www.ipcc.ch/SPM6avr07.pdf.

152. Christian Aid, *Human Tide: The Real Migration Crisis*, May 2007. See http://www.christianaid.org.uk/Images/human_tide3__tcm15-23335.pdf.

153. See Sen. Dick Durbin and Chuck Hagel, "Global Warming Poses Serious Threat To National Security" March 28, 2007, http://www.durbin.senate.gov/record.cfm?id=271514.

154. Ibid.

155. Amanda Griscom Little, "Will global warming threaten national security?" See http://www.salon.com/news/feature/2007/04/09/muckraker/print.html.

156. Julian Borger, "Darfur Conflict Heralds Era of Wars Triggered by Climate Change, UN Report Warns," *The Guardian*, June 23, 2007. http://www.guardian.co.uk/environment/2007/jun/23/sudan.climatechange.

157. CNA Corporation. *National Security and the Threat of Climate Change*. See http://securityandclimate.cna.org/. (CNAC is a non-profit research organization that operates the Center for Naval Analyses http://www.cna.org/about/cna/ and the Institute for Public Research http://www.cna.org/about/ipr/

158. CNA Corporation, *National Security and the Threat of Climate Change*, p. 10.

159. Ibid.

160. Julian Borger, "Darfur Conflict Heralds Era of Wars Triggered by Climate Change, UN Report Warns," *The Guardian*, June 23, 2007. http://www.guardian.co.uk/environment/2007/jun/23/sudan.climatechange.

161. Ibid.

162. Ibid.

163. Suzanne Goldenberg, "Nearly Half of Americans Believe Climate Change Threat Is Exaggerated," *March 11, 2010 by The Guardian/UK*. http://www.commondreams.org/headline/2010/03/11-9.

164. Jason Linkins, "'Climategate' Debunking Gets Less Coverage Than Original Trumped-Up Scandal," July 12, 2010. http://www.huffingtonpost.com/2010/07/12/climategate-debunking-get_n_642980.html?ir=Green.

165. Jason Linkins. See also John M. Broder, "Researcher on Climate Is Cleared in Inquiry" *The New York Times*, February 3, 2010. http://www.nytimes.com/2010/02/04/science/earth/04climate.html?_r=2. The study by Union of Concerned Scientists, Other reports refuting "Climate Gate" are "Debunking Misinformation About Stolen Climate Emails in the 'Climategate' Manufactured Controversy" is at http://www.ucsusa.org/global_warming/science_and_impacts/global_warming_contrarians/debunking-misinformation-stolen-emails-climategate.html. Dr. Nafeez Mosaddeq Ahmed, "The Real ClimateGate: Getting Over the Nonexistent 'Email Scandal'" (Part 1) November 13, 2010, http://www.truth-out.org/the-real-climategate-getting-over-nonexistent-email-scandal-part-165012. The parliamentary science and technology select committee is discussed in James Randerson, "Climate researchers' 'secrecy' criticised – but MPs say science remains intact" *The Guardian*, March 31, 2010. http://www.guardian.co.uk/environment/2010/mar/31/climate-mails-inquiry-jones-cleared. The Lord Oxburgh report is discussed at David Adam, "Scientists cleared of malpractice in UEA's hacked emails inquiry" *The Guardian*, April 14, 2010 http://www.guardian.co.uk/environment/2010/apr/14/oxburgh-uea-cleared-malpractice. The Muir Russell report is discussed at David Adam, "Climategate scientists cleared of manipulating data on global warming" *The Guardian*, July 8, 2010 http://www.guardian.co.uk/environment/2010/jul/08/muir-russell-climategate-climate-science. Also "Britain: Inquiry Finds No Distortion of Climate Data" New York Times April 14, 2010 By Reuters http://www.nytimes.com/2010/04/15/science/earth/15briefs-Britbrf.html?emc=tnt&tntemail0=y. See also Bradford Plumer, "'Climategate' Was Overblown" *The New Republic*, April 15, 2010 http://www.nonesoblind.org/blog/?p=6445. Some of the key conclusions:-—We saw no evidence of any deliberate scientific malpractice in any of the work of the Climatic Research Unit and had it been there we believe that it is likely that we would have detected it. Rather we found a small group of dedicated if slightly disorganized researchers who were ill-prepared for being the focus of public attention. As with many small research groups their internal procedures were rather informal. … ---After reading publications and interviewing the senior staff of CRU [Climatic Research Unit] in depth, we are satisfied that the CRU tree-ring work has been carried out with integrity, and that allegations of deliberate misrepresentation and unjustified selection of data are not valid.–We believe that CRU did a public service of great value by carrying out much time-consuming meticulous work on temperature records at a time when it was unfashionable and attracted the interest of a rather small section of the scientific community. CRU has been among the leaders in international efforts to determining the overall uncertainty in the derived temperature records and where work is best focused to improve them.

166. John Stauber and Sheldon Rampton, "The Public Relations Industry's Secret War on Activists." http://mediafilter.org/CAQ/caq55.prwar.html.

167. "How to Talk to a Climate Skeptic: Responses to the most common skeptical arguments on global warming" A Grist Special Series, http://gristmill.grist.org/skeptics.

168. Joe Romm, "Science: CO2 levels haven't been this high for 15 million years, when it was 5° to 10°F warmer and seas were 75 to 120 feet higher." http://climateprogress.org/2009/10/18/science-co2-levels-havent-been-this-high-for-15-million-years-when-it-was-5%c2%b0-to-10%c2%b0f-warmer-and-seas-were-75-to-120-feet-higher-we-have-shown-that-this-dramatic-rise-in-sea-level-i/.

169. Chris Mooney, "Why Republicans Distrust Mainstream Climate Science," http://www.huffingtonpost.com/chris-mooney/why-republicans-distrust-_b_44254.html.

170. Raymond Colitt, "World Must Fix Climate in Less than 10 years: UN" November 27, 2007, Reuters. http://www.enn.com/climate/article/25936.

171. For the statement and a list of people that signed it, see http://www.christiansandclimate.org/statement.

172. Laurie Goodstein, "Evangelical Leaders Join Global Warming Initiative," *The New York Times*, February 8, 2006, http://www.nytimes.com/2006/02/08/national/08warm.html. See also The Evangelical Climate Initiative, at www.christiansandclimate.org.

173. Paul Krugman, "Enemy of the Planet," *New York Times*, April 17, 2006, http://select.nytimes.com/2006/04/17/opinion/17krugman.html.

174. Union of Concerned Scientists, *Smoke, Mirrors and Hot Air: How ExxonMobil Uses Big Tobacco's Tactics to Manufacture Uncertainty on Climate Science*, January 2007. See http://www.ucsusa.org/assets/documents/global_warming/exxon_report.pdf.

175. Union of Concerned Scientists, *Smoke, Mirrors and Hot Air*, p. 1.

176. Union of Concerned Scientists, *Smoke, Mirrors and Hot Air*, p. 7.

177. Gelbspan, *The Heat Is On*, p. 86.

178. Ibid.

179. Union of Concerned Scientists, *Smoke, Mirrors and Hot Air*, p. 10.

180. Ibid, p. 11.

181. Union of Concerned Scientists, *Smoke, Mirrors and Hot Air*, p. 12. For instance, one of the main "skeptics" is Patrick Michaels, a meteorologist who edited a 2005 book, *Shattered Consensus: the True State of Global Warming*. Patrick Michaels belongs to at least 11 organizations funded by ExxonMobil, including the Marshall Institute.

182. Mooney, Chris. "Blinded By Science How 'Balanced' Coverage Lets the Scientific Fringe Hijack Reality," *Columbia Journalism Review*, November/December 2004. See http://findarticles.com/p/articles/mi_qa3613/is_200411/ai_n9460969.

183. Gelbspan, "Straddling Solutions and Survivalism."

184. Dr. Nafeez Mosaddeq Ahmed, "The Real ClimateGate: Getting Over the Nonexistent 'Email Scandal'" (Part 1) November 13, 2010, http://www.truth-out.org/the-real-climategate-getting-over-nonexistent-email-scandal-part-165012. See also http://uk.ibtimes.com/articles/20100929/big-oil-spends-halfbillion-dollars-oppose-climate-legislation.htm.

185. "Sallie Baliunas & Willie Soon" Part of the *A Rundown of the Skeptics & Deniers* series, from Logical Science. http://www.logicalscience.com/skeptics/skeptics.htm.

186. Sallie Baliunas and Willie Soon work for the Harvard-Smithsonian Center for Astrophysics. Soon is a paid consultant and Baliunas is a senior scientist for the George C. Marshall Institute, a think tank partially funded by Exxon Mobil, Olin Chemical, Gulf Oil Corporation, and the White Star Oil Company. The Marshall Institute opposes limits on carbon dioxide emissions and disputes human's role in global climate change. Baliunas and Soon are also contributing editors to World Climate Report, which is run by Patrick Michaels and funded by the Western Fuels Association. Baliunas has also written papers for the Oil and Tobacco funded Heartland Institute. Baliunas works for numerous think tanks that have a conflict of interest due to funding by big oil, yet she claims it is the mainstream scientists that are twisting the science for monetary gain. In an interview with ABC news she said: "It's the money! . . . If scientists and researchers were coming out releasing reports that global warming has little to do with man, and most to do with just how the planet works, there wouldn't be as much money to study it." "Sallie Baliunas & Willie Soon" Part of the *A Rundown of the Skeptics & Deniers* series, from Logical Science. http://www.logicalscience.com/skeptics/Sallie-Baliunas.html.

187. Union of Concerned Scientists, *Smoke, Mirrors and Hot Air*, p. 15. (The peer-reviewed rebuttal to the Soon-Baliunas paper is M. E. Mann et al., "On past temperatures and anomalous late-20th century warmth," *EOS Transactions, AGU* 84(27), p. 256, http://stephenschneider.stanford.edu/Publications/PDF_Papers/EosForum2003_revised20June.pdf.)

188. Dr. Nafeez Mosaddeq Ahmed, "The Real ClimateGate: Getting Over the Nonexistent 'Email Scandal'".

189. TCS Newsflash: "Important Global Warming Study Audited -- Numerous Errors Found," E-Wire, Press Release of October 27, 2003. See http://www.ewire.com/display.cfm/Wire_ID/1856.

190. Nick Schulz, Op-Ed, "Researchers question key global-warming study" *USA Today*, October 28, 2003. See http://www.usatoday.com/news/opinion/editorials/2003-10-28-schulz_x.htm.

191. Ibid.

192. James Inhofe, Speech on the floor of the U.S. Senate, October 29, 2003. http://inhofe.senate.gov/pressreleases/climate.htm.

193. Ibid.

194. Dr. Nafeez Mosaddeq Ahmed, "The Real ClimateGate: Getting Over the Nonexistent 'Email Scandal'" (Part 1) November 13, 2010, http://www.truth-out.org/the-real-climategate-getting-over-nonexistent-email-scandal-part-165012. The study is at http://www.centerforinquiry.net/opp/news/senate_minority_report_on_global_warming_not_credible/

195. Naomi Oreskes, "Beyond the ivory tower: The scientific consensus on climate Change," *Science*, December 3, 2004. (Quoted in UCS. *Smoke, Mirrors and Hot Air*, p. 30.)

196. Ibid.

197. Gelbspan, "Straddling Solutions and Survivalism."

198. Jeff Fleischer, "Interview with Robert F. Kennedy on his book *Crimes Against Nature*," in *MotherJones*, October 7, 2004. See http://www.motherjones.com/news/qa/2004/10/09_402.html.

199. James Hansen, "Political Interference with Government Climate Change Science," Testimony to Committee on Oversight and Government Reform, United States House of Representatives, March 19, 2007, See http://www.columbia.edu/~jeh1/20070319105800-43018.pdf, p. 4. See Mark Bowen, *Censoring Science: Inside the Political Attack on Dr. James Hansen and the Truth of Global Warming, New York, NY,* Dutton Adult, 2007.

200. Ibid, p. 6.

201. Ibid, p. 6.

202. Ibid.

203. Union of Concerned Scientists. *Smoke, Mirrors and Hot Air*, p. 21.

204. Ibid, p. 20.

205. Ibid. p. 21.

206. Union of Concerned Scientists, *Atmosphere of Pressure: Political Interference in Federal Climate Science*. See http://www.ucsusa.org/scientific_integrity/interference/atmosphere-of-pressure.html. It is not just global warming information that is being suppressed. Another UCS study on scientific integrity lists instances of other problems; one of them surrounds the medication Vioxx. In September 2004, the pharmaceutical company Merck pulled Vioxx from the market after clinical trials showed that those taking the drug had an increased risk of heart attack and stroke. Subsequently, David Graham, associate science director of the Office of Drug Safety at the Food and Drug Administration (FDA) who was working on a separate agency review of Vioxx, alleged that FDA officials tried to suppress and delay the results of his work, which drew similar conclusions. Graham claimed that the agency was "virtually incapable of protecting America" and that "the review and clearance process had been turned into a battleground, full of contention and intimidation because our managers, the people who fill out our performance evaluations, had created a system where it was taking a great risk to stand firm in our scientific beliefs." His study indicates that between 88,000 and 139,000 people had heart attacks or strokes and about 55,000 died as a result of taking the pain medication. And yet managers at the FDA wanted to suppress this information. For this additional study, see http://www.ucsusa.org/assets/documents/scientific_integrity/Scientific-Integrity-at-Risk-FDA.pdf.

207. Union of Concerned Scientists, *Atmosphere of Pressure: Political Interference in Federal Climate Science*, p. 3.

208. Gelbspan, *Boiling Point*, p. 144.

209. Union of Concerned Scientists, *Smoke, Mirrors and Hot Air*, p. 10.

210. Elizabeth Kolbert, "Annals of Science—The Climate Of Man," *New Yorker*. This was a three-part series on April 25, May 2, and May 9, 2005, Part I http://www.newyorker.com/archive/2005/04/25/050425fa_fact3, Part II, http://www.newyorker.com/archive/2005/05/02/050502fa_fact3,PartIII, http://www.newyorker.com/archive/2005/05/09/050509a_fact3. It has subsequently been collected and expanded into a book, *Field Notes from a Catastrophe: Man, Nature, and Climate Change*. Bloomsbury, 2006.

Chapter 2

1. Joanna Macy, "The Great Turning as Compass and Lens: What it means to be alive at a moment of global crisis and possibility," Yes! Magazine, Summer 2006. Italics added. http://www.yesmagazine.org/issues/5000-years-of-empire/the-great-turning-as-compass-and-lens

2. Gleick, James. *Chaos: Making a New Science*., p. 16.

3. Ibid. p. 16.

4. Buchanan, *Ubiquity: Why Catastrophes Happen*, pp. 17–18.

5. Ibid, p. 39.

6. National Academy of Sciences. *Abrupt Climate Changes*, p. 14.

7. John Briggs, and F. David Peat. *Turbulent Mirror: An Illustrated Guide to Chaos Theory and the Science of Wholeness*, p. 134.

8. Ajit Jain, editor, *Solidarity*, pp. 7–19.

9. Fritjof Capra, *Web of Life*, p. 72.

10. Fritjof Capra, "Ecology and Community," p. 1. See http://www.ecoliteracy.org/pdf/community.pdf.

11. Capra, "Ecology and Community," pp. 7–8.

12. David Korten, *Agenda for a New Economy*, p.107.

Chapter 3

1. George Lakoff, *The Political Mind*, p. xv.

2. George Lakoff, *Thinking Points: Communicating Our American Values and Vision*, pp. 31 and 35.

3. Lakoff, *The Political Mind*, p. xv.

4. Ibid, p. xv.

5. Ibid, p. xv.

6. These ideas are from *Moral Politics* and *Don't think of an Elephant*, and the Rockridge Institute.

7. Ibid.

8. From "Characteristics of the Renaissance" in *A Guide to the Study of Literature*.

9. Murray Bookchin, *Toward an Ecological Society*, p. 29.

10. Quoted in David Korten, *The Great Turning*, p. 3.

11. Ibid, p. 29.

12. Korten, *The Great Turning*, p. 39.

13. Arnie Cooper, "David Korten on Putting an End to Global Competition," *The Sun* Interview September 2007, Issue 381. http://www.thesunmagazine.org/issues/381.

14. Ibid.

15. Korten, *The Great Turning*, p. 34.

16. Ibid.

17. David Korten, *To Change the Future, Change the Story*, http://thegreatturning.net/PDF/Change%20the%20Story%208.pdf.

18. David Korten, "5,000 Years of Empire: The Great Turning: From Empire to Earth Community," *Yes! Magazine*, Summer 2006 http://www.yesmagazine.org/article.asp?ID=1463.

19. Ibid.

20. Lakoff, *Moral Politics*, p. 2.

21. Ibid.

22. Alfie Kohn, *No Contest: the Case against Competition*, p. 149.

23. Ibid. pp. 183–184.

24. Peter F. Drucker, *The Post-Capitalist Society*, p. 93.
25. Ibid, pp. 108–109.
26. Herman Daly, *Common Good,* p. 204.
27. Fritjof Capra, *Web of Life*, p. 227.
28. Ibid, p. 231.
29. Ibid.
30. Ibid, p. 232. See also "The Evolutionary Idea," http://www.edge.org/documents/ThirdCulture/g-Pt.1Intro.html the Introduction to *The Third Culture: Beyond the Scientific Revolution* by John Brockman (Simon & Schuster, 1995). Full text at http://www.edge.org/documents/ThirdCulture/d-Contents.html.
31. Fritjof Capra, "Ecology and Community," pp. 7–8, italics added. See http://www.ecoliteracy.org/pdf/community.pdf.
32. Elizabeth Fox-Genovese, *Feminism without Illusions: a Critique of Individualism*, p. 57.
33. Ibid, p. 57.
34. David Korten, "To Change the Future, Change the Story." The Earth Charter is at http://www.earthcharterusa.org/ and http://www.earthcharter.org/.
35. Korten, *The Great Turning*, p. 33.
36. David Korten, "We Are Hard-Wired to Care and Connect," *Yes! Magazine,* Fall 2008. See http://www.yesmagazine.org/article.asp?id=2848.
37. Ibid.
38. Ibid.
39. Ibid.
40. Korten, *The Great Turning*, p. 33.
41. Ibid, p. 36.
42. Jeremy Rifkin's Foundation on Economic Trends, http://www.foet.org/JeremyRifkin.htm.
43. Rifkin, *The Empathic Civilization,* pp. 9–10.
44. Ibid, p.10.
45. Ibid, pp. 15–16.
46. Ibid, p. 12.
47. Ibid.
48. Ibid.
49. Ibid, p. 42.
50. Ibid, p. 3.

Chapter 4

1. David Korten, "Imperial Politics, Christianity and the True Jesus: Reflections on the 2004 Election," http://www.yesmagazine.org/happiness/imperial-politics-christianity-and-the-true-jesus-reflections-on-the-2004-election .
2. Quoted in Korten, *The Great Turning,* p. 257.
3. Milton Rokeach, *The Open and Closed Mind: Investigations into the Nature of Belief Systems and Personality Systems*, p. 15.
4. Ibid, p. 6.
5. Ibid, p. 45
6. Ibid, p. 17.
7. Ibid, p. 48. Rokeach says there are layers to our beliefs. The first layer of belief is the content of primary beliefs—those that encompass our basic beliefs about physical reality, including, in particular, whether the world is friendly or not, whether parental or authority figures are loving or judgmental, whether people are to be trusted or feared, and whether the future is safe or frightening. It also includes beliefs about self-identity, self worth, autonomy or dependence (p. 41). The next "layer" of beliefs begins to define the structure apart from the content. Rokeach defines time perspective as one measure of openness: a more open system uses a broad time perspective including the past, present and future, whereas a more closed system overemphasizes one time dimension, often the far distant future (as in apocalyptic religions or Marxist historical determinism) (p. 45).
8. Joanna Macy, "The Great Turning," Interview, *Yes! Magazine*, Spring 2000. http://www.yesmagazine.org/article.asp?ID=333.
9. Personal communication.
10. Marcus Borg, "An Emerging Way," Chapter 1 in Michael Schwartzentruber, ed., *The Emerging Christian Way: Thoughts, Stories, and Wisdom for a Faith of Transformation,* pp. 13–14.
11. Ibid, p. 17.
12. Ibid, p. 18.
13. Ibid, pp. 21–22, 25.
14. Ibid.
15. Ibid, pp. 19–20.
16. Marcus Borg, *Jesus A New Vision: Spirit, Culture, and the Life of Discipleship*, p. 110.
17. William Herzog, in *Parables as Subversive Speech* says, Day laborers fall into a class of people in advanced agrarian societies know as "the expendables." Those forced into the expendable class fell below a subsistence level of living and were eventually forced to beg before dying from malnutrition. Westminster/John Knox press, Louisville: 1994, pp. 88–90.
18. Marcus Borg, *Meeting Jesus Again for the First Time: The Historical Jesus and the Heart of Contemporary Faith*, p. 51.
19. Marcus Borg, *Jesus A New Vision,* pp. 83–86.
20. Borg, *Meeting Jesus*, p. 52.
21. Ibid, pp. 55–56.
22. Ibid, p. 86.
23. Ibid, pp. 3, 39.
24. Ibid, p. 49.
25. Borg, *Jesus A New Vision*, p. 2.
26. Ibid, p. 5.
27. Ibid, p. 6.

28. Borg, *Meeting Jesus*, p. 119.
29. Ibid, p. 128.
30. Ibid, p. 7.
31. Ibid, p. 47.
32. Borg, *Jesus A New Vision*, p. 8.
33. David Ray Griffin, *God and Religion*, p. xiv.

Chapter 5

1. Herman Daly and John Cobb Jr., *For the Common Good: Redirecting the Economy toward Community, the Environment, and a Sustainable Future*, p. 13.
2. Elizabeth Martinez and Arnoldo García, "What is 'Neo-Liberalism'? A Brief Definition," Updated: February 26th, 2000. See
3. http://www.globalexchange.org/campaigns/econ101/neoliberalDefined.html.
4. John Gray, False Dawn: The Delusions of Global Capitalism, The New Press, 1998, p. 1.
5. Martinez and García, "What is 'Neo-Liberalism'? A Brief Definition."
6. Susan George, "A Short History of Neoliberalism," Speech at Conference on Economic Sovereignty in a Globalising World, Bangkok, March 24–26, 1999. See http://www.globalexchange.org/campaigns/econ101/neoliberalism.html.
7. Paul Krugman, "Greed Is Bad," *New York Times,* June 4, 2002, http://nytimes.com/2002/06/04/opinion/04KRUG.html.
8. Ibid.
9. George, "A Short History of Neo-liberalism."
10. Krugman, "Greed Is Bad."
11. Martinez and García, "What is 'Neo-Liberalism'? A Brief Definition."
12. George, "A Short History of Neo-liberalism."
13. Ibid.
14. Luke 12:48 and George, "A Short History of Neo-liberalism."
15. George, "A Short History of Neo-liberalism."
16. Daly and Cobb, *For the Common Good*, p. 99.
17. Ibid, p. 113.
18. Ibid, p. 113.
19. Ibid, pp. 194, 110.
20. Ibid, p. 192.
21. Robert Weissman and Harvey Rosenfield, "Sold Out: How Wall Street and Washington Betrayed America." See http://www.wallstreetwatch.org/reports/sold_out.pdf (large document). The executive summary is at http://www.wallstreetwatch.org/reports/executive_summary.pdf. The report introduction by Harvey Rosenfield is at http://www.wallstreetwatch.org/reports/part2.pdf. Highlights of the report's campaign contribution and lobbyist data are at http://www.wallstreetwatch.org/reports/part2.pdf.
22. Ibid.
23. Frontline "The Long Demise of Glass-Steagall: A chronology tracing the life of the Glass-Steagall Act,"" http://www.pbs.org/wgbh/pages/frontline/shows/wallstreet/weill/demise.html.
24. Ibid.
25. Joseph E. Stiglitz, "Capitalist Fools," ***VANITY FAIR,*** January 2009. http://www.truth-out.org/121008R or http://www.vanityfair.com/magazine/2009/01/stiglitz200901.
26. Ibid.
27. Ellen Brown, "Borrowing from Peter to Pay Paul: The Wall Street Ponzi Scheme Called Fractional Reserve Banking," December 29, 2008. See http://www.webofdebt.com/articles/ponzi.php.
28. Ibid.
29. Weissman and Rosenfield, "Sold Out: How Wall Street and Washington Betrayed America."
30. Stiglitz, "Capitalist Fools."
31. Weissman and Rosenfield, "Sold Out: How Wall Street and Washington Betrayed America."
32. Stiglitz, "Capitalist Fools."
33. Weissman and Rosenfield, "Sold Out: How Wall Street and Washington Betrayed America."
34. Warren Buffett, Chairman, Berkshire Hathaway, Report to Shareholders, February 21, 2003. See http://www.berkshirehathaway.com/letters.
35. Stiglitz, "Capitalist Fools."
36. Ibid.
37. Ellen Brown, "It's The Derivatives, Stupid! Why Fannie, Freddie and AIG All Had to Be Bailed Out," September 18, 2008. See http://www.webofdebt.com/articles/its_the_derivatives.php.
38. Ellen Brown, "The Collapse of a 300 Year Ponzi Scheme: The Real Debate Is Crony Socialism or Financial Sovereignty," October 16, 2008. See http://www.webofdebt.com/articles/modest_proposal.php.
39. William G. Gale and Peter R. Orszag, "The Cost of Tax Cuts," *Minneapolis Star-Tribune,* September 19, 2004,
40. http://www.brookings.edu/articles/2004/0919useconomics_gale.aspx.
41. Stiglitz, "Capitalist Fools."
42. Ibid.
43. David Korten, "Beyond the Bailout: Agenda for a New Economy," *Yes! Magazine,* Winter 2009. http://www.yesmagazine.org/article.asp?ID=3050.
44. Stiglitz, "Capitalist Fools."
45. Linda J. Bilmes and Joseph E. Stiglitz, "The Iraq War Will Cost Us $3 Trillion, and Much More," *Washington Post*, Sunday, March 9, 2008 http://www.washingtonpost.com/wp-dyn/content/article/2008/03/07/AR2008030702846.html, they are co-authors of *The Three Trillion Dollar War:*

The True Cost of the Iraq Conflict.

46. Mark Harris, "As Long As We're Talking About Socialism" April 28, 2009. See http://www.commondreams.org/view/2009/04/28-5.
47. Rana Foroohar, "May the Best Theory Win," *Newsweek,* February 1, 2010 issue, http://www.newsweek.com/id/232111
48. Ibid.
49. Ibid.
50. Ibid.

Chapter 6

1. See Naomi Wolf, "Fascist America, in 10 Easy Steps," The Guardian/UK, April 24, 2007. http://www.commondreams.org/archive/2007/04/24/708. *Laurence W. Britt, "Fascism Anyone?" Free Inquiry Magazine, Vol. 23, No. 2. See* http://www.secularhumanism.org/library/fi/britt_23_2.htm. **Thom Hartmann, Book Review of** *They Thought They Were Free: The Germans, 1938-45,* by Milton Mayer. See http://blog.buzzflash.com/hartmann/10019. Thom Hartmann, "Reclaiming the Issues: Islamic or Republican Fascism?" August 28, 2006. See http://www.commondreams.org/views06/0828-23.htm.
2. Thom Hartmann, "The Ghost of Vice President Wallace Warns: 'It Can Happen Here'" July 19, 2004 *CommonDreams.org* http://www.commondreams.org/views04/0719-15.htm.
3. Ibid.
4. Laurence W. Britt, "Fascism Anyone?" *Free Inquiry Magazine, Vol. 23, No. 2.* http://www.secularhumanism.org/library/fi/britt_23_2.htm.
5. Hartmann, "The Ghost of Vice President Wallace Warns."
6. Ibid.
7. The John Birch Society. http://www.jbs.org/node/740. The website editor said this.
8. See Thom Hartmann, *Unequal Protection.* He gives a history of how this change occurred.
9. Gwen Braxton and Mary Zepernick, "Racism and Global Corporations," in the "Organizing Packet for the Campaign to Abolish Corporate Personhood," by the Women's International League for Peace and Freedom. See http://www.wilpf.org/corp/ACP/acp-packet.pdf. And http://www.wilpf.org/docs/ccp/corp/ACP/CP_article%2Btimeline.pdf.
10. Molly Morgan, "Abolish Corporate Personhood," Women's International League for Peace and Freedom. This speech is available for anyone to give. See p. 5f, http://www.wilpf.org/CPOWER_abolish.
11. "Abolish Corporate Personhood, *US section of the Women's International League for Peace and*

    Freedom (WILPF). http://www.wilpf.org/docs/ccp/corp/ACP/CP_article%2Btimeline.pdf.
12. Buckly v. Valeo (1976), First National Bank of Boston v. Bellotti (1978), Central Hudson Gas & Electric v. Public Service Commission of New York (1980), and Pacific Gas & Electric v. Public Utilities Commission (1986).
13. David C. Korten, *Living Economies for a Living Planet Part II: Corporate Pathology and the Suicide Economy.* See http://www.pcdf.org/Living_Economies/I_Intro.htm.
14. Morgan, "Abolish Corporate Personhood."
15. Herman E. Daly, "When Protectionism Is a Good Thing," PCD Forum Column #67, January 25, 1994. See http://www.pcdf.org/1994.htm and http://globalpolicy.org/component/content/article/209-bwi-wto/42749.html.
16. Daly, "When Protectionism Is a Good Thing.
17. Vandana Shiva, "The Polarised World of Globalisation," May 28, 2005 (A response to Thomas Friedman's Flat Earth Hypothesis). See http://www.zmag.org/sustainers/content/2005-05/27shiva.cfm.
18. See "Trading Democracy" on Bill Moyers' NOW program. Transcript is at http://www.pbs.org/now/printable/transcript_tdfull_print.html.
19. Ibid.
20. Noreen Hertz, *The Silent Takeover*, p. 92.
21. Sam Ferguson, Jason Leopold, and Kyle Berlin, "Supreme Court Shreds Campaign-Finance Laws, Lifts Corporate Spending Restrictions" January 21, 2010 http://www.truthout.org/supreme-court-decision-radically-overhauls-campaign-finance-laws-favor-corporations56261.
22. Ibid.
23. Ruth Marcus, "Court's campaign finance decision a case of shoddy scholarship" January 23, 2010. http://www.washingtonpost.com/wp-dyn/content/article/2010/01/22/AR2010012203897.html.
24. Greg Palast, "The Supreme Court Just Handed Anyone, Including bin Laden or the Chinese Government, Control of Our Democracy" *AlterNet* January 22, 2010, http://www.truthout.org/the-supreme-court-just-handed-anyone-including-bin-laden-or-chinese-govt-control-our-democracy56332.
25. Ibid.
26. Michael Waldman, "Campaign Finance Ruling Reflects Supreme Court's Growing Audacity" *January 22, 2010. The Washington Post* http://www.commondreams.org/view/2010/01/22-10.
27. Adam Liptak, "Justices, 5-4, Reject Corporate Spending Limit" January 21, 2010 *New York Times.* http://www.nytimes.com/2010/01/22/us/politics/22scotus.html.
28. Paul Krugman, "In Media Res," *New York Times,* November 29, 2002. http://www.nytimes.com/2002/11/29/opinion/29KRUG.html.
29. *Merging Media,* April 26 2002.See http://www.pbs.org/newshour/bb/media/jan-june02/merging_6-25.html.
30. Telecommunications Act of 1996. See http://www.fcc.gov/telecom.html.
31. Media Ownership and Deregulation, http://www.journalism.org/node/1751.
32. John Nichols, "As Not Seen on TV," February 24, 2003. See http://www.thenation.com/doc.mhtml?i=20030224&s=nichols.

33. "Who owns the media?" http://www.stopbigmedia.com/chart.php.

34. D.C. Circuit Court decision, Meredith Corp. v. FCC. The full history of the Fairness Doctrine is at http://www.museum.tv/archives/etv/F/htmlF/fairnessdoct/fairnessdoct.htm. For a more complete history of Media Regulations, see NOW, on PBS http://www.pbs.org/now/politics/mediatimeline.html.

35. Ross Gelbspan, "Straddling Solutions and Survivalism," lecture at Massachusetts Climate Action Network Conference, November 19, 2006. See http://www.heatisonline.org/contentserver/objecthandlers/index.cfm?id=6161&method=full.

36. Bill Moyers, "Is the Fourth Estate a Fifth Column," In These Times, July 11, 2008. See http://www.truthout.org/article/is-fourth-estate-a-fifth-column.

37. Ibid.

38. FAIR (Fairness & Accuracy In Reporting) "Media Blackout on Single-Payer Healthcare," http://www.fair.org, March 7, 2009. http://www.commondreams.org/view/2009/03/07-6.

39. Ben H. Bagdikian, *The Media Monopoly,* 6th ed., p. 216.

40. Ibid, p. 138.

41. David Cromwell, *The Propaganda Model: An Overview.* See http://www.medialens.org/frameset_articles.html. An extract from the "Spotlight on the Media," Chapter 3 of *Private Planet.* See http://www.private-planet.com.

42. Michael Copps, "Copps: Broadband and Media Threats 'Joined at the Hip,'" from *Federal Communications Commission*, August 3, 2007, by http://www.freepress.net/news/25181. and http://www.savetheinternet.com.

43. Paul Krugman, "Digital Robber Barons?" *New York Times* op-ed, December 6, 2002. http://www.nytimes.com/2002/12/06/opinion/06KRUG.html.

44. See http://www.jubileeusa.org/start.htm. In the Bible, ancient Israel was instructed by God to celebrate a jubilee—a year of the Lord—every 50 years. During this year, slaves were set free, land was redistributed, and debts were forgiven.

45. Soren Ambrose, "Multilateral Debt: The Unbearable Burden," *Foreign Policy in Focus,* Vol. 6, No. 37, November 2001. See http://www.fpif.org/briefs/vol6/v6n37multidebt.html.

46. Rick Rowden, "A World of Debt," *The American Prospect,* July 16, 2001. http://www.prospect.org/cs/articles?article=a_world_of_debt.

47. Ibid.

48. John Cavanagh and Jerry Mander, *Alternatives to Economic Globalization,,* p. 37.

49. Rowden, "A World of Debt."

50. Ibid.

51. Sada Diarra, Statement of the Jubilee Movement International For Economic and Social Justice http://www.jubileeresearch.org/jmi/jmi-policies/statement_bamako_english.htm.

52. 50 Years Is Enough. See http://50years.org. These demands have been formulated by the 50 Years Is Enough Network, a U.S.-based coalition of more than 200 organizations committed to the fundamental transformation of the IMF and World Bank, through consultations with the members of its South Council (representing economic justice organizations in 13 countries in Asia, Africa, Latin America, and the Caribbean) and others.

53. These organizations include Interamerican Regional Organization of Workers (ORIT, which includes national labor federations, such as the AFL-CIO and the Brazilian CUT) the Latin American Congress of Rural Organizations (CLOC), the Mexican Action Network on Free Trade (RMALC), ART and REBRIP, Convergence of People's Movements of the Americas (COMPA), Grito de los Excluidos, and Friends of the Earth Latin America.

54. Alternatives for the Americas. See http://www.globalexchange.org/campaigns/alternatives/americas or http://www.asc-hsa.org/pdf/altering2.pdf.

55. Deborah James, "A Decade's Struggle Ends in Victory," *Global Exchange*, May 15, 2005. See http://www.globalexchange.org/campaigns/ftaa/3070.html.

56. http://www.globalexchange.org/campaigns/alternatives/americas, and http://www.asc-hsa.org/pdf/altering2.pdf.

57. Friends of the Earth International, "What is Ecological Debt?" See http://www.foei.org/publications/ecodebt/statement.html.

58. Andrew Simms, "Ecological Debt: Global Warming's Victims Could Take U.S. to Court," *The International Herald Tribune,* August 7, 2001. See http://www.iht.com/articles/2001/08/07/edsimms_ed2_.php.

59. Thom Hartmann, *Unequal Protection: The Rise of Corporate Dominance and the Theft of Human Rights.*

60. Margery Kelly, "Enron and the Myths of Runaway Capitalism," *Business Ethics,* February 20, 2002. http://www.alternet.org/story/12447.

61. Ibid, pp. 98–99.

62. Ibid, p. 58, Italics added.

63. Cavanagh and Mander, *Alternatives to Economic Globalization,* pp. 140–141.

64. Margaret Kelly, *The Divine Right of Capital: Dethroning the Corporate Aristocracy,* p. 29.

65. Ibid, p. 24.

66. Ibid, p. 96.

67. Ibid, p. 2.

68. Ibid, p. xi.

69. Peter Drucker, *The Post-Capitalist Society,* p. 92.

70. Ibid, p. 77.

71. Paul Krugman, "For Richer," *New York Times Magazine,* October 20, 2002. See http://www.nytimes.com/2002/10/20/magazine/for-richer.html.

72. Ibid.

73. Ibid.

74. Ibid.

75. Robert H. Dahl, *A Preface to Economic Democracy*, p. 162.

76. Ibid, pp. 162–163.
77. Quoted in David Korten, "Revitalizing Democracy, Transforming Society" Keynote Address to Revitalizing Democracy Conference, Duluth, July 28–30, 2002. http://www.davidkorten.org/revitalizingdemocracy.
78. Elizabeth Debold, "The Business of Saving the World," *What is Enlightenment Magazine,* March–May 2005, p. 71. See http://www.wie.org/j28/business.asp. See also http://www.wie.org/business.

Chapter 7

1. Lionel Hurst, "Small Island Countries' Dialogue on Water and Climate," Address to the World Water Forum, Kyoto, Japan, March 16–23, 2003.
2. George Lakoff, *Whose Freedom,* pp. 119–120.
3. Ibid. p. 121.
4. Ibid, pp. 43, 129.
5. Ibid, p. 23.
6. George Lakoff, *Don't Think of an Elephant,* p. 16.
7. Ibid, p. 28.
8. Ibid, pp. 15–16.
9. George Lakoff, *The Political Mind*, pp. 47–48.
10. Ibid, pp. 47–48.
11. Bill McKibben, "The Christian Paradox," *Harper's,* September 15, 2005, at http://www.harpers.org/ExcerptTheChristianParadox.html. See also http://www.billmckibben.com/.
12. Doug Pibel and Sarah van Gelder, "Health Care: It's What Ails Us," *Yes Magazine*, posted Jul 19, 2006. http://www.yesmagazine.org/issues/health-care-for-all/health-care-its-what-ails-us. This article says, "Single payer" systems can cover their entire populations at half of our per capita cost. The United States spends by far the most on health care per person—more than twice as much as Europe, Canada, and Japan which all have some version of national health insurance. Yet we are near the bottom in nearly every measure of our health. The World Health Organization (WHO) ranks the U.S. health care system 37th of 190 countries, well below most of Europe, and trailing Chile and Costa Rica. The United States does even worse in the WHO rankings of performance on level of health—a stunning 72nd. Life expectancy in the U.S. is shorter than in 27 other countries; the U.S. ties with Hungary, Malta, Poland, and Slovakia for infant mortality—ahead of only Latvia among industrialized nations. For information on the myths and realities of health care, see John R. Battista, M.D. and Justine McCabe, Ph.D., "The Case For Single Payer, Universal Health Care For The United States" *June 4, 1999.* http://cthealth.server101.com/the_case_for_universal_health_care_in_the_united_states.htm.
13. Ibid, p
14. Lakoff, *Whose Freedom*, p. 56, 65–66, 140–141.

    Lester Brown, *Plan B 4.0: Mobilizing to Save Civilization.* pp. 134–135, 137.
15. Ibid, p. 190.
16. *State of the World 2000*, p. 16.
17. Population Clocks. See http://www.census.gov/ipc/www/popclockworld.html.
18. *The World at Six Billion.* http://www.un.org/esa/population/publications/sixbillion/sixbilpart1.pdf. World Abortion Policies 2007 provides the most up-to-date, accurate and objective information available on the legal status of induced abortion for the 195 Member and non-Member States of the United Nations. http://www.un.org/esa/population/publications/2007_Abortion_Policies_Chart/2007AbortionPolicies_wallchart.htm
19. 2002 Earth Policy Institute. See http://www.envirosagainstwar.org/sayno/UNEP-Toepher.pdf. UNDP poverty page: www.undp.org/poverty. Describes UNDP poverty alleviation programs and initiatives, publications, and good practices and facilitates poverty and related discussions and events. World Bank poverty website: www.worldbank.org/poverty.
20. "World Population to Exceed 9 Billion By 2050," http://www.un.org/esa/population/publications/wpp2008/pressrelease.pdf.
21. Ibid, p. 3.
22. Richard Cincotta, Global Security Brief #2, Youth Bulge, Unemployment Raise Risks of Civil Conflict, *Worldwatch* Global Security Briefs. http://www.worldwatch.org/node/76.
23. Ibid.
24. Cristina Page, "What Pro-Choice Presidential Candidates Should Be Saying," *Huffington Post*, August 14, 2007. See http://www.commondreams.org/archive/2007/08/14/3167/.
25. Interview with Cristina Page, author of *How the Pro-Choice Movement Saved America*: *Freedom, Politics and the War on Sex*, http://www.truthout.org/article/did-pro-choice-movement-save-america.
26. Page, *How the Pro-choice Movement Saved America.*
27. "Birth Control Foe to Head Family Planning" Oct. 18, 2007, http://www.cbsnews.com/stories/2007/10/18/health/main3380290.shtml.
28. Page, *How the Pro-choice Movement Saved America*, p. 29.
29. Ibid, p. 96.
30. Page, "The Fight to Stop Contraception," Published on Friday, June 6, 2008, by *The Capital Times* (Madison). See http://www.commondreams.org/archive/2008/06/04/9455/.
31. Page, *How the Pro-choice Movement Saved America.*
32. ACLU's Reproductive Freedom Project. See http://www.aclu.org/images/asset_upload_file576_29402.pdf.
33. Laurie Goodstein, "Disowning Conservative Politics, Evangelical Pastor Rattles Flock," *New York Times,* July 30, 2006. http://www.nytimes.com/2006/07/30/us/30pastor.html.
34. Brown, *Plan B 4.0.: Mobilizing to Save Civilization.* p. 211
35. Dave Lindorff, "Your Tax Dollars at War: More Than 53% of Your Tax Payment Goes to the Military" *April 13,*

2010 http://www.commondreams.org/view/2010/04/13-4. See War Resisters League Pie Chart Flyer – "Where Your Income Tax Money Really Goes," http://www.warresisters.org/federalpiechart. See also Dean Baker "Massive Defense Spending Leads to Job Loss" November 11, 2009 truthout http://www.truthout.org/1109097

36. Dave Lindorff, "Your Tax Dollars at War See also Carl Conetta, PDA (Project on Defense Alternatives) Briefing Report #20, January 18, 2010, "An Undisciplined Defense: Understanding the $2 Trillion Surge in US Defense Spending" Executive Summary is at http://www.comw.org/pda/fulltext/1001PDABR20exsum.pdf, and the Full Report is at http://www.comw.org/pda/fulltext/1001PDABR20.pdf.

37. Carl Conetta, "How Much Does the Pentagon Need? A Quick Guide to US Defense Spending," Project on Defense Alternatives February 15, 2010. http://www.comw.org/pda/1002BudgetSurge.html.

38. Ibid.

39. Ibid.

40. Ibid.

41. Tom Engelhardt, "Believe It or Not (2010 Imperial Edition) --- U.S. War-Fighting Numbers to Knock Your Socks Off" *April 7, 2010 by* TomDispatch.com http://www.commondreams.org/view/2010/04/07

42. The Editors, *Monthly Review* "U.S. Military Bases and Empire" http://www.monthlyreview.org/0302editr.htm

43. Ibid.

44. Ibid.

45. Wikipedia "Hijackers in the September 11 attacks," http://en.wikipedia.org/wiki/Hijackers_in_the_September_11_attacks. Of the other 4 hijackers, two were from the United Arab Emirates, one from Lebanon, and one from Egypt.

46. The Editors, *Monthly Review* "U.S. Military Bases and Empire" http://www.monthlyreview.org/0302editr.htm

47. Ibid.

48. Carl Conetta, "A Prisoner to Primacy." Project on Defense Alternatives, Briefing Memo #43 February 5, 2008. http://www.comw.org/pda/0802bm43.html.

49. Ibid.

50. *Chalmers Johnson,* "737 U.S. Military Bases = Global Empire" Excerpted from Chalmers Johnson's book, *Nemesis: The Last Days of the American Republic, February 19, 2007* http://www.alternet.org/story/47998?page=entire.

51. *Ibid.*

52. Conetta, "A Prisoner to Primacy."

53. Carl Conetta, "Dissuading China and Fighting the 'Long War'" *World Policy Journal,* Summer 2006. http://www.comw.org/pda/fulltext/06conetta.pdf.

54. Ibid.

55. Conetta, "A Prisoner to Primacy."

56. Conetta, "Dissuading China and Fighting the 'Long War.'"

57. Ibid.

58. Ibid.

59. Carl Conetta, "Pyrrhus on the Potomac: How America's Post-9/11 Wars have Undermined US National Security" Project on Defense Alternatives, Briefing Report #18 September 05, 2006. http://www.comw.org/pda/fulltext/0609br18.pdf.

60. The Editors, "U.S. Military Bases and Empire" *Monthly Review* http://www.monthlyreview.org/0302editr.htm. See testimony by the Unocal Corporation to the House Committee on International Relations in February 1998 (reprinted as "A New Silk Road: Proposed Pipeline in Afghanistan" in *Monthly Review*, December 2001, http://www.serendipity.li/wot/wsap212982.htm.) The history of Unocal's Central Asian pipeline project is also discussed in detail in Ahmed Rashid, *Taliban* (New Haven: Yale University Press, 2000), 151-80.

61. Transcript/film clip of Thomas Friedman's Addicted to Oil documentary, Jun 30, 2006 (*From the CalCars-News archive*). http://www.calcars.org/calcars-news/462.html

62. Ibid.

63. Ibid.

64. Ibid.

65. Ibid.

66. Conetta, "Dissuading China and Fighting the 'Long War.'"

67. Harry Mika and Howard Zehr, "Kaleidoscope of Justice" and "Restorative Justice Signposts." See http://www.bostontheological.org/rec/rj.htm.

68. Ibid.

69. Kay Pranis, "Building Community Support for Restorative Justice—Principles and Strategies" by Pranis is Restorative Justice Planner, Minnesota Department of Corrections, See http://www.restorativepractices.org/library/building.html. For links to Restorative Justice sites, see http://rjp.umn.edu/links.html. Some define it with a religious basis as the Boston Theological Institute does; see http://www.bostontheological.org/programs/transforming_and_restorative_justice.htm and "Four Terms that Stand Together: Reconciliation – Forgiveness –Restorative Justice– the Church," at http://www.bostontheological.org/programs/four-terms.pdf. Tom Cavanagh, "What is Restorative Justice?" http://www.bostontheological.org/programs/transforming_and_restorative_justice.htm .See See also *Schooling for Peace: Using* "Restorative Justice and School Violence: Building Theory and Practice," at http://www.realjustice.org/library/morrison_bullying.html.

70. David Miers and Jolien Willemsens, editors, *Mapping Restorative Justice: Developments in 25 European Countries,* 2006. See http://www.restorativejustice.org/editions/2006/july06/mapping.

71. The Center for Restorative Justice at Suffolk University. See http://www.suffolk.edu/research/6953.html.

72. Cavanagh, "What is Restorative Justice?"

73. Harry Mika and Howard Zehr,"Kaleidoscope of Justice" and "Restorative Justice Signposts." See http://www.bostontheological.org/rec/rj.htm.

74. See http://webmail.bostontheological.org/reconciliation_

peace_and_justice.html.

75. John Jay College of Criminal Justice: An international leader in educating for justice, see http://www.jjay.cuny.edu/.

76. 'Stop Snitching' Movement Confounding Criminal Justice May 8, 2008 Listen to the Story News & Notes May 8, 2008. National Public Radio. http://www.npr.org/templates/story/story.php?storyId=90280108

77. Ibid.

78. Suzanne Smalley, "Always on My Mind David Kennedy's obsession with drug dealers has made him a highly sought-after criminologist." *Newsweek* Feb 9, 2009. http://www.newsweek.com/2009/01/30/always-on-my-mind.html.

Chapter 8

1. David Korten, *Investing in Planetary Health and Human Survival*. http://www.pcdf.org/2000/Social%20Investment%20Forum.htm.

2. Herman E. Daly, "When Protectionism Is A Good Thing," The People-Centered Development Forum PCDForum Column #67, January 25, 1994 http://www.pcdf.org/1994/67daly.htm.

3. David C. Korten, *Agenda for a New Economy: From Phantom Wealth to Real Wealth,* p. 33. There is a new edition available from http://www.davidkorten.org/NewEconomyBook.

4. Bernard Lietaer, Dr. Robert Ulanowicz, and Dr. Sally Goerner , "White Paper on All the Options for Managing a Systemic Bank Crisis," http://www.lietaer.com/images/White_Paper_on_Systemic_Banking_Crises_final.pdf, p. 13.

5. Ibid.

6. Ibid, p. 14.

7. Ibid, p. 14.

8. Ibid, p. 14.

9. Ibid, p. 14.

10. Ibid, p. 14.

11. Ibid, p. 15.

12. Ibid, p. 15.

13. Ibid, p. 16.

14. Paul Hawken, "Natural Capitalism," *Mother Jones*, March/April 1997. http://motherjones.com/politics/1997/03/natural-capitalism

15. Lester R. Brown, *Plan B: Rescuing a Planet under Stress and a Civilization in Trouble*, Chapter 11. http://www.earth-policy.org/Books/Eco_contents.htm.

16. Ibid.

17. Lester Brown, Janet Larsen, Jonathan G. Dorn, and Frances Moore, *Time for Plan B: Cutting Carbon Emissions 80 Percent by 2020.* The full report is available in pdf without endnotes http://www.earthpolicy.org/Books/PB3/80by2020.pdf or with endnotes http://www.earthpolicy.org/Books/PB3/80by2020notes.pdf

18. Robert Kennedy, Jr., in "Global Warming: A Real Solution," *Rolling Stone Magazine*. June 18, 2007. http://www.rollingstone.com/politics/story/15051506/global_warming_a_real_solution.

19. Ibid.

20. Ibid.

21. George Lakoff, "Oil and Betrayal in Iraq," *Truthout Perspective*, September 20, 2007. http://blog.buzzflash.com/contributors/1315

22. James Hansen, "Copenhagen has given us the chance to face climate change with honesty," *The Observer,* December 27, 2009. http://www.guardian.co.uk/environment/2009/dec/27/james-hansen-copenhagen-agreement-opportunities.

23. Ibid.

24. James Hansen, "G-8 Failure Reflects U.S. Failure on Climate Change," July 9, 2009. http://www.huffingtonpost.com/dr-james-hansen/g-8-failure-reflects-us-f_b_228597.html. See H.R. 2454**:** American Clean Energy and Security Act of 2009, http://www.opencongress.org/bill/111-h2454/show.

25. Lisa Lerer, "Senate Democrats to W.H.: Drop Cap-and-Trade" December 27, 2009, Politico.com. http://www.commondreams.org/headline/2009/12/27-0.

26. Steve Connor, "Warning: Oil Supplies Are Running Out Fast" *August 3, 2009 The Independent/UK* http://www.commondreams.org/headline/2009/08/03.

27. Jeff Rubin, *Why Your World is about to Get a Whole Lot Smaller: Oil and the End of Globalization,* p. 88.

28. Ibid, p. 38.

29. Ibid, p. 46.

30. Ibid, p. 42.

31. Ibid, pp. 48, 50.

32. Robert Kunzig, "The Canadian Oil Boom" *National Geographic*, March 2009. http://ngm.nationalgeographic.com/2009/03/canadian-oil-sands/kunzig-text.

33. Ibid, p. 42.

34. Rubin, *Why Your World is about to Get a Whole Lot Smaller,* p. 35.

35. Ibid, pp. 66–68.

36. Hansen, "G-8 Failure Reflects U.S. Failure on Climate Change."

37. Ibid, pp. 158–166.

38. Ibid, pp. 158–166.

39. Ibid, pp. 169–170.

40. Ibid, pp. 174–175.

41. Ibid, p. 176.

42. Ibid, p. 21.

43. Bernard Lietaer,*The Future of Money* p. 144.

44. Herman Daly, "Globalization and Its Discontents." *Philosophy and Public Policy Quarterly*, Vol. 21, No. 2/3, Spring/Summer 2001. See http://www.publicpolicy.umd.edu/IPPP/reports/Spring-Summer%20Vol21%202001/221056.pdf.

45. Bernard Lietaer, pp.59–60, using Harlan Cleveland.

Leadership and the Information Revolution, Minneapolis: World Academy of Art and Science, 1970, and *The Knowledge Executive: Leadership in Information Society,* NY: Truman Tally Books/E. P. Dutton, 1985, Howard Rheingold, *Virtual Reality and Virtual Community,* N.Y.: Harper Perennial, 1993

46. Ibid, pp. 532–533.

47. Ibid, p. 528.

48. Jeremy Rifkin, *The Empathic Civilization: the Race to Global Consciousness in a World in Crisis,* pp. 528–529.

49. Ibid, pp. 531–532.

50. Ibid, p. 529.

51. Ibid, pp. 530–531.

52. Ibid, p. 534.

53. Rifkin, *The Empathic Civilization*, p. 533.

54. Ibid, p. 534. See also *Wikinomics*, by Don Tapscott and Anthony D. Williams.

55. Ibid, pp. 531–532.

56. Ibid, p. 543.

57. Herman Daly, *Beyond Growth*, p. 18,

58. Daly, "Globalization and Its Discontents," p. 19.

59. Ibid.

60. The first three are from Herman Daly, "Globalization and Its Discontents," p. 19, *Philosophy and Public Policy Quarterly*, Vol. 21, No. 2/3, Spring/Summer 2001. The rest are from John Cavanagh and Jerry Mander, *Alternatives to Economic Globalization.* p. 19.

61. John Cavanagh, Jerry Mander, *Alternatives to Economic Globalization,* p. 28.

62. Ibid, p. 28.

63. Ibid, p. 28.

64. Ibid, p. 29.

65. Daly, "When Protectionism Is A Good Thing."

66. Ibid.

67. Cavanagh and Mander, *Alternatives to Economic Globalization* p. *19*

68. David Korten "Living Wealth: Better Than Money" *Yes! Magazine* Fall 2007. http://www.yesmagazine.org/article.asp?ID=1834

69. Cavanagh and Mander, *Alternatives to Economic Globalization,* p. 107

70. Ibid, p. 107

71. Ibid, p. 108

72. Michael Schuman, *Going Local: Creating Self-Reliant Communities in a Global Age*, pp. 49–52.

73. Cavanagh and Mander, *Alternatives to Economic Globalization,* p. 25

74. In *Cities and the Wealth of Nations,* Jane Jacobs distinguishes between import substitution and import replacement. Import substitution is when a region replaces imports with goods it already produces. Import replacement is when a region replaces imports with goods it can now make for itself. See http://www.planning.unc.edu/courses/261/drucker/.

75. Shuman, *Going Local,* pp. 83–90.

76. Ted Howard, "Ownership Matters" *Yes!* Magazine Spring 1999. At http://www.yesmagazine.org/article.asp?ID=776.

77. David Korten, "Economies for Life" *Yes!* Fall 2002 p. 14 http://www.yesmagazine.org/issues/living-economies/economies-for-life.

78. Howard, "Ownership Matters."

79. Business Alliance for Local Living Economies (BALLE) http://www.livingeconomies.org/

80. Korten, "Economies for Life."

81. Korten, *Agenda for a New Economy*, p. 109.

82. Ibid.

83. Herman Daly, "The Steady-State Economy: Postmodern Alternatives to Growthmania" in *Spirituality*, David Ray Griffin, ed., p. 110.

84. Herman E. Daly, "Economics in a Full World," *Scientific American,* September 2005, http://www.beyondpeak.org/files/ScientificAmerican_Daly_05.pdf.

85. Robert H. Nelson, *Economics as Religion*, pp. xv, 43.

86. Ibid, p. 65.

87. Ibid, p. 68.

88. Daly, *Beyond Growth.*

89. Genuine Progress Indicator: What's Wrong With The GDP As A Measure Of Progress http://www.rprogress.org/projects/gpi/

90. Global Footprint Network. http://www.footprintnetwork.org/en/index.php/GFN/page/ecological_debtors_and_creditors/.

91. GPI- http://www.rprogress.org/sustainability_indicators/genuine_progress_indicator.htm. Download report: *The Genuine Progress Indicator 2006* at http://www.rprogress.org/publications/2007/GPI%202006.pdf. They also have community and regional indicators at http://www.rprogress.org/sustainability_indicators/about_sustainability_indicators.htm.

92. Environmental Justice and Climate Change Initiative, http://www.ejcc.org. See J. Andrew Hoerner and Nia Robinson, "A Climate of Change: African Americans, Global Warming, and a Just Climate Policy in the U.S." Text at http://www.rprogress.org/publications/2008/climateofchange.pdf.

93. The New Economics Foundation, Centre for Well-being, www.neweconomics.org/programs/well-being and http://www.happyplanetindex.org/learn/.

94. Korten, *Agenda for a New Economy*, p. 99.

95. Alan Durning, *How Much Is Enough*, New York: W.W. Norton & Company, 1992. pp. 25 and 41.

96. "Rethinking the Good Life" http://www.worldwatch.org/features/consumption/sow/trendsfacts/2004/09/15/

97. Daly, "Economics In a Full World."

98. Ibid.

99. Quoted in "Boulding, Kenneth Ewart." In: *Encyclopedia of Earth*. Eds. Cutler J. Cleveland (Washington, D.C.: Environmental Information Coalition, National Council for Science and the Environment). http://www.eoearth.org/article/Boulding,_Kenneth_Ewart.

100. Herman Daly, *Steady-State Economics*. Chapter 5: A Catechism of Growth Fallacies. First published in 1977, this volume caused a sensation because of Daly's radical view that "enough is best." Today, his ideas are recognized as the key to sustainable development, and Steady-State Economics is universally acknowledged as the leading book on the economics of sustainability. There is also *Steady-State Economics:* Second Edition with New Essays 1991. http://www.worldpolicy.org/globalrights/environment/daly.html.

101. Ibid.

102. Daly, *Steady-State Economics*. Article was in *Fortune*, 1976, p. 116.

103. Daly, "Economics in a Full World."

104. Herman Daly "Five policy Recommendations for a Sustainable Economy" http://www.feasta.org/documents/feastareview/daly2.htm.

105. Herman Daly, "Towards a New Economics: Questioning Growth" http://www.worldpolicy.org/globalrights/ecoindex.html. See also Daly, Herman E. and Kenneth N. Townsend "Valuing the Earth: Economics, Ecology, Ethics" (1993) http://www.worldpolicy.org/globalrights/ecoindex.html, and Daly, Herman. *Steady-State Economics* Chapter 5: A Catechism of Growth Fallacies, http://eebweb.arizona.edu/courses/Ecol206/DalyHermanSteady-StateEconomics.pdf.

106. Daly, "Towards a New Economics: Questioning Growth."

107. Daly, *Beyond Growth*, p. 1

108. Bertrand "In Praise of Idleness." Quoted in Daly "Towards a New Economics: Questioning Growth" http://www.worldpolicy.org/globalrights/ecoindex.html.

109. Herman Daly, *Steady-State Economics,* Chapter 5.

110. Ibid.

111. Ibid.

112. David Korten, "Economics for Life" Yes Magazine, Fall 2002.

113. Daly, *Beyond Growth*, p. 7

114. Korten, *Agenda for a New Economy*, p. 10

Chapter 9

1. Sarah Van Gelder, "Dialogue: Beyond Greed & Scarcity." *YES! Magazine*! Spring 1997, Interview with Bernard Lietaer. http://www.yesmagazine.org/issues/money-print-your-own/beyond-greed-and-scarcity.

2. Ibid.

3. See Bernard Lietaer's website: http://www.transaction.net

4. David Korten, *Agenda for a New Economy*, p. 1.

5. Roger Langrick, *A Monetary System for the New Millennium* http://www.worldtrans.org/whole/monetarysystem.html.

6. Ibid.

7. Ibid.

8. Korten, *Agenda for a New Economy*, p. 23.

9. Ellen Brown, "Borrowing from Peter to Pay Paul: The Wall Street Ponzi Scheme Called Fractional Reserve Banking" December 29, 2008. www.webofdebt.com/articles/ponzi.php. Ellen Brown's websites are www.webofdebt.com, www.ellenbrown.com, and www.public-banking.com.

10. Ellen Brown, "Dollar Deception: How Banks Secretly Create Money," July 3rd, 2007. http://www.webofdebt.com/articles/dollar-deception.php. Quoting Board of Governors of the Federal Reserve, "M3 Money Stock" (discontinued series), http://research.stlouisfed.org/fred2/data/M3SL.txt.

11. Brown, "Borrowing from Peter to Pay Paul."

12. Bernard Lietaer, *The Future of Money,* p. 54. The full text now online at http://www.transaction.net/money/book.

13. Ibid.

14. Brown, "Borrowing from Peter to Pay Paul."

15. Ibid.

16. Ellen Brown, *Web of Debt,* p. 14.

17. Ellen Brown, "Sustainable Government: Banking For A "New" New Deal" December 8th, 2008. www.webofdebt.com/articles/newdeal.php

18. Ibid.

19. Brown, *Web of Debt*, p. 7.

20. Ellen Brown, "Putting the "Federal" Back in the Federal Reserve," July 26, 2008. http://www.webofdebt.com/articles/federal-back-fr.php.

21. Ibid.

22. Robert Owen, *The Federal Reserve Act* (1919); "Who Was Philander Knox?" www.worldnewsstand.net/history/PhilanderKnox.htm. (1999). Quoted in Ellen Brown, "Borrowing from Peter to Pay Paul."

23. Brown, "Putting the 'Federal' Back in the Federal Reserve."

24. Brown, *Web of Debt*, p. 24.

25. Brown, "Putting the 'Federal' Back in the Federal Reserve" (see also Ellen Brown, "The Secret Bailout of JPMorgan," May 13, 2008, www.webofdebt.com/articles; and "What's the Difference Between Lehman Brothers and Bear Stearns?" June 14, 2008, http://www.webofdebt.com/articles/lehman_brothers.php.)

26. Matt Taibbi, "The Big Takeover: The Global Economic Crisis isn't about Money—It's about Power. How Wall Street Insiders Are using the Bailout to Stage a Revolution," March 19, 2009, http://www.rollingstone.com/politics/story/26793903/the_big_takeover

27. David Korten, "Beyond the Bailout: Agenda for a New Economy" *Yes! Magazine,* Winter 2009, *http://www.yesmagazine.org/article.asp?ID=3050*

28. Brown, "Dollar Deception: How Banks Secretly Create Money."

29. Bernard Lietaer, Dr. Robert Ulanowicz , and Dr. Sally

Goerner, "White Paper on All the Options for Managing a Systemic Bank Crisis" http://www.lietaer.com/images/White_Paper_on_Systemic_Banking_Crises_final.pdf. p. 5.

30. Brown, "Borrowing from Peter to Pay Paul."
31. Korten, *Agenda for a New Economy,* p. 139.
32. Ibid, p. 142.
33. Ibid, p. 143.
34. Ellen Brown, "Revive Lincoln's Monetary Policy: an Open Letter to President Obama," *YES! Magazine* April 9, 2009. http://www.yesmagazine.org/article.asp?ID=3394
35. Ibid.
36. Ellen Brown, "Bank On It: How Cash-Starved States Can Create Their Own Credit." http://www.yesmagazine.org/article.asp?ID=3361. For the North Dakota bank, see www.banknd.nd.gov. See also Ellen Brown, "The Growing Movement for Publicly Owned Banks" March 18, 2010 *YES! Magazine*. http://www.commondreams.org/headline/2010/03/18-7.
37. Matthew Cardinale, "US States Consider Starting Their Own Banks" April 30, 2010, *Inter Press Service,* http://www.truthout.org/us-states-consider-starting-their-own-banks59084.
38. Lietaer, *The Future of Money,* p. 116.
39. Ibid, pp. 145–146.
40. Ron Morrison, "Keynes without Debt," www.prosperityuk.com/prosperity/articles/keynes.html, April 2004. Quoted in Ellen Brown, "Borrowing from Peter to Pay Paul."
41. Lietaer, "White Paper," p. 17.
42. Lietaer, *The Future of Money,* pp. 33 and 41.
43. Lietaer, *The Future of Money,* pp. 33 and 41. Lietaer says money has sometimes played three other roles in the past and happens to play them today as well:
    1) A store of value. Historically, this has only rarely been the case. For example, the word capital derives from the Latin capus, capitis, which means head, and referred to heads of cattle just as is still done today in Texas or among the Tutsi in Africa: "He is worth 1,000 head." Another example: in Egypt through the Late Classical period, and in Europe throughout the Middle Ages and until the late eighteenth Century, wealth was stored mainly in land and its gradual improvements.
    2) A tool for speculative profit, most emphatically today when more than 95 percent of all currency transactions in the world are motivated by speculation, and less than 5 percent are for trades of goods and services. This has been systematically possible only since August 1972, when President Nixon created the floating currency nonsystem we now have.
    3) A tool of empire. The control by the former Soviet Union of the external trade of Comecon countries via the "convertible ruble" is a recent example.
    4) These changes have lead to the destruction of the environment, as well as a very unstable economic system and a drive to eliminate as many jobs as possible. Bernard Lietaer's essay, "Community Currencies: A New Tool for the 21st Century."
44. Ibid, p. 5.
45. Richard Douthwaite, *Short Circuit, Strengthening Local Economies for Security in an Unstable World,* Lilliput Press, 1996. Also at http://www.feasta.org/documents/shortcircuit/index.html?sc4/share.html
46. Michael Shuman, *Going Local: Creating Self-Reliant Communities in a Global Age,* p. 50.
47. Edgar Cahn, "Money That Builds Connection" *YES! Magazine,* Spring 1997, http://cms.yesmagazine.org/issues/money-print-your-own/889.
48. Time Banks, http://www.timedollar.org/
49. Lietaer, "Beyond Greed and Scarcity."
50. Ibid.
51. Ibid.
52. Lietaer, "White Paper."
53. Ibid, p. 18.
54. Ibid, p. 19.
55. Ibid, p. 16.
56. Ibid, p. 20.
57. Ibid, p. 21.
58. Ibid, p. 24.
59. Ibid, p. 22.
60. Citizens for Tax Justice, "The Bush Tax Cuts Cost Two and a Half Times as Much as the House Democrats' Health Care Proposal," September 8, 2009, http://www.ctj.org/pdf/bushtaxcutsvshealthcare.pdf.

Chapter 10

1. Joanna Macy, "The Great Turning," Interview, *Yes! Magazine,* Spring 2000.
2. Ibid.
3. Ibid.
4. Thomas Berry, "An Ecologically Sensitive Spirituality," *Earth Ethics,* Fall 1996. http://center1.com/earth-ethics-1996-fall.html.
5. Thomas Moore, *Care of the Soul,* p. xi.
6. Ibid, p. xiv.
7. Ibid, p. 32.
8. Ibid, p. 168.
9. Ibid, p. 304.
10. Ibid, pp. 252–253.
11. Ibid, p. 17.
12. Matthew Fox, *The Coming of the Cosmic Christ,* p. 79.
13. Ibid, p. 239.
14. Suzi Gablik, *The Reenchantment of Art,* p. 1.
15. Ibid, p. 42.

16. Ibid, p. 44.

17. Leslie Kaufman, "Darwin Foes Add Warming to Targets," *New York Times*, March 3, 2010.

18. Morris Berman, *Reenchantment of the World*, p. 10.

19. Carolyn Merchant, *The Death of Nature: Women, Ecology, and the Scientific Revolution*, p. 193.

20. Ibid, p. 30.

21. Ibid, p. 3.

22. David Ray Griffin, ed., "Introduction," *The Reenchantment of Science: Postmodern Proposals*, p. 11.

23. Griffin and Smith, *Primordial Truth*, p. 19.

24. Ibid.

25. Griffin, *The Reenchantment of Science: Postmodern Proposals*, p. 11.

26. Griffin, *God and Religion*, p. 57.

27. Griffin, "Peace and the Postmodern Paradigm," *Spirituality*, pp. 148 and 20.

28. Griffin, *God and Religion*, p. 16.

29. Lynn White, "The Historical Roots of our Ecological Crisis," pp. 24–25. http://www.uvm.edu/~gflomenh/ENV-NGO-PA395/articles/Lynn-White.pdf.

30. White, p. 52.

31. White, p. 12.

32. Gerald O. Barney, *Global 2000 Revisited: What Shall We Do?* Arlington, VA: The Millennium Institute, 1993, p. 5.

33. Michael J. Cohen, "Naturally Attracted: Connecting with Michael J. Cohen." A free, 45-minute online DVD. See http://www.imdb.com/video/wab/vi2114192153/ Project NatureConnect: Educating, Counseling and Healing with Nature. See http://www.ecopsych.com/. Natural Attraction Ecology, The Pure Art and Science of Creating Balanced, Therapeutic Relationships: A Practical Application of Spirituality. See http://www.naturalattractionecology.com/. *The Children & Nature Network (C&NN) working to reconnect children with nature. See* www.childrenandnature.org.

34. Richard Louv. See http://richardlouv.com/. *The Children & Nature Network* keeps track of Regional Campaigns and Initiatives in the children and nature movement. See http://www.childrenandnature.org/movement/.

35. Griffin, *Reenchantment of Science*, p. 11.

36. Ibid, p. 25.

37. Ibid, p. 26.

38. Rosemary Radford Ruether, *Sexism and God-Talk: Toward a Feminist Theology*, p. 53.

39. Mary Condren. Interview. See http://www.tallgirlshorts.net/marymary/marycondren.html.

40. Mary Condren, *The Serpent and the Goddess*, p. 13.

41. Ibid. p. 11.

42. Ibid, p. 121.

43. Ibid, p. 22.

44. Ibid, p. 121.

45. Ibid, pp. 21 and 24.

46. Ibid, p. 40.

47. Ibid, p. 59.

48. Andrew Lawless, "Sexuality, Sin, and Sacrifice: Deconstructing the Patriarchy," an interview with Dr. Mary Condren. See http://www.threemonkeysonline.com/als/_mary_condren_serpent_goddess_feminism.html.

49. Condren, *The Serpent and the Goddess*, p. 60.

50. Ibid, p. 81.

51. Ibid, p. 36.

52. Ibid, p. 152.

53. Ibid, p. 79.

54. Ibid, p. 84.

55. Ibid, p. 94.

56. Ibid, p. 87.

57. Ibid, pp. 111–112.

58. Ibid, p. 178.

59. Ibid, p. 179.

60. Ibid, p. 236.

61. Leonard Shlain, *The Alphabet Versus the Goddess*, pp. 4–5.

62. Shlain, p. 19. Although some people say the concept of the left and right brain is a broad oversimplification, Shlain says "certain facts remain beyond dispute. If a right-handed person has a major stroke in the controlling left hemisphere, with few exceptions, a catastrophic deficit of speech, right-sided muscle paralysis and/or dysfunction in abstract thinking will occur. Conversely, damage to the right brain will impair the afflicted person's ability to solve spatial problems, recognize faces, appreciate music, besides paralyzing the left side of the body," p. 18.

63. Ibid. p. 5.

64. Ibid, p. 7.

65. Ibid, p. 6.

66. Ibid, pp. 295–296.

67. Ibid, p. 302.

68. Ibid, p. 337.

69. Ibid, p. 338.

70. Ibid, p. 322.

71. Ibid, p. 342.

72. Ibid, p. 361.

73. Ibid, p. 364.

74. Ibid, pp. 370–371.

75. Ibid, p. 373.

76. Ibid, p. 374.

77. Ibid, p. 375.

78. Ibid, p. 408.

79. Ibid, p. 442. Italics in the original.

80. Ibid, p. 443.
81. Ibid, p. 444.

Chapter 11

1. The Alban Institute is an independent center of learning and leadership development with a focus on congregations. Located in greater Washington, D.C., Alban is a not-for-profit membership organization that develops and shares knowledge through consulting, publishing, research, and education programs. The Alban Institute was founded in 1974 by Loren Mead as a major resource for American congregations facing the challenges of a changing society. http://www.alban.org/
2. Speech given by John Cobb, Jr., Naperville, Illinois, May 8, 1995.
3. Ibid, p. 95.
4. Loren Mead, in *The Once and Future Church: Reinventing the Congregation for a New Mission Frontier,* p. 11.
5. Ibid, p. 10.
6. Ibid, p. 14.
7. Ibid, p. 15.
8. Ibid, p. 22.
9. Ibid, p. 17.
10. Ibid, p. 17.
11. Peter Hodgson, *Revisioning the Church: Ecclesial Freedom in the New Paradigm,* pp. 15–16.
12. Mead, p. 63.
13. Ibid, p. 18.
14. Ibid, p. 43.
15. Ibid, p. xx,
16. Ibid, p. 87.
17. Ibid, pp. 133–134, gives charts of membership in 5 denominations, first in terms of actual numbers and then as a % of population.
18. Ibid, p. 16.
19. Ibid, p. 23.
20. Ibid, p. x.
21. Ibid, p. 22.
22. Ibid, p. 32.
23. Ibid, p. 54.
24. Peter Senge, *The Fifth Discipline*, p. 266.
25. Ibid, p. 206.
26. Ibid, p. 221.
27. Mead, p. 368.

Chapter 12

1. Bill McKibben, "Warning on Warming," http://www.nybooks.com/articles/19981.
2. Lester Brown, Janet Larsen, Jonathan G. Dorn, and Frances Moore, *Time for Plan B: Cutting Carbon Emissions 80 Percent by 2020* The full report is available in pdf without endnotes http://www.earthpolicy.org/Books/PB3/80by2020.pdf or with endnotes. http://www.earthpolicy.org/Books/PB3/80by2020notes.pdf
3. Al Gore, "A Generational Challenge to Repower America" http://www.wecansolveit.org/pages/al_gore_a_generational_challenge_to_repower_america/
4. Jim Hansen, "Climate Change: On the Edge" Italics added. http://www.commondreams.org/views06/0217-22.htm
5. MSNBC staff and news service reports, May 4, 2007, http://www.msnbc.msn.com/id/18466213/
6. Coal Moratorium NOW http://cmnow.org/. The Sierra Club has information on all the proposed coal plants across the country at http://www.sierraclub.org/environmentallaw/coal/plantlist.asp.
7. James Hansen, "Why We Can't Wait," *The Nation*, May 7, 2007, http://www.thenation.com/article/why-we-cant-wait.
8. Jonathan G. Dorn, "The End of an Era: Closing the Door on Building New Coal-fired Power Plants in America" Plan B Updates, March 31, 2009. http://www.earth-policy.org/index.php?/plan_b_updates/2009/update81
9. Matthew L. Wald, "Citing Global Warming, Kansas Denies Plant Permit" October 20, 2007, *New York Times*. http://www.nytimes.com/2007/10/20/business/20plant.html.
10. Union of Concerned Scientists, "Coal vs. Wind," http://www.ucsusa.org/clean_energy/coalvswind/c01.html
11. Ibid.
12. Ibid.
13. Ibid.
14. Lester R. Brown, "U.S. Moving Toward Ban on New Coal-Fired Power Plants," February 14, 2008 - 2 http://www.earth-policy.org/Updates/2008/Update70.htm
15. Ibid.
16. Jeffery Ball, "Wall Street Shows Skepticism Over Coal -Banks Push Utilities To Plan for Impact Of Emissions Caps" *Wall Street Journal* February 4, 2008. http://online.wsj.com/article/SB120209079624339759.html?mod=us_business_whats_news.
17. Ibid.
18. Sierra Club, "Factsheet: The Dirty Truth about Coal," http://www.sierraclub.org/coal/downloads/0508-coal-report-fact-sheet.pdf.
19. Marilyn Berlin Snell, "Can Coal Be Clean?" Sierra Club January/February 2007. Sierra Club's Coal website, http://www.sierraclub.org/sierra/200701/coal.asp.
20. Ben Block, "U.S. Environmental Groups Divided on 'Clean Coal'" March 19, 2008, http://www.worldwatch.org/node/5654
21. Ibid.
22. Brown, "U.S. Moving Toward Ban on New Coal-Fired Power Plants."

23. Ibid. and *The Future of Coal* at http://web.mit.edu/coal/

24. Ben Block, "U.S. Environmental Groups Divided on 'Clean Coal," March 19, 2008, http://www.worldwatch.org/node/5654.

25. Andrew C. Revkin "A 'Bold' Step to Capture an Elusive Gas Falters" *New York Times*, February 3, 2008, http://www.nytimes.com/2008/02/03/weekinreview/03revk.html.

26. Brown, "U.S. Moving Toward Ban on New Coal-Fired Power Plants," See http://fossil.energy.gov/news/techlines/2008/08003-DOE_Announces_Restructured_FutureG.html

27. Revkin "A 'Bold' Step to Capture an Elusive Gas Falters"

28. Sierra Club, "100 Coal Plants Unplugged," http://www.sierraclub.org/coal/100plants/ and http://www.geni.org/globalenergy/library/technical-articles/generation/climate-change/sierraclub/100-coal-plants-prevented-or-abandoned/index.shtml.

29. Lester R. Brown, "The Flawed Economics of Nuclear Power" Plan B Updates, October 28, 2008. http://www.earth-policy.org/index.php?/plan_b_updates/2008/update78

30. Amory B. Lovins, Imran Sheikh, and Alex Markevich, "Forget Nuclear" http://www.rmi.org/sitepages/pid467.php.

31. *Helen Caldicott,* "Nuclear power is the problem, not a solution" April 13, 2005 http://www.theaustralian.news.com.au/common/story_page/0,5744,12835747%5E12332,00.html

32. David Biello, "Atomic Weight: Balancing the Risks and Rewards of a Power Source." *Scientific American* January 27, 2009. http://www.scientificamerican.com/article.cfm?id=nuclear-power-plant-safety.

33. Ibid.

34. Ibid.

35. Ibid.

36. Ibid.

37. *Environment News Service (ENS)* "Chernobyl Radiation Killed Nearly One Million People: New Book" *April 26, 2010.* http://www.commondreams.org/headline/2010/04/26. Alexey V. Yablokov (Center for Russian Environmental Policy, Moscow, Russia), Vassily B. Nesterenko, and Alexey V. Nesterenko (Institute of Radiation Safety, Minsk, Belarus). Consulting Editor Janette D. Sherman-Nevinger (Environmental Institute, Western Michigan University, Kalamazoo, Michigan) *Chernobyl: Consequences of the Catastrophe for People and the Environment.* December 2009. The book is available from the New York Academy of Sciences, http://www.nyas.org/Publications/Annals/Detail.aspx?cid=f3f3bd16-51ba-4d7b-a086-753f44b3bfc1.

38. Ibid.

39. Ibid.

40. Ibid.

41. Ibid.

42. David Biello, "Spent Nuclear Fuel" *Scientific American,* January 28, 2009. http://www.scientificamerican.com/article.cfm?id=nuclear-waste-lethal-trash-or-renewable-energy-source

43. Brown, "The Flawed Economics of Nuclear Power."

44. Ibid.

45. Suzanne Struglinski "Yucca 'flunked' Messages show DOE knew that rock alone wouldn't isolate waste." *Las Vegas SUN.* April 22, 2005 http://www.lasvegassun.com/sunbin/stories/text/2005/apr/22/518648054.html

46. Ibid.

47. Keay Davidson, "Last-minute glitches in Yucca nuclear-waste burial plan" *The San Francisco Chronicle* 11/17/2003. http://www.sfgate.com/cgi-bin/article.cgi?file=/chronicle/archive/2003/11/17/MNGB633I4B1.DTL

48. Sierra Club, "Deadly Nuclear Waste Transport," http://www.sierraclub.org/nuclearwaste/yucca_factsheet.asp

49. Ibid.

50. Biello, "Spent Nuclear Fuel."

51. Ibid.

52. Brown, "The Flawed Economics of Nuclear Power."

53. David Biello Will the Nuclear Power "Renaissance" Ever Reach Critical Mass?" *Scientific American* News, May 21, 2009. http://www.scientificamerican.com/article.cfm?id=will-nuclear-power-reach-critical-mass.

54. Biello "Spent Nuclear Fuel"

55. Biello, "Atomic Weight"

56. Biello "Spent Nuclear Fuel"

57. David Biello, "Reactivating Nuclear Reactors for the Fight against Climate Change" *Scientific American* January 27, 2009. http://www.scientificamerican.com/article.cfm?id=reactivating-nuclear-reactors-to-fight-climate-change.

58. Physicians for Social Responsibility "Dirty, Dangerous and Expensive: The Truth About Nuclear Power." http://www.psr.org/resources/nuclear-power-factsheet.html.

59. Julio Godoy, "European Heat Wave Shows Limits of Nuclear Energy" OneWorld.net http://us.oneworld.net/article/view/137077/1/ and http://www.commondreams.org/headlines06/0728-06.htm

60. Physicians for Social Responsibility "Dirty, Dangerous and Expensive.

61. Ibid.

62. Mark Z. Jacobson, "Review of Solutions to Global Warming, Air Pollution, and Energy Security," http://www.stanford.edu/group/efmh/jacobson/0810EnergySeminar.pdf.

63. Dave Reed, "Return of the Nuclear Salesman" http://www.rmi.org/sitepages/pid642.php.

64. Amory Lovins, "How Not to Parachute More Cats" https://www.rmi.org/images/PDFs/AboutRMI/G96-01_ParaCats.pdf

65. Ibid.

66. Brown, *Plan B 3.0: Mobilizing to Save Civilization,* p. 214.

67. Lovins, Sheikh, and Markevich, "Forget Nuclear."

68. Brown "The Flawed Economics of Nuclear Power"

69. Jonathan Fahey, "How NRG Energy Wants to Revive Nuclear Industry," *Forbes Magazine* December 14, 2009. http://www.forbes.com/forbes/2009/1214/energy-power-nrg-energy-wants-revive-nuclear-industry.html.

70. Michael Grunwald, "Nuclear's Comeback: Still No Energy Panacea," *Time Magazine*, Dec. 31, 2008. http://www.time.com/time/magazine/article/0,9171,1869203,00.html. And Biello, "Reactivating Nuclear Reactors.

71. Mark Cooper "All Risk, No Reward for Taxpayers and Ratepayers" November 2009 Issue Brief. http://www.vermontlaw.edu/Documents/11_03_09_Cooper-All_Risk_No_reward__Issue_Brief_1.pdf

72. Biello, "Reactivating Nuclear Reactors."

73. Cooper "All Risk, No Reward for Taxpayers and Ratepayers"

74. Ibid.

75. Physicians for Social Responsibility, Safe Energy-Economics. http://www.psr.org/safe-energy/economics.html.

76. Brown "The Flawed Economics of Nuclear Power"

77. *Ryan Alexander* "The real cost of new U.S. nuclear reactors" *Bulletin of the Atomic Scientists, August 21, 2009.* http://www.thebulletin.org/web-edition/op-eds/the-real-cost-of-new-us-nuclear-reactors.

78. Physicians for Social Responsibility "Dirty, Dangerous and Expensive."

79. Physicians for Social Responsibility, Safe Energy-Economics. http://www.psr.org/safe-energy/economics.html.

80. Grunwald, "Nuclear's Comeback: Still No Energy Panacea,"

81. Joe Romm, "Exclusive analysis, Part 1: The staggering cost of new nuclear power" *January 5, 2009.* http://climateprogress.org/2009/01/05/study-cost-risks-new-nuclear-power-plants/.

82. Brown "The Flawed Economics of Nuclear Power."

83. Ibid.

84. Ibid.

85. Ibid.

86. Biello, "Reactivating Nuclear Reactors."

87. Steve Connor, "Warning: Oil Supplies Are Running Out Fast" *August 3, 2009 The Independent/UK* http://www.commondreams.org/headline/2009/08/03.

88. Christopher Flavin "Over the Peak" *Worldwatch*, January/February 2006. https://www.worldwatch.org/node/4030.

89. Ibid.

90. Connor, "Warning: Oil Supplies Are Running Out Fast."

91. Flavin "Over the Peak."

92. "Steep Decline In Oil Production Brings Risk of War and Unrest, Says New Study" October 22, 2007 The Guardian/UK http://www.commondreams.org/archive/2007/10/22/4737/.

93. Ibid.

94. Ibid.

95. Ibid.

96. Ibid.

97. Michael T. Klare "Entering the Tough Oil Era" TomDispatch.com August 16, 2007, http://www.truthout.org/docs_2006/081607D.shtml.

98. Connor, "Warning: Oil Supplies Are Running Out Fast."

99. Ibid.

100. Amory B. and L. Hunter Lovins. "The Nuclear Option Revisited" *The Los Angeles Times* July 8 2001, Also at http://www.commondreams.org/views01/0708-05.htm.

101. Amory B. Lovins and L. Hunter Lovins "Fool's Gold in Alaska" *Foreign Affairs*, July/August 2001. p. 78. http://www.rmi.org/images/PDFs/Energy/E01-04_FoolsGoldAnnot.pdf.

102. Robert F. Kennedy Jr. "Early 1980s: Better Gas Mileage, Greater Security, a Stronger Economy" November 12, 2007, http://www.fuel-efficient-vehicles.org/energy-news/?p=382#more-382.

103. Ibid.

104. Amory B. Lovins "Powerful Ways to Save and Replace Oil Have Been Quietly Emerging for 30 Years" http://www.rmi.org/sitepages/pid151.php.

105. http://www.sierraclub.org/globalwarming/cleanenergy/efficiency.asp.

106. Brown, *Plan B 2.0,* p. 185.

107. Brown, *Time for Plan B.*

108. Ibid.

109. Architecture 2030. http://www.architecture2030.org/2030_challenge/index.html.

110. Jeremy Rifkin, *The Empathic Civilization: the Race to Global Consciousness in a World in Crisis,* p. 519.

111. See http://www.eere.energy.gov/consumer/your_home/designing_remodeling/index.cfm/mytopic=10360.

112. Michael Shuman, in *Going Local: Creating Self-Reliant Communities in a Global Age*, p. 65.

113. Brown, *Plan B 2.0.* Chapter 10.

114. Brown, *Time for Plan B.*

115. Ibid.

116. Energy Efficient Lighting at http://www.eartheasy.com/live_energyeff_lighting.htm.

117. "Small steps, big energy savings" November 26, 2007 *The Los Angeles Times*, http://www.recycled-nergy.com/news-room/news/small_steps_big_energy_savings.

118. Ibid.

119. Seth Dunn and Christopher Flavin, in "Sizing up Micropower," *State of the World 2000* p. p. 143.

120. Amory Lovins, *Small Is Profitable introduces engineering and financial practitioners, business managers and strategists, public policymakers, designers, and interested*

citizens to the new value opportunities presented by considering these economic benefits. It also provides a basic introduction to key concepts from such disciplines as electrical engineering, power system planning, and financial economics. (www.smallisprofitable.org)

121. *State of the World 1999* p. 30.

122. Seth Dunn and Christopher Flavin, "Sizing up Micropower" *State of the World 2000* p. 144.

123. Michael T. Klare "Entering the Tough Oil Era" TomDispatch.com, August 16, 2007 http://www.truthout.org/docs_2006/081607D.shtml

124. Ibid.

125. Ibid.

126. Jeremy Rifkin, "Hydrogen: Empowering the People" *The Nation*, December 23, 2002, http://www.thenation.com/doc/20021223/rifkin.

127. Ibid, p. 38.

128. Repower America, "National Unified Smart Grid" http://www.repoweramerica.org/elements/national-unified-smart-grid/.

129. Ibid.

130. Ibid.

131. From http://www.repoweramerica.org/us/opportunity/clean-cars/, See Mark Z Jacobson. "Review of Solutions to Global Warming, Air Pollution, and Energy Security." http://www.stanford.edu/group/efmh/jacobson/revsolglobwarmairpol.htm.

132. Rifkin, *The Empathic Civilization,* p. 524.

133. *State of the World 2003,* pp. 91 and 110.

134. Brown, *Time for Plan B.*

135. Kate Gordon, Julian L. Wong, and J.T. McLain "Out of the Running? How Germany, Spain, and China Are Seizing the Energy Opportunity and Why the United States Risks Getting Left Behind," Center for American Progress, March 2010. http://www.americanprogress.org/issues/2010/03/pdf/out_of_running.pdf.

136. Brown, *Time for Plan B.*

137. Ibid.

138. Brown, *Plan B 2.0.* Chapter 10.

139. Brown, *Time for Plan B.*

140. Lester R. Brown "Europe Leading World into Age of Wind Energy," Eco-Economy Update 2004-6 http://www.earth-policy.org/Updates/Update37.htm .

141. Brown, *Time for Plan B.*

142. *State of the World 1999* p. 36.

143. Ibid, p. 89.

144. Lester R. Brown, "The Short Path to Oil Independence: Gas-Electric Hybrids and Wind Power Offer Winning Combination," *Plan B* Updates, October 13, 2004. http://www.earth-policy.org/Updates/Update43.htm

145. Lester Brown, *Time for Plan B.*

146. Lester Brown, Janet Larsen, Jonathan G. Dorn, and Frances Moore, *Time for Plan B: Cutting Carbon Emissions 80 Percent by 2020* The full report is available in pdf without endnotes http://www.earthpolicy.org/Books/PB3/80by2020.pdf or with endnotes. http://www.earthpolicy.org/Books/PB3/80by2020notes.pdf. Algeria, now a leading oil exporter, has plans to develop 6,000 megawatts of solar thermal electric generating capacity for export to Europe via undersea cable. A project on that scale could meet the household electricity demand of a country the size of Portugal. Solar rooftop water and space heaters will also play a major role in cutting CO2 emissions in the Plan B economy, with a 2020 installation goal of more than 1 million thermal megawatts. In China, some 40 million rooftop solar water heaters have been installed in recent years, both in cities and in villages, for as little as $200 each. Collectively they harness energy equal to the output of 54 coal-fired power plants. The Chinese government aims to more than double the current 124 million square meters of rooftop solar water heaters to 300 million square meters by 2020. The European Solar Thermal Industry Federation's goal is even higher: by 2020 they are calling for 500 million square meters of solar water and space heaters, or one square meter for every European. (Israel now leads the world on a per person basis, at 0.74 square meters.) Achieving China's and Europe's goals, while ramping up installations in the United States, Japan, and the rest of the world, would capture enough solar energy to equal the output of 690 coal-fired power plants.

147. "Solar Thermal Energy," http://en.wikipedia.org/wiki/Solar_thermal_energy

148. "Renewable Generation" http://www.repoweramerica.org/elements/renewable-generation/

149. Ibid.

150. Lester R. Brown, *Time for Plan B.*

151. Ibid.

152. Ibid.

153. John J. Berger, *Charging Ahead: The Business of Renewable Energy and What It Means for America.* (New York: Henry Holt, 1997, p. 55. quoted in *Going Local* p.68.

154. *State of the World 2000,* p. 149.

155. Noah Buhayar "Getting a (Firm) Grip on Renewables" http://www.rmi.org/sitepages/pid518.php

156. Lovins, Sheikh, and Markevich, "Forget Nuclear" http://www.rmi.org/sitepages/pid467.php

157. Ibid.

158. *Time for Plan B.* Iceland currently heats close to 90 percent of its homes with energy from the earth. In the Philippines, 25 percent of electricity comes from geothermal power plants. In El Salvador the figure is 22 percent. Other countries rich in geothermal energy are those bordering the Pacific in the so-called Ring of Fire, including Chile, Peru, Mexico, the United States, Canada, Russia, China, Japan, Indonesia, and Australia, as well as the countries along the Great Rift Valley of Africa and those around the Eastern Mediterranean.

159. "Renewable Generation" http://www.repoweramerica.org/elements/renewable-generation/.

160. http://www.eere.energy.gov/consumer/your_home/space_heating_cooling/index.cfm/mytopic=12640.

161. Lester R. Brown, *Time for Plan B.*

162. The Report of the World Commission on Dams, *Dams and Development: A New Framework for Decision-Making*, http://www.dams.org/report/wcd_overview.htm. p. 198.

163. *State of the World 2004* http://www.worldwatch.org/topics/consumption/sow/trendsfacts/2004/03/03/.

164. International Rivers Network http://www.irn.org/.

165. Lester R. Brown, *Time for Plan B.*

166. Matthew R. Simmons, "At Risk: The Sustainability Of Oil And Gas" AON Annual Energy Insurance Symposium, January 14, 2010, Houston, TX. http://www.simmonsco-intl.com/files/AON%20Annual%20Energy%20Insurance%20BW.pdf.

167. Lovins and Lovins "Fool's Gold in Alaska."

168. Robert F. Kennedy Jr. "Early 1980s: Better Gas Mileage."

169. Ibid.

170. Lovins and Lovins "Fool's Gold in Alaska"

171. Mark Fischetti, "Why Not a 40-MPG SUV? Technology exists to double gas guzzlers' fuel efficiency. So what's the holdup?" November 2002. http://www.technologyreview.com/Energy/12989/.

172. Ibid, p. 40.

173. Ibid, p. 46.

174. Quoted in Lester R. Brown, *Plan B 2.0,* Chapter 11. Designing Sustainable Cities, from Richard Register, "Losing the World, One Environmental Victory at a Time—And a Way to Solve That Problem," (Oakland, CA: Ecocity Builders, Inc., 31 August 2005); Richard Register, *Ecocities: Building Cities in Balance with Nature* (Berkeley, CA: Berkeley Hill Books, 2002).

175. Ibid, pp. 43–44.

176. Brown, *Plan B 3.0,* Designing Cities for People. http://www.earth-policy.org/Books/PB3/PB3ch10_ss7.htm.

177. Ibid, p. 46.

178. Ibid, p. 47.

179. Ibid.

180. National Complete Streets Coalition, http://www.completestreets.org/.

181. Brown, *Plan B 3.0.*

182. Ibid, p. 48.

183. Ibid, p. 55.

184. Ibid, p. 49.

185. Ibid.

186. Hawken, *Natural Capitalism* p. 24.

187. Ibid, p. 24.

188. Ibid, p. 24.

189. Ibid, pp. 97–98. And deriving hydrogen from natural gas for the initial generation of fuel cells would allow the early stages of a hydrogen economy to be based on the extensive natural gas pipelines and other equipment already in place. Small-scale reforming units that convert natural gas into hydrogen could be placed in homes, office buildings, and service stations. One near-term solution to the "chicken-and-egg" infrastructure problem in many countries would be to use small-scale natural gas reformers at fueling stations, relying on existing natural gas pipelines to distribute the fuel.

190. Hawken, *Natural Capitalism* p. 24.

191. Ibid, p. 25.

192. The Hypercar at the Rocky Mountain Institute, http://move.rmi.org/markets-in-motion/case-studies/automotive/hypercar.html, and http://move.rmi.org/markets-in-motion/case-studies.

193. Hawken, *Natural Capitalism* p. 38.

194. Ibid, pp. 31–32.

195. Ibid, p. 32.

196. Amory B. Lovins, "How to get real Security" Whole Earth Fall 2002 p. 8-16. See www.rmi.org, or www.wholeearth.com. .

197. Hawken, *Natural Capitalism* p. 27.

198. Ibid, p. 35.

199. Ibid, p. 27.

200. Ibid, pp. 34–35.

201. Lester R. Brown, "The Short Path to Oil Independence: Gas-Electric Hybrids and Wind Power Offer Winning Combination" *Plan B* Updates, October 13, 2004. http://www.earth-policy.org/Updates/Update43.htm

202. Ibid.

203. Fred Pearce, "Fuels gold: Big risks of the Biofuel Revolution" *New Scientist* September 25, 2006. http://environment.newscientist.com/channel/earth/mg19125701.200-fuels-gold-big-risks-of-the-biofuel-revolution.html.

204. George Monbiot "A Lethal Solution" Guardian March 27 2007. http://www.monbiot.com/archives/2007/03/27/a-lethal-solution/

205. Monbiot "A Lethal Solution" and Wetlands International, 8th December 2006. Bio-fuel less sustainable than realized. http://www.wetlands.org/news.aspx?ID=804eddfb-4492-4749-85a9-5db67c2f1bb812. And http://www.biofuelwatch.org.uk/resources.php#2007Jan31

206. Fred Pearce, "Fuels gold: Big risks of the biofuel revolution"

207. David Roberts, "Alternative Energy Guru Reflects on Policy, Fuels" Grist.org. July 26, 2007. A conversation with energy guru Amory Lovins. http://www.grist.org/news/maindish/2007/07/26/lovins/index.html.

208. Frances Cerra Whittelsey "Bio-Hope, Bio-Hype: A users' guide to biofuels" September/October 2007, *Sierra Club Magazine*. http://www.sierraclub.org/sierra/200709/bio.asp.

209. Ross Gelbspan, "Why we need to worry about global warming, now" San Jose Mercury News, Perspective Section, April 9, 2006. http://www.heatisonline.org/contentserver/

objecthandlers/index.cfm?id=5896&method=full. The size of the fund was calculated by energy policy specialists at the Tellus Institute.

210. Ibid.

211. Ross Gelbspan http://www.heatisonline.org/contentserver/objecthandlers/index.cfm?id=6320&method=full

212. George Monbiot. "Giving up on two degrees" *Guardian*, May 1, 2007, http://www.monbiot.com/archives/2007/05/01/1058/#more-1058 .

213. Bill McKibben, "Global Warning" New York Review of Books http://www.nybooks.com/articles/19981.

214. Ross Gelbspan, *The Boiling Point*, p. 134; 137.

Chapter 13

1. "Just the Facts," *Yes! Magazine* Spring 2009, p. 24.

2. *Worldwatch magazine*, September/October 2000. Press Releases September 1, 2000, http://www.worldwatch.org/press/news/2000/09/01/.

3. U.N. Food and Agriculture Organization, *Livestock's Long Shadow: Environmental Issues and Options.* http://www.fao.org/docrep/010/a0701e/a0701e00.HTM.

4. Michael Pollan, "Farmer in Chief" *New York Times Magazine*, October 12, 2008. http://www.nytimes.com/2008/10/12/magazine/12policy-t.html.

5. Lester R. Brown, "The Oil Intensity of Food" Book Bytes, June 25, 2009 Adapted from *Plan B 3.0: Mobilizing to Save Civilization*, Chapter 2, "Deteriorating Oil and Food Security."

6. Ibid.

7. Ibid.

8. Alan Thein Durning "Matters of Scale—The Price of Beef" *World Watch Magazine*, July/August, 1994. http://www.worldwatch.org/node/791

9. Brown, "The Oil Intensity of Food."

10. Ibid.

11. Ibid.

12. "Hamburger Connection Fuels Amazon Destruction" issued by the Center for International Forestry Research (CIFOR) http://www.cifor.cgiar.org/publications/pdf_files/media/Amazon.pdf.

13. Rattan Lal, Testimony for a Senate hearing on July 8, 2003. "Concerning Soil Carbon Sequestration by Agricultural and Forestry Land Uses to Mitigate Climate Change," http://epw.senate.gov/108th/Lal_070803.htm.

14. Fred Pearce Science, "Ground-breaking solutions to global warming," *The Independent*, (London), Dec 8, 2000. http://business.highbeam.com/4341/article-1G1-10912527/highest-reading-scientists-grapple-greenhouse-gases.

15. Paul Hawkin, *Natural Capitalism,* p. 205.

16. Testimony of Rattan Lal,

17. Ibid.

18. Ibid.

19. "Can Soils Soak Up Greenhouse Gases?" ABC Broadcast 16/12/00, Summary http://www.abc.net.au/rn/science/earth/stories/s226590.htm

20. "Global Warming: Does the Solution Lie in the Soil?" http://www.fairfaxcounty.gov/nvswcd/newsletter/globalwarming.htm

21. "Can Soils Soak Up Greenhouse Gases?" ABC Broadcast 16/12/00, Summary http://www.abc.net.au/rn/science/earth/stories/s226590.htm

22. Ibid.

23. "Global Warming: Does the Solution Lie in the Soil?"

24. Eric Holt-Gimenez "Food Rebellions: 7 Steps to Solving the Food Crisis" April 16, 2009 *YES! Magazine*, Spring 2009. http://www.yesmagazine.org/issues/food-for-everyone/food-rebellions-7-steps-to-solving-the-food-crisis.

25. Brown, *Plan B 4.0*, p. 4.

26. Lester R. Brown, "Could Food Shortages Bring Down Civilization?" *Scientific American Magazine*, April 22, 2009. http://www.scientificamerican.com/article.cfm?id=civilization-food-shortages.

27. Neil Macfarquhar, "Experts Worry as Population and Hunger Grow." *New York Times*, October 21, 2009. http://www.nytimes.com/2009/10/22/world/22food.html, See also http://www.fao.org/hunger/en/.

28. Brown, "Could Food Shortages Bring Down Civilization?"

29. Center for International Forestry Research (CIFOR), "Hamburger Connection Fuels Amazon Destruction." http://www.cifor.cgiar.org/publications/pdf_files/media/Amazon.pdf.

30. Lester Brown, *Plan B 4.0*, p. 5, http://www.earthpolicy.org/images/uploads/book_files/pb4book.pdf.

31. Brown, "Could Food Shortages Bring Down Civilization?"

32. Ibid.

33. Brown, *Eco-economy,* p. 148 .

34. Lester R. Brown. "World Grain Harvest Falling Short by 54 Million Tons Water Shortages Contributing to Shortfall," November 23, 2001-3. http://www.mail-archive.com/bdnow@envirolink.org/msg07660.html.

35. Lester R. Brown, "Could Food Shortages Bring Down Civilization?"

36. Ibid.

37. Ibid.

38. Ibid.

39. Lester R. Brown, "The Copenhagen Conference on Food Security" Earth Policy Release November 10, 2009. http://www.earthpolicy.org/index.php?/plan_b_updates/2009/update84

40. Brown, "Could Food Shortages Bring Down Civilization?"

41. Brown, "World Grain Harvest Falling Short"

42. From The Committee for Sustainable Agriculture CSA Web site http://www.csa-efc.org/organic.html.

43. From The Committee for Sustainable Agriculture Web site

http://www.csa-efc.org/organic.html.

44. Richard Douthwaite, *Short Circuit*, p. 253.

45. Brown, "Could Food Shortages Bring Down Civilization?"

46. Brown, *Plan B 4.0,* p. 4.

47. Lester R. Brown "China Losing War With Advancing Deserts" Update 26: August 5, 2003-6 Earth Policy Institute

48. Ibid.

49. Lester R. Brown, "Rising Temperatures & Falling Water Tables Raising Food Prices, Record World Grain Harvest Shortfall of 83 Million Tons in Prospect" August 21, 2002-12, Earth Policy Institute http://earth-policy.org/Updates/Update16.htm

50. Sierra Club- "Sprawl" http://www.sierraclub.org/sprawl/.

51. Gary Gardner *Shrinking Fields: Cropland Loss in a World of Eight Billion,* Worldwatch.

52. Ted C. Fishman, "The Chinese Century" *New York Times Magazine,* July 4, 2004, http://query.nytimes.com/gst/fullpage.html?res=9C04E2DC1438F937A35754C0A9629C8B63.

53. Lester Brown "Paving the Planet: Cars and Crops Competing for Land," http://www.earth-policy.org/index.php?/plan_b_updates/2000/alert12.

54. Stephen Leahy "Climate Change: Food Supply Hangs in the Balance" *Inter Press Service* October 02 2009 at http://www.truthout.org/100309G

55. Ibid.

56. Brown, "The Copenhagen Conference on Food Security."

57. Janet Larsen "Other Fish n The Sea, But For How Long?" http://www.earth-policy.org/Updates/Update25.htm

58. WWF, "Mediterranean bluefin tuna stocks collapsing now as fishing season opens" *The Independent*, April 14, 2009. http://www.panda.org/?162001/Mediterranean-bluefin-tuna-stocks-collapsing-now-as-fishing-season-opens.

59. Martin Hickman "Revealed: the bid to corner world's bluefin tuna market" *The Independent*, June 3, 2009. http://www.independent.co.uk/environment/nature/revealed-the-bid-to-corner-worlds-bluefin-tuna-market-1695479.html

60. "The Issue: Snapper Limits Mulled." *Sun Sentinel*, July 28, 2009. http://articles.sun-sentinel.com/2009-07-28/news/0907270280_1_snapper-bottom-fishing-fishermen.

61. Johann Hari. "Could We be the Generation that Runs out of Fish?"*The Independent*, *June 5 2009.* http://www.independent.co.uk/opinion/commentators/johann-hari/johann-hari-could-we-be-the-generation-that-runs-out-of-fish-1697247.html.

62. Pew Ocean s Commission, http://www.pewoceans.org/oceans/

63. Jeremy Hance "Using fish as livestock feed threatens global fisheries" November 18, 2009. http://news.mongabay.com/2009/1118-hance_fishmeal.html

64. Ibid.

65. "Just the Facts" Yes Magazine Spring 2009. P. 24.

66. Brown, "Could Food Shortages Bring Down Civilization?"

67. Ibid.

68. "Food and the poor" *The Economist,* Apr 17th 2008. http://www.economist.com/world/international/displaystory.cfm?story_id=11049284%20

69. Chris Hedges "Food Is Power and the Powerful Are Poisoning Us" *September 7, 2009 by TruthDig.com* http://www.commondreams.org/view/2009/09/07-3

70. Brandon Keim "Food Riots Begin: Will You Go Vegetarian?" April 21, 2008 http://blog.wired.com/wiredscience/2008/04/food-riots-begi.html#previouspost and Lester R. Brown "World Facing Huge New Challenge on Food Front: Business-as-Usual Not a Viable Option" Plan B Updates April 16, 2008 http://www.earth-policy.org/index.php?/plan_b_updates/2008/update72. For examples of food price unrest see www.earth-policy.org/datacenter/xls/update72_1.xls

71. Lester R. Brown "World Facing Huge New Challenge on Food Front: Business-as-Usual Not a Viable Option" Plan B Updates April 16, 2008. http://www.earth-policy.org/index.php?/plan_b_updates/2008/update72

72. "Food and the poor" *The Economist* Apr 17th 2008. http://www.economist.com/world/international/displaystory.cfm?story_id=11049284%20.

73. Salena Tramel "Why Food Sovereignty Is the New Food Security" *August 12, 2009* http://www.commondreams.org/view/2009/08/12-14

74. Ibid.

75. Cavanagh, *Alternatives to Economic Globalization*, p. 172.

76. Ibid.

77. "Agribusiness Concentration, Not Low Prices, is Behind Global Farm Crisis," Worldwatch Institute, Press Release September 1, 2000. http://www.worldwatch.org/press/news/2000/09/01/.

78. Elisabeth Rosenthal "As More Eat Meat, a Bid to Cut Emissions" *New York Times* December 4, 2008.

79. "Rearing cattle produces more greenhouse gases than driving cars, UN report warns," http://www.un.org/apps/news/story.asp?NewsID=20772&Cr=global&Cr1=environm.

80. Jeremy Rifkin, *Los Angeles Times*, May 27, 2002, Quoted in "Is Meat Sustainable?" by The Editors, *Worldwatch*, July/August 2004, p. 15, http://www.worldwatch.org/node/549.

81. Robert Goodland and Jeff Anhang, "Livestock and Climate Change" *Worldwatch,* Nov/Dec 2009. http://www.worldwatch.org/node/6294.

82. Goodland and Anhang, "Livestock and Climate Change." See also, Robert Goodland, "Livestock and Climate change": Critical Comments and Responses. *Worldwatch*, March/April 2010, http://blogs.worldwatch.org/revolt/livestock/.

83. Jeremy Rifkin, *Los Angeles Times*, May 27, 2002, Quoted in *Worldwatch*, July/August 2004, p. 15.

84. Michael Pollan "Unhappy Meals" *New York Times* January 28, 2007. http://www.nytimes.com/2007/01/28/magazine/28nutritionism.t.html.

85. Antibiotics in agriculture are not needed to fight against foot-and-mouth and "mad cow" diseases. Antibiotics -- which

are effective against bacteria and other microorganisms -- are not used to treat either foot-and-mouth or "mad cow" disease. Foot-and-mouth disease is caused by a virus, a class of organisms against which antibiotics are ineffective. "Mad cow" [Bovine Spongiform Encephalopathy (BSE)], a poorly understood disease, is thought to be caused by "prions" -- a type of protein, and is transmitted by feeding meat and bone meal from BSE-afflicted animals. BSE is not caused by microorganisms and is not treated with antibiotics. From Union of Concerned Scientists.

86. Pollan, "Farmer in Chief."

87. Ibid.

88. Jessica Roy, "Jonathan Safran Foer's Beef with Factory Farms" http://mobile.salon.com/books/int/2009/11/06/jonathan_safran_foer/index.html

89. Abundant Life Seed Foundation is a charitable organization that promotes the genetic diversity of plants through the use of open-pollinated seed. They educate seed growers and support a seed distribution network. The Abundant Life farmers grow about 200 varieties of open-pollinated seeds a year at the Abundant Life Seed Garden in historic Port Townsend, Washington. You can join to support these growouts, build an educational database, research additional rare varieties, assist new seed growers, and supply seed to people in need through the World Seed Fund. http://www.abundantlifeseed.org/programs.htm.

90. Douthwaite, *Short Circuit*, p. 271

91. Ibid, p. 275.

92. Ibid, p. 255.

93. Ibid, p. 254–255.

94. Ibid, p. 270.

95. Ibid, pp. 270–271.

96. Ibid, p. 271.

97. Marwaan Macan-Markar, "Neglect of Farming Led to Rice Crisis," *April 26, 2008 by Inter Press Service* http://www.commondreams.org/archive/2008/04/26/8546.

98. Karen Kaplan "A 'time bomb' for world wheat crop," *Los Angeles Times*. June 14, 2009. http://articles.latimes.com/2009/jun/14/science/sci-wheat-rust14.

99. Ibid.

100. Norman E. Borlaug "Stem Rust Never Sleeps" *New York Times* April 26, 2008. http://www.nytimes.com/2008/04/26/opinion/26borlaug.html?ref=wheat.

101. Karen Kaplan "A 'time bomb' for world wheat crop." .

102. See Food First and Pesticide Action Network "The Case for a GM-Free Sustainable World" http://www.foodfirst.org/progs/global/ge/isl/summary.

103. Cavanagh, *Alternatives to Economic Globalization* p. 85.

104. Ibid, pp. 85–86.

105. How the FATT Affects You http://www.globalexchange.org/campaigns/ftaa/affects.html.

106. Cavanagh *Alternatives to Economic Globalization,* p. 85.

107. Vandana Shiva in *Alternatives to Economic Globalization*, p. 87-88

108. Ibid.

109. Ibid.

110. Ibid.

111. Vandana Shiva, "War Against Nature and the People of the South" In Sarah Anderson ed. *Views from the South: The Effects of Globalization and the WTO on Third World Counties.* San Francisco: International Forum on Globalization, 2000. Quoted in *Alternatives to Economic Globalization*, p. 88.

112. Cavanagh, *Alternatives to Economic Globalization*, p. 96.

113. Ibid. pp. 88–89.

114. From Food First and Pesticide Action Network "The Case for a GM-Free Sustainable World" http://www.foodfirst.org/progs/global/ge/isl/summary.

115. Ibid.

116. Editorial, "Do Seed Companies Control GM Crop Research?" *Scientific American,* August 13, 2009 http://www.scientificamerican.com/article.cfm?id=do-seed-companies-control-gm-crop-research.

117. Ibid.

118. Ibid.

119. Brown, "Could Food Shortages Bring Down Civilization?"

120. Michael Pollan "No Bar Code" *Mother Jones* May/June 2006 Issue. http://www.motherjones.com/environment/2006/05/no-bar-code.

121. From The Committee for Sustainable Agriculture Web site http://www.csa-efc.org/organic.html.

122. Ibid.

123. From the Union of Concerned Scientists, http://www.ucsusa.org/agriculture/index.html.

124. Alison Benjamin, "Fears for Crops as Shock Figures from America Show Scale of Bee Catastrophe" May 2, 2010 *The Observer/UK*, http://www.commondreams.org/headline/2010/05/02-1.

125. Ibid.

126. Ibid.

127. Ibid.

128. Ibid.

129. Pollan "No Bar Code"

130. Ibid.

131. Chris Hedges "Food Is Power and the Powerful Are Poisoning Us" *September 7, 2009 by TruthDig.com* http://www.commondreams.org/view/2009/09/07-3.

132. http://www.sare.org/csa/index.htm--has a list of community supported agriculture- by state and zip code.

133. Community Supported Agriculture, http://www.umassvegetable.org/food_farming_systems/csa/.

134. UMass Extension Agroecology Program. http://www.umass.edu/umext/csa/about.html.

135. See Community Food Security Coalition at http://

foodsecurity.org/faq.htm.

136. Lester R. Brown "The Localization of Agriculture," Book Byte December 1, 2009. http://www.earthpolicy.org/index.php?/book_bytes/2009/pb4ch09_ss5.

137. "Urban Agriculture- A Global Initiative" from City Farmer at http://www.cityfarmer.org/unbrochure.html.

138. Brian Halweil, *Home Grown: The Case for Local Food In A Global Market,* Worldwatch Institute, November 2002. http://www.worldwatch.org/node/827, p. 34.

139. Ibid, pp. 34–35.

140. Ibid, p. 35.

141. Ibid, p. 36.

142. Ibid, p. 35.

143. Ibid, pp. 35.

144. Ibid, pp. 59–60.

145. http://www.cardinalgroup.ca/grhcc/main.htm

146. Halweil, *Home Grown*, p. 51.

147. Food Security, Farming, and the FTAA and WTO, http://www.globalexchange.org/campaigns/ftaa/FTAAWTOAgriculture.html.

148. Clara Jeffery "Michael Pollan Fixes Dinner" (Extended Interview) *Mother Jones* March 2009. http://www.motherjones.com/environment/2009/03/michael-pollan-fixes-dinner-extended-interview .

149. Ibid.

150. Ibid.

151. Ibid.

152. Ibid.

153. Claire Hope Cummings, "The Good Food Revolution" *Yes! Magazine*, Spring 2009, http://www.yesmagazine.org/issues/food-for-everyone/the-good-food-revolution.

154. Pollan, "Farmer in Chief."

155. Ibid.

156. Cavanagh, *Alternatives to Economic Globalization*, p. 172.

157. "Agribusiness Concentration, Not Low Prices, is Behind Global Farm Crisis," Worldwatch Institute Press Release, September 1, 2000. http://www.worldwatch.org/press/news/2000/09/01/

Chapter 14

1. "Moving Toward a Less Consumptive Economy" *State of the World 2004.* http://www.worldwatch.org/features/consumption/sow/trendsfacts/2004/05/05/.

2. Lester R. Brown, "Throwing Out the Throwaway Economy" Book Bytes August 20, 2009 Adapted from Chapter 1, and Chapter 6, in, *Plan B 3.0: Mobilizing to Save Civilization* www.earthpolicy.org/index.php?/books/pb3.

3. Lester R. Brown, *Plan B 3.0: Mobilizing to Save Civilization* From Chapter 11 http://www.earth-policy.org/Books/PB3/PB3ch11_ss6.htm.

4. GrassRoots Recycling Network, http://www.grrn.org.

5. *State of the World 2004* and http://www.worldwatch.org/features/consumption/sow/trendsfacts/2004/05/05/

6. Sam Cole and Brian Ladd, "Who's Already Started on the Road to Zero Waste?" GRRN's Zero Waste Campaign, http://www.ecocycle.org/zero/road.cfm.

7. Ibid.

8. Ibid.

9. See research at http://www-new.dbe.bee.qut.edu.au/research/postgraduate_research.shtml.

10. See an article on Reverse Logistics, in the April 12, 1999, edition of *Information Week* available online at http://www.informationweek.com/729/logistics.htm. See also The Center for Clean Products and Clean Technologies at http://eerc.ra.utk.edu/ccpct/index.html

11. *State of the World 2004.* See also http://www.worldwatch.org/features/consumption/sow/trendsfacts/2004/05/05/

12. Michael Jessen, "The Ripple Effect of Zero Waste" http://www.yourplanetearth.org/article.pl?sid=01/05/29/1816244

13. Janine Benyus, *Biomimicry: Innovation Inspired by Nature.*

14. Ibid.

15. Ibid.

16. Reliable Prosperity. http://www.reliableprosperity.net/index.html.

17. Paul Hawken, *Natural Capitalism,* p. 16.

18. Jeremy Rifkin, *The Empathic Civilization: the Race to Global Consciousness in a World in Crisis,* p. 537.

19. Ibid, p. 539.

20. Ibid, p. 539.

21. Ibid, p. 538.

22. William McDonough and Michael Braungart, "The NEXT Industrial Revolution" *Atlantic Monthly.* http://www.theatlantic.com/magazine/archive/1998/10/the-next-industrial-revolution/4695/

23. Ibid.

24. Elizabeth Becker, "White House Undermined Chemical Tests, Report Says," *The New York Times*, April 2, 2004.

25. Our Stolen Future, "Why endocrine disruption challenges current approaches to regulation of chemicals." http://www.ourstolenfuture.org/Basics/challenge.htm

26. John Cavanagh and Jerry Mander *Alternatives to Economic Globalization* p. 77.

27. Science and Environmental Health Network , http://www.sehn.org/index.html.

28. "What's Needed for Strong REDD Policy?" http://www.ucsusa.org/global_warming/solutions/recognizing-forests-role-in-climate-change.html

29. WWF http://www.panda.org/about_wwf/ http://www.panda.org/about_wwf/what_we_do/forests/problems/forest_crime.cfm

30. WWF - http://www.panda.org/about_wwf/what_we_do/forests/problems/index.cfm.

31. Lester Brown, Janet Larsen, Jonathan G. Dorn, and Frances Moore, *Time for Plan B: Cutting Carbon Emissions 80 Percent by 2020* The full report is available in pdf without endnotes http://www.earthpolicy.org/Books/PB3/80by2020.pdf or with endnotes. http://www.earthpolicy.org/Books/PB3/80by2020notes.pdf

32. The Earth Island Institute's ReThink Paper campaign See http://www.earthisland.org/ei/paper/aboutrtp.html.

33. Meghan Clancy-Hepburn. "Agricultural Residues: A Promising Alternative to Virgin Wood Fiber." http://www.rca-info.org/alts/meghanhayes.html.

34. Ibid.

35. Rethink paper "KENAF The Clean Paper Crop An Ecological Alternative to Virgin Wood-Based Paper."

36. Ted Williams, "Hemp! Legalize It!" *Audubon Magazine*. http://magazine.audubon.org/incite/incite9911.html.

37. Rethink paper "KENAF The Clean Paper Crop An Ecological Alternative to Virgin Wood-Based Paper."

38. Ross Gelbspan "Straddling Solutions and Survivalism" (Talk at Mass. Climate Action Network Conference: Nov. 19 2006). http://www.heatisonline.org/contentserver/objecthandlers/index.cfm?id=6161&method=full.

39. Working Group III Report: Mitigation of Climate Change, Summary for Policymakers, p. 29. http://www.ipcc.ch/SPM040507.pdf.

40. Ross Gelbspan, *Boiling Point*, p. 134.

41. See http://www.usmayors.org/climateprotection/climateChange.asp

42. Eban Goodstein, "Climate Change: What the World Needs Now Is. . . Politics" *Worldwatch* January/February 2006. http://www.worldwatch.org/node/3865, or http://www.amazon.com/Climate-change-world-needs-politics/dp/B000E5KTL6.

43. David Korten, "The Great Turning: From Empire to Earth Community" *Yes! Magazine*, Summer 2006. http://www.yesmagazine.org/article.asp?ID=1463

Chapter 15

1. Samuel Bowles and Herbert Gintis, *Democracy and Capitalism: Property, Community and the Contradictions of Modern Social Thought*, p. 125.

2. Ronald Heifetz, *Leadership without Easy Answers*, p. 35.

3. Ibid, p. 37.

4. Ibid, p. 235.

5. Ibid, p. 37.

6. Ibid, p. 241.

7. Ibid, p. 38.

8. Thomas Rochon, *Mobilizing for Peace: The Antinuclear Movements in Western Europe*. p. 2.

9. Sarah Ruth van Gelder, "A Culture Gets Creative," an interview with Paul Ray and Sherry Anderson, Yes! Magazine. Winter 2001, Italics added. Ray and Anderson are authors of *The Cultural Creatives: How 50 Million People Are Changing the World*, Harmony Books. See http://www.culturalcreatives.org.

10. Ibid.

11. Ibid.

12. John Briggs and F. David Peat, *Turbulent Mirror: An Illustrated Guide to Chaos Theory and the Science of Wholeness*, p. 180.

13. Peter Senge, *The Fifth Discipline: The Art and Practice of the Learning Organization*, p. 206.

14. Ibid, p. 155.

15. Berkana Institute, http://www.berkana.org/resources/irresistiblefuture.html and http://www.wiserearth.org/group/berkana. See also *Turning to One Another: Simple Conversations to Restore Hope to the Future,* by Margaret Wheatley.

16. Thomas Prugh, "Idiocy and Sustainability," *Yes! Magazine* Winter 2001. See http://www.yesmagazine.org/article.asp?ID=397

17. Ibid.

18. Fred Kofman and Peter M. Senge, "Communities of Commitment: The Heart of Learning Organizations" in Chawla and Renesch's (ed) *Learning Organizations: Developing Cultures for Tomorrow's Workplace,* pp. 14–43, http://leeds-faculty.colorado.edu/larsenk/learnorg/kof_sen.html.

19. The Co-Intelligence Institute, http://www.co-intelligence.org/links.html, and http://www.democracyinnovations.org/

20. Systems Thinker, "The World Café Living Knowledge through Conversations That Matter" and Mini-Manual for Conversation Café Hosts, http://www.theworldcafe.com/.

21. http://www.co-intelligence.org/P-worldcafe.html http://www.co-intelligence.org/P-worldcafe2.html. See http://www.theworldcafe.com/.

22. Systems Thinker, "The World Café Living Knowledge through Conversations That Matter." Italics added. www.conversationcafe.org/docs/HostManual.doc.

23. Ibid.

24. Ibid.

25. Harrison Owen, "Opening Space for Emerging Order," http://www.openspaceworld.com/brief_history.htm. See also Harrison Owen, *Open Space Technology: A User's Guide.*

26. Ibid.

27. Understanding and Using Co-Intelligence, http://www.co-intelligence.org/CIcontents.html

28. Howard Rheingold, *Smart Mobs: the Next Social Revolution,* http://www.smartmobs.com/book/.

29. Ibid.

30. Ibid.

Chapter 16

1. Ron Morrison, "Keynes without Debt," *Prosperity,* April, 2004. www.prosperityuk.com/prosperity/articles/keynes.html.

Index

A

B

D

E

F

G

H

I

J

K

L

Q

R

S

W

Y

Z

CPSIA information can be obtained at www.ICGtesting.com
Printed in the USA
LVOW121529120212

268317LV00006B/34/P

9 780983 261308